Infrared Spectroscopy of Symmetric and Spherical Top Molecules for Space Observation 1

Infrared Spectroscopy Set

coordinated by
Pierre Richard Dahoo and Azzedine Lakhlifi

Volume 3

Infrared Spectroscopy of Symmetric and Spherical Top Molecules for Space Observation 1

Pierre Richard Dahoo
Azzedine Lakhlifi

iSTE

WILEY

First published 2021 in Great Britain and the United States by ISTE Ltd and John Wiley & Sons, Inc.

ISTE Ltd
27-37 St George's Road
London SW19 4EU
UK

www.iste.co.uk

John Wiley & Sons, Inc.
111 River Street
Hoboken, NJ 07030
USA

www.wiley.com

Library of Congress Control Number: 2020951963

British Library Cataloguing-in-Publication Data
A CIP record for this book is available from the British Library
ISBN 978-1-78630-568-8

Contents

Foreword

Before the early 1950s, the immensity of the cosmos seemed essentially cold and barren. The launch of Sputnik 1 on October 4, 1957, marked the beginning of the space age, and soon afterwards automated space explorers were sent through the solar system and orbiting terrestrial observatories were launched. At the same time, the development of advanced detection techniques and increasingly powerful telescopes contributed to the unprecedented rapid expansion of terrestrial observatories. All these developments led to a radical transformation of our view of both planetary atmospheres and interstellar medium, revealing an unexpected molecular abundance and giving birth to a new discipline: astrochemistry.

This transformation relies heavily on the use of light as a messenger, providing information on the composition of these media by taking advantage of the available range of wavelengths. Indeed, although Newton understood the decomposition of visible light as early as 1666, it was not until 1800 that Herschel broadened our view with the discovery of infrared radiation, followed shortly by Ritter's discovery of ultraviolet radiation, and then Röntgen's 1895 discovery of X-rays. Then, Maxwell unified these various radiations in his electromagnetic field theory. Starting with the 1920s, the development of quantum physics by Heisenberg, Schrödinger, Pauli, Dirac, etc. brought the tools to understand why chemical elements (atoms and molecules) left a discrete signature, in the form of absorption or emission lines in the electromagnetic spectra, an observation that Wollaston and Fraunhofer had made over a century earlier. This laid the basis for the use of spectroscopy as an essential analytical tool.

Nevertheless, since molecules mainly absorb short wavelength radiation, infrared and microwave radiations, largely absorbed by the Earth's atmosphere, the advent of space observatories was decisive for astrochemistry. However, observatories in the microwave range were also set up at the ground level, at high altitude and very dry regions (ALMA), or airborne (stratospheric balloons, SOFIA). The astrophysicists now have a significant array of instruments available that are dedicated to spectroscopy, which obviously includes spectrometers airborne by space probes heading for Venus (Venus Express), Mars (Mars Express, TGO, etc.), Jupiter (Juno), Saturn (Cassini-Huygens), comets (Rosetta), etc. An exciting future lies ahead in this field, given the terrestrial observatories (E-ELT, etc.), space observatories (JWST, WFIRST) and various planetary missions.

Moreover, the discovery in 1995 by Mayor and Quéloz (Nobel Prize in Physics 2019) of the first exoplanet, and the exponential number of discoveries of these celestial bodies ever since, opened the way for conducting the very first spectroscopic studies of their atmospheres. The coming new space instruments (JWST, ARIEL) will even be partially or entirely dedicated to the spectroscopic characterization of these extrasolar planets, the target being the research of biosignatures.

However, using all these instruments and their large amounts of data requires significant upstream experimental and theoretical laboratory work, in order to record and model the spectra of many molecules. This may involve relatively complex organic molecules as well as simple molecules. Indeed, contrary to a preconceived idea, there is definitely still not enough modeling of the infrared spectrum of "small" molecules (CO_2, H_2O, CH_4, NH_3, etc.) in the current databases. In fact, the data needed by planetologists nowadays need to cover "extreme" conditions that were to a little extent, if at all, studied in the laboratory, that is: very high temperatures and pressures, molecules in confined environments, etc.

The aim of this book is to review the theoretical knowledge required for understanding and modeling the spectra of two molecules that are essential in planetology, ammonia (NH_3) and methane (CH_4), and to provide the tools for their spectroscopic study in a confined environment, such as the clathrates.

Vincent BOUDON
Research Director (CNRS – French National Center for Scientific Research)
Laboratoire interdisciplinaire Carnot de Bourgogne (ICB)
January 2021

Preface

Infrared (IR) spectroscopic analysis is of fundamental interest for understanding the physics of the atmosphere and planets, as well as for the study of observable molecules in astrophysics. In the space field of space sciences or exploration, observations conducted by means of instruments at the ground level, or airborne by space probes or space telescopes, contribute to the discoveries that drive science forward or participate in observational sciences. The resulting database enables the confirmation or clarification of theoretical predictions that improve our understanding of the physical and chemical phenomena and processes in the surrounding environments. These observations are facilitated by technological advances and progresses both in the implementation of detection systems and in the analysis conducted by increasingly high-performance computers using ever better controlled statistical methods or theoretical models for the analysis of observational data. A recent example of black hole observation is worth mentioning:

> The Event Horizon Telescope (EHT) is a large telescope array consisting of a global network of radio telescopes and the EHT project combines data from several very-long-baseline interferometry (VLBI) stations around Earth with angular resolution sufficient to observe objects of the size of a supermassive black hole's event horizon.

The authors collaborating on the EHT project apply the methods that rely on the interference of electromagnetic waves using an array of instruments located at various sites on the Earth's surface. The methods applied are increasingly sophisticated and require international collaborations for data collection, analysis and development in order to reveal the observed phenomenon.

The diversity of discoveries contributes to advances in the field of astrophysics and cosmology, building a better understanding of the phenomena at the origin of the universe and addressing the nature and distribution of its constituents that are currently considered to be composed of below 5% visible matter, about 25% dark matter and 70% dark energy, which is responsible for a force that repels gravity and is believed to contribute to the expansion of the universe. Thanks to the analyzed data from space observations, astronomers and physicists have to improve the theoretical approaches, among which the following are worth mentioning: the cosmological model and Einstein's equation in the general theory of relativity or the geometric theory of gravitation published in 1915 [EIN 15], and baryogenesis as an interpretation of the predominance of matter over antimatter [SAK 67]. Similarly, using automated and connected instruments, such as that of the Mars Perseverance Rover 2020 which was successfully launched on July 30, 2020, planetary exploration programs pave the way for observations and data analysis whose interpretation requires theoretical models adapted to various ranges of the electromagnetic spectrum, including IR spectroscopy, which is the focus of this volume.

Theoretical and experimental spectroscopies contribute to the development of methods and devices for the observation and analysis of spectra corresponding to chemical species, molecules, radicals and ions in specific environments, as shown in Volumes 1 and 2 of the set *Infrared Spectroscopy* [DAH 17, DAH 19]. In the IR range, various types of instruments can be used for space observation, in order to detect molecules or chemical species (ions, radicals, macromolecules, nanocages, etc.) present in the atmosphere of planets, including the Earth, and their satellites, as well as in interstellar media, comets or exoplanets.

In molecular physics, the rovibrational energy levels of a molecule are calculated using the molecular Hamiltonian established by Wilson and Howard [WIL 36], reformulated by Darling and Dennison [DAR 40] and later simplified by Watson [WAT 68]. The contact transformation introduced by Van Vleck [VAN 29] is applied to determine the levels of vibration–rotation energy at different orders of approximation. The formalism of this method is described in Volumes 1 and 2, and its application is illustrated on diatomic and triatomic molecules. It was initially applied by Schaffer *et al.* [SCH 39a] and further improved by Nielsen, Amat and Goldsmith [AMA 57a, AMA 57b, AMA 58, GOL 56, GOL 57, NIE 51]. This method makes it possible to regroup the Hamiltonian in interacting polyads with the application of unit transformations, leading to a matrix form in blocks which is easier to diagonalize. Based on the method proposed by Watson [WAT 67, WAT 68b, WAT 68c] for the theoretical study of isolated states, Flaud and Camy-Peyret [FLA 81, FLA 90] showed that symmetry properties of nonlinear triatomic molecules (H_2O, O_3, etc.) can be used to build unitary transformations, leading to the transformation of the initial Hamiltonian into interacting blocks that could be linked to the experimental observations of interacting vibrational levels. Using similar methods developed in the field of IR spectroscopy, a study of ammonia and methane molecules in molecular physics was conducted by the research groups at the CNRS (French National Center for Scientific Research) laboratories in Paris and Dijon. The results obtained by these groups were shared in the PhD theses conducted in this field. The theses referred to in this volume are included in the references section. In particular, it is worth mentioning the theses of Moret-Bailly, Tarrago, Champion, Gherissi, Loëte, Coudert, Coquart, Boudon, Gabart and Permogorov [MOR 61,TAR 65, CHA 78, GHE 79, LOË 84, COU 86, COQ 94, BOU 95, GAB 96, PER 96], which are dedicated to the analysis of gas phase spectra and cover both cold and hot bands or combinations thereof. In this work, which is dedicated to the effects of an environment on the spectrum of an NH_3 or CH_4 molecule, as an example of symmetric or spherical tops, the cold bands are only addressed in situations where the degree of rotation is hindered either by a weak coupling (rotational motions limited to weak values of J) or a strong coupling (librational motions) depending on the type of environment, as studied in the theses of Abouaf-Marguin, Dubost, Gauthier, Boissel, Lakhlifi, Brosset and Dahoo [ABO 73, DUB 75, GAU 80, BOI 85, LAK 87, BRO 93, DAH 96].

As mentioned in the preface to Volumes 1 and 2, the application of methods and tools of theoretical spectroscopy initially developed in molecular spectroscopy for the gas phase and adapted to environments in which the motion of the considered molecule is perturbed makes it possible to not only determine the structure of chemical species (in the gas, liquid or solid phase), but to also identify the species (atoms, molecules, molecular fragments, radicals, etc.) in various environments (nanocavities, media containing various species, ice surface, dust surface, etc.). The species themselves can be used as probes in order to characterize the environment (temperature, pressure, composition) and determine its nature based on theoretical models developed for the analysis of corresponding data.

This book describes the theoretical methods developed in the framework of fundamental research for the interpretation of the spectra of ammonia molecules, which are characterized by a ternary axis of symmetry, from observed spectra in the IR range when these molecules are subjected to an environment in which the temperature and pressure modify their IR gas phase spectra or in nanocages. It describes the theoretical models for the study of ammonia and methane molecules in these media based on the theoretical models elaborated for the gas phase. The modification of the IR spectra of these molecules can also be interpreted, such as the shift of the band centers or the modification of the rovibrational spectrum in nanocages or on surfaces.

This book is intended for students at master and doctoral levels, teaching academics and researchers, astronomers and astrophysicists who analyze the data derived from the interaction between electromagnetic radiation and matter in the IR range, in order to identify the chemical species and their environments.

This book (Volume 3), which will be followed by a complementary book on applications (Volume 4), presents, to both beginners and experts in the field of spectroscopy, the methods developed through theoretical models using group theory for rigid and non-rigid molecules characterized by the tunneling effect phenomenon and large amplitude motions. Simulations of these models enable the analysis of IR data and the identification of molecules based on transitions and profiles not only in the gas phase from fundamental bands but also when they are forced to evolve in an environment that can be a nanocage or a surface.

Ammonia and methane molecules are members of the class of the molecules with a ternary and quaternary axis of symmetry. The first part is organized into two chapters concerning the symmetry and the use of group theory for the study of symmetric and spherical tops, as well as a review of the line profile of molecules that evolve in nanocages or on surfaces. Then the theoretical models elaborated for the study of vibration–rotation spectra of polyatomic molecules in the gas phase are briefly recalled as developed in molecular physics. They allow the study of both cold and hot bands, as well as the combination bands, and have been extensively described in the books devoted to the spectroscopic study in the gas phase.

In this volume, the theoretical description only covers the study of cold bands, whose spectra result from transitions from the fundamental vibration–rotation level. The IR spectroscopic study concerns the libration or hindered rotation movements that result from a strong coupling with the environment and the constrained rotational movements when the coupling with the environment is weak (rotation at low J).

The theoretical part resulted from molecular physics was partly inspired by the molecular physics lectures given in the 2nd year of master degree studies by G. Amat at the UPMC and those of J.M. Flaud and C. Camy-Peyret in the master degree of advanced studies, “Laser and Matter”, at the UPSUD, and the research works conducted at the CNRS laboratories (Paris, Orsay, Dijon, Grenoble and Reims), which are available particularly in the form of (PhD) theses on spectroscopic studies of symmetric or spherical top molecules in the IR range. The theoretical models specific to research work on the effects of an environment have been developed, particularly in the group of molecular physics in Besançon (L. Galatry, D. Robert, J. Bonamy, L. Bonamy, C. Girardet, A. Lahklifi, etc.), in order to analyze observations on molecules subjected to pressure or isolated in condensed phase media in collaboration with researchers from Paris and Orsay (L. Abouaf, H. Dubost, B. Gauthier, J.P. Boissel, P.R. Dahoo, etc.)

These models are applied to study the molecules subjected to interactions in various media, whose effects manifest particularly at the nanometer scale, which modifies the profile of the IR spectra of these molecules. The theoretical inclusion model or the extended model proposed by Lakhlifi and Dahoo was explained in Volumes 1 and 2, and certain programs for the numerical calculation were described in these volumes, mainly in Volume 2. They are used to calculate the IR spectra of molecules in nanocages.

Chapter 1 describes the main tools and methods developed in group theory and tensor algebra for their applications and IR spectroscopy. The finite and continuous point symmetry groups allow for classifying the energy levels associated with electronic, vibrational and rotational degrees of freedom by their symmetries connected to an irreducible representation of the group. The exchange of identical nuclei in a molecule, where tunneling inversion can be observed through experimentation, is studied as part of the permutation–inversion group.

In Chapter 2, group theory methods are applied to determine the symmetry groups of symmetric top molecules with reference to the NH_3 molecule and the spherical top molecules with reference to the CH_4 molecule. The isotopologues (the hydrogen atom H is replaced by the deuterium atom D) of these molecules are also studied from the point of view of symmetry groups. The symmetry properties of NH_3 and CH_4 molecules and their isotopic varieties are presented in this chapter and their geometric symmetry groups are determined. It is shown that the loss of a symmetry element modifies the symmetry group from a more symmetric group to a less symmetric one, and consequently modifies the observable IR spectra. The theory of permutation–inversion groups is also applied to determine the corresponding CNPI groups.

In Chapter 3, the group theoretical methods are applied to determine the functions that generate irreducible representations of the symmetry group of symmetric top molecules and spherical top molecules by referring to NH_3 and CH_4 molecules and to their isotopes for various degrees of freedom, either electronic, vibrational, rotational and electron and nuclear spins. An approximate Hamiltonian is used, which makes it possible to express the total wave function as a product of wave functions by neglecting the vibration–rotation interaction. The statistical weights of the levels are calculated by applying the generic formula given by Landau, and the inversion phenomena is described in the molecular symmetry group following the approach of Bunker and Jensen. The various expressions used to calculate a line profile observable in a spectrum, as well as the method used to determine the selection rules in general, are recalled. By neglecting the couplings, the product form of the total wave function that describes the molecule in the zeroth-order approximation and its symmetry properties are taken into account. The objective is to identify the types of symmetry of the energy levels of each type of degree of freedom, and to determine the allowed transitions that give the selection rules in absorption, emission or diffusion spectroscopy. These transitions are the fingerprint of the molecule when it interacts with light in the studies of IR absorption, emission spectroscopy and Raman spectroscopy.

In Chapter 4, the theoretical model developed for the study of IR spectra that result from transitions between the levels of vibration–rotation energy of symmetric top molecules is described with reference to the NH_3 molecule. The resolution of the Schrödinger equation gives eigenvalues of the molecular system that depend on the degrees of freedom of the nuclei and of the electrons in the molecule that do not have simple analytical solutions. An approximate Hamiltonian is built in the Born and Oppenheimer approximation in order to decouple the rapid motion of electrons from that of nuclei. For each electronic state, the Hamiltonian of nuclei can be used for the study of vibration–rotation motions of the molecule. The various steps in the calculation of the vibration–rotation energy levels are explained in the applications of quantum mechanics applied to molecules. The decoupling of the translation motion from the other degrees of freedom is achieved by studying the vibration–rotation motion in a reference frame attached to the equilibrium configuration of the molecule (mobile reference frame) whose origin coincides with its center of mass (Eckart–Sayvetz conditions). The vibration is then studied in the approximation of low amplitudes. A partial decoupling is achieved between the vibration and rotation motions of the molecule. The Van Vleck contact method is used for splitting the Hamiltonian into blocks, in order to study the IR spectra of the molecules. The transitions between the energy levels resulting from the interaction between the dipole moment or the induced dipole moment of the molecule and the IR electromagnetic radiation lead respectively to the absorption and/or emission spectra of the molecule or to the Raman spectra of the molecule.

In Chapter 5, the theoretical model developed by the spectroscopic group of Dijon in France is illustrated with reference to the methane molecule (CH_4), a spherical top, in the gas phase. The theoretical study of this molecule is not easy. In our opinion, the Dijon approach appears to be the most complete, though it relies on a difficult mathematical formalism. Group theory methods are applied within the framework of the tensor formalism developed in Dijon. More than one chapter would be needed in the case of spherical tops, in order to tackle the various theoretical models elaborated for the calculation of the energy levels and the transitions between these levels. An outline of the model is given to study methane spectroscopy in the gas phase. Based on this theoretical model, applied to the lowest level in the vibration–rotation expansion, the effect of an environment on the spectroscopic characteristics of methane CH_4 can be studied when it is embedded in various media, based on the symmetry considerations as presented in the first two chapters. Its application is illustrated on the dipole moment for the IR spectroscopy transitions of methane in the gas phase.

Pierre Richard DAHOO
Professor at University of Versailles St-Quentin
Azzedine LAKHLIFI
Senior Lecturer at University of Franche-Comté
January 2021

1

Group Theory in Infrared Spectroscopy

This chapter reviews the use of group theory methods for the study of a molecule using infrared (IR) spectroscopy. The application of group theory involves the use of character tables in the theory of representation of a point symmetry group associated with a molecule. The focus is on the elements of the theory of representation, which are used in the following chapters for the study of ammonia (NH_3) molecules (an example of a symmetric top), and methane (CH_4) (an example of a spherical top), and their isotopologues. Elements of the tensor operator algebra are also reviewed for the study of CH_4 within the framework of tensor formalism. The permutation-inversion group CNPI is also presented with reference to NH_3, in order to distinguish between the symmetry group based on the geometry of the molecule, i.e. on its symmetry elements, and the permutation group of the identical nuclei.

1.1. Introduction

The point symmetry group of a molecule reflects the geometric form of the molecule. It is well adapted to the Hamiltonian of Schrödinger's equation, the energy operator which is the sum of the kinetic energy operator and the potential energy operator of the molecule, since it is built on space variables of the position of the electrons and nuclei. When the transformations involving identical nuclei exchanges are taken into account, the point symmetry group is not adapted to the study of the symmetries of the wave function which includes quantum variables related to the spins of identical nuclei when describing a vibration–rotation state for a given electronic state. The identical particle permutation group, which is the symmetric group S_n (not to be confused with the point group S_{2n}) of permutations of n identical particles of order $n!$, is more appropriate.

Molecules characterized by a potential energy function with several minima are not rigid molecules. To take into account the inversion in the case of semi-rigid molecules such as NH_3 and its isotopic varieties, a group of higher order that has the irreducible representations needed to list the irreducible representations used for the classification of the molecule wave functions depending on the Hamiltonian symmetry properties has been used. This group is built from the permutation-inversion group that takes into account situations in which inversion can be observed through identical nuclei exchange in the molecule [BUN 98, HOU 62, HOU 63, HOU 65, LON 63].

1.2. The point-symmetry group of a molecule

The point-group theory applies to rigid molecules, in the sense that the motions of nuclei during a vibration are described by low-amplitude motions and the rotation of the molecule as a whole is described within the framework of rigid rotators (see section 2.2). The construction of the point group relies on the geometric form of the molecule. The operators associated with the elements of a point symmetry group l of a molecule generally commute with one part of the exact or full Hamiltonian (energy operator of an effective Hamiltonian) of the molecule (at the zeroth-order approximation). The focus is on the space position variables of the electrons and nuclei and on the Schrödinger equation for vibration–rotation for a given electronic state.

In this case, the complete nuclear permutation-inversion (CNPI) group [BUN 98, HOU 62, HOU 63, HOU 65, LON 63] is built from a permutation-inversion group (see section 1.6). The elements of this group are the permutations of identical nuclei of a molecule (group of permutation S_n) and the group built on the operator whose effect is the inversion of all the coordinates of nuclei and electrons. This group operates on the exchange of nuclei independently of the observable geometric form. There is a direct link between the symmetry properties of the Hamiltonian and the operators of this group, while the link with the operators of the point symmetry group is indirect. It is worth noting that in some textbooks dedicated to group theory applications, the construction of double symmetry groups is explicitly presented [BUN 98, HAM 62, LAN 75, WIL 55] by building a higher-order group by tensor product of the point symmetry group with a second-order group built on the space coordinate inversion operator. This approach was initially proposed by H. Bethe [BET 29], who used group theory to interpret the degeneracy lifting of atomic orbitals of atoms inserted in crystallographic sites. His work was based on the effect on spin variable symmetry linked to the irreducible representation of a double group. For the structures that were not addressed by Bethe's and Opechowski's approach

[OPE 40], which showed that the classes of a double group G^* built with G, a subgroup of the three-dimensional rotation group SO(3), could be less than twice the number of G classes.

1.2.1. *Symmetry operations and symmetry elements of a molecule*

A symmetry or isometric operation that leaves distances and angles unchanged is an operation that brings a physical object from a given orientation to a new orientation that is equivalent to the former one. In the case of a molecule, the equilibrium configuration of the molecule is invariant under the symmetry operations associated with the symmetry elements of the molecule and therefore the molecule is brought to coincide with itself. There are two types of transformations: rotations of the three-dimensional space corresponding to the transformations of Euler r angles in the rotational space (direct operations) and other transformations such as inversion, reflections with respect to a plane, rotations followed by an inversion, rotations–reflections that reverse the direction of the mobile 3D reference frame (inverse operations).

The possible symmetry elements of a molecule that contribute to building a point group are rotation axes (operations termed as "direct"), inverse rotation axes, planes of reflection symmetry and an inversion center. There are five possible types of symmetry operations for a molecule (see Figure 1.1 and Table 1.1):

– identity operation E, equivalent to a rotation by 2π about C_n, ($C_z(\varphi = 2\pi)$) with $n = 1$ or an absence of operation ($C_z(\varphi = 0)$). The molecules remain unchanged under the action of E; E belongs to all the symmetry groups;

– a rotation by $2\pi/n$ about C_n, the rotation axis of order n; for a linear molecule, there are infinite possible rotations by angle φ about the internuclear axis z, the axis $C_z(\varphi)$ including the identity operation E; for $n = 2$ the molecule is an asymmetric top (H_2O); for $n = 3$, the molecule is a symmetric top (NH_3) if there is only one axis of symmetry of order 3; for $n = 4$ and because of the existence of four axes of symmetry of order 3, the CH_4 molecule is a spherical top. It is worth noting that it is a classification based on rigid rotators (see section 2.4) and on the symmetry about an axis. In general, if n is not a prime number, there is an integer k ($n = kp$), such that the p axes C_n^k are confounded with axis C_n. If there is an axis C_n with n even and a symmetry plane σ perpendicular to C_n, there is also a symmetry operation of inversion i;

– a rotation by $2\pi/n$ about the improper axis S_n, followed by the reflection operation in a plane perpendicular to the (rotation–reflection) axis; for a linear molecule, there is an infinity of improper rotations $S_z(\varphi)$ (rotation by an angle φ

about axis z followed by a reflection operation in a plane perpendicular to the axis at the center of the bond), including the reflection operation with respect to the plane xy (rotation by an angle 0 about the axis z and inversion S_2 with respect to the center of the bond); for a symmetric linear molecule, there is an infinity of rotations $C_t(\pi)$ by an angle π about axes t perpendicular to z and passing through the origin. If n is even, the axes C_{2n} are merged with axes S_n, ($S_n = C_{2n}$). If n is odd, the axes C_n are merged with axis S_n and there is a symmetry plane σ that is perpendicular to the axis S_n, ($S_n = C_n \otimes \sigma$). There is also the operation of inversion symmetry i. Generally, if n is odd and the axis is of type S_{2n}, then i exists;

– the symmetry planes σ_v, v stand for vertical (containing the highest-order rotation axis), σ_h, h stands for horizontal (perpendicular to the highest-order rotation axis) or σ_d, d stands for diagonal. For a linear symmetric molecule, there is an infinity of reflections $\sigma_{zt}(\sigma_v)$ with respect to zt planes containing the axis z and making an arbitrary angle with the zx plane; a molecule is invariant after the reversal of coordinates with respect to the symmetry plane σ;

– inversion i with respect to a center of symmetry of the molecule; for a linear symmetric molecule, S_2 corresponds to the inversion with respect to the center of the bond. Molecules that are invariant to this symmetry operation have an inversion center.

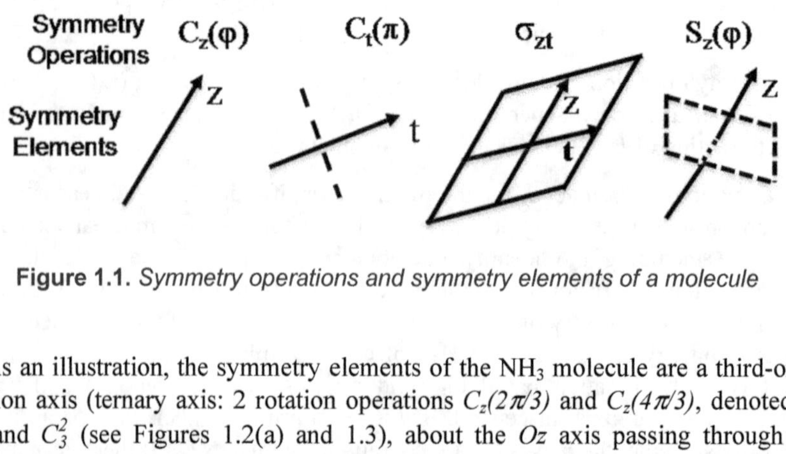

Figure 1.1. *Symmetry operations and symmetry elements of a molecule*

As an illustration, the symmetry elements of the NH_3 molecule are a third-order rotation axis (ternary axis: 2 rotation operations $C_z(2\pi/3)$ and $C_z(4\pi/3)$, denoted by C_3^1, and C_3^2 (see Figures 1.2(a) and 1.3), about the Oz axis passing through the nitrogen atom and three reflection symmetry planes $\sigma(0)$, $\sigma(2\pi/3)$ and $\sigma(4\pi/3)$, denoted by σ_v, σ_v' and σ_v'' (see Figures 1.2(b) and 1.3). If the identity operation E $(C_z(0))$ is added to these operations, the overlapping group of NH_3 and ND_3 molecules, denoted by C_{3v}, can be built (see Table 1.2). Generally, the Oz axis is placed along the highest-order rotational axis.

Symmetry operation	Herman–Mauguin	Schoenflies
Identity	1	C_1 or E
Rotation (2π/n)	n	C_n
Mirror	m	σ, C_s, S_1
Inversion	$\bar{1}$	i, C_i, S_2
Rotation-reflection	-	S_{2n}
Rotation-inversion	$\bar{n}$	-

Table 1.1. *Schoenflies and Hermann–Mauguin notations*

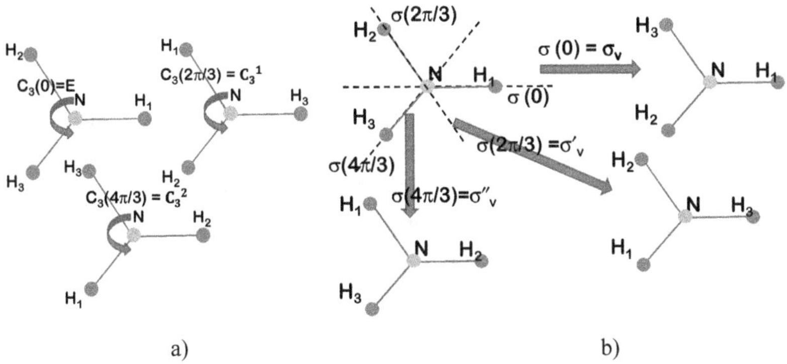

Figure 1.2. *Equivalent symmetry operations of plane rotation and reflection overlapping the equilibrium configurations of the NH_3 molecule. For a color version of this figure, see www.iste.co.uk/dahoo/infrared.zip*

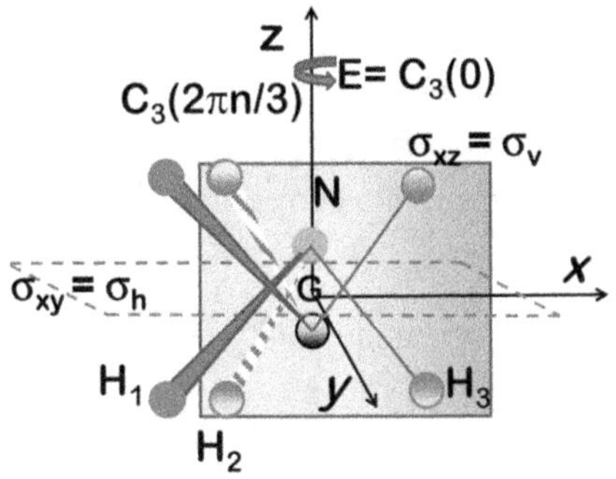

Figure 1.3. *Symmetry elements of the NH_3 molecule and its reversed equilibrium configuration. For a color version of this figure, see www.iste.co.uk/dahoo/infrared.zip*

C_{3v}	E	C_3^1	C_3^2	σ_v	σ_v'	σ_v''
E	E	C_3^1	C_3^2	σ_v	σ_v'	σ_v''
C_3^1	C_3^1	C_3^2	E	σ_v''	σ_v	σ_v'
C_3^2	C_3^2	E	C_3^1	σ_v'	σ_v''	σ_v
σ_v	σ_v	σ_v'	σ_v''	E	C_3^1	C_3^2
σ_v'	σ_v'	σ_v''	σ_v	C_3^2	E	C_3^1
σ_v''	σ_v''	σ_v	σ_v'	C_3^1	C_3^2	E

Table 1.2. *(Cayley) group C_{3v} multiplication table*

1.2.2. *Point symmetry group and laws of composition*

The set of symmetry operations of a molecule forms a point group (that leaves a point, invariant) in the mathematical sense [AMA 80a, BAR 88, BRA 72, BUN 98, CAM 88, COT 63, CRO 63, DEC 77, HAM 62, HER 45, LAN 75, MAR 83, MIX 72, TIN 03, WIG 59, WIL 55, WIL 80], and the symmetry elements can be regrouped into classes (symmetry operations that are equivalent: for NH_3, the two rotation operations $C_z(2\pi/3)$ and $C_z(4\pi/3)$, denoted by C_3^1 and C_3^2 (Figures 1.2(a) and 1.3), about the Oz axis passing through the nitrogen atom form a class, and the three planes of reflection symmetry $\sigma(0)$, $\sigma(2\pi/3)$ and $\sigma(4\pi/3)$ (Figures 1.2(b) and 1.3) form a second class.

The set of symmetry operations O_j constitutes a group if the following conditions are met:

– the product $O_k = O_iO_j$ of two operations of the set belongs to the set;

– the product is associative $O_i(O_jO_k) = (O_iO_j)O_k$;

– there is an operation E referred to as an identity such that $EO_i = O_iE = O_i$ (neutral element);

– each operation O_i can be put into correspondence with an operation O_i^{-1} known as the inverse of O_i, such that $O_iO_i^{-1} = O_i^{-1}O_i = E$.

If additionally $O_iO_j = O_jO_i$ (commutativity property), the group is referred to as "abelian".

The number of elements of a finite group is known as the "order of the group".

If two elements O and O' of a group G of order n are such that $O' = O_iOO_i^{-1}$ for any $O_i \in G$, the elements O and O' are conjugate. If O_i scans throughout group G,

n elements O' are obtained. The set of distinct elements O' constitutes a class of conjugate elements (or equivalent class). The number h of elements is known as the order of the class. Any finite group can be decomposed into a certain number of classes of conjugate elements. For example, the group C_{3v} (Table 1.2) can be decomposed into three classes, the identity E, the two rotations $C_3(2\pi/3)$ or C_3^1 and $C_3(4\pi/3) = C_3(-2\pi/3)$ or $C_3^2 = C_3^{-1}$, and the three reflection operations σ_v, σ_v' and σ_v''. The order of this group is $n = 6$.

According to the Schoenflies notation (see Tables 1.1 and 1.3; Figure 1.4) used in molecular physics and in spectroscopy, the point groups are systematically denoted by C_n, S_{2n} or D_n (C for cyclic group, S for *spiegel*, which is the German word for mirror, D for dihedron) in relation, respectively, to the main symmetry axis of order n (a single one) that corresponds to the highest rotation order, the presence of a symmetry plane perpendicular to the main axis or of n axes of order 2 perpendicular to the main axis. For intermediate angles, the rotation axis $C_n(2p\pi/n)$ is denoted by C_n^p.

The indices v, h and d are added after the letter n in the presence of symmetry planes, v stands for vertical when the plane (σ_v) contains the main axis, h stands for horizontal when the plane (σ_h) is perpendicular to the main rotational axis, and d for the vertical planes containing the main bisector axis (σ_d) of axes of order 2 (C_{nv}, C_{nh}, D_{nh}, D_{nd}).

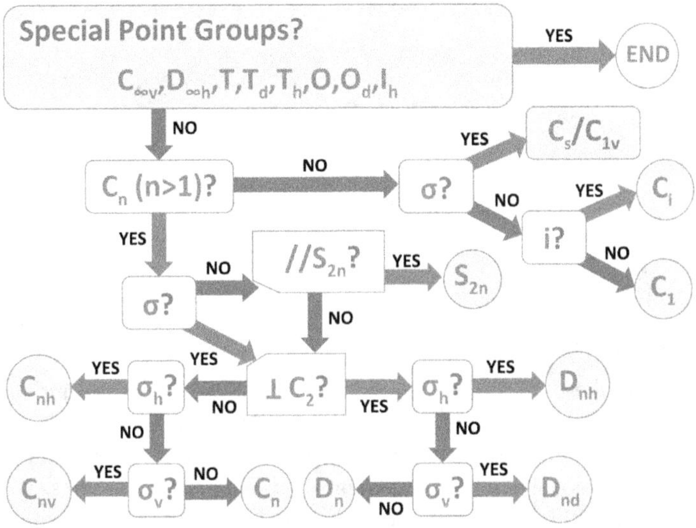

Figure 1.4. *Symmetry operations and flow diagram of point groups according to the Schoenflies notation. For a color version of this figure, see www.iste.co.uk/dahoo/infrared.zip*

A second convention, referred to as Hermann–Mauguin (see Tables 1.1 and 1.2), is preferred in crystallography: the cyclic groups are denoted by their order *n* (rotational symmetry axis of order *n*), a symmetry plane is denoted by *m* and the group is denoted by *n/m* ($C_{2v} \equiv 2m$); if the plane is perpendicular to the symmetry axis, then *n/m* ($C_{2h} \equiv 2/m$) is used; the presence of a symmetry center is indicated by a bar above the order of the group ($C_i = S_2 \equiv \bar{1}$, $S_4 \equiv \bar{4}$, $S_6 = C_i \otimes C_3 \equiv \bar{3}$). For example, $O_h \equiv m3m$, $D_{3h} \equiv \bar{3}m2$, etc. (see Tables 1.1 and 1.3).

Crystal system	Number of point groups	Herman–Mauguin	Schoenflies
Triclinic	2	$1, \bar{1}$	C_1, C_i
Monoclinic	3	2, m, 2/m	C_2, C_s, C_{2h}
Orthorhombic	3	222, 2mm, mmm	D_2, C_{2v}, D_{2h},
Rhombohedral	5	3, 32, 3m $\bar{3}$, $\bar{3}$m	C_3, D_3, C_{3v} S_6, D_{3d}
Hexagonal	7	6, 622, 6mm, 6/m, $\bar{6}$, $\bar{6}$2m, 6/mmm	C_6, D_6, C_{6v}, C_{6h} C_{3h}, D_{3h}, D_{6h}
Tetragonal	7	4, $\bar{4}$, 422, 4mm, $\bar{4}$2m, 4/m, 4/mmm	C_4, S_4, D_4, C_{4v}, D_{2d}, C_{4h}, D_{4h}
Cubic	5	23, m3, 432, $\bar{4}$3m, m3m	T, T_h, O T_d, O_h

Table 1.3. *The point groups of 32 classes of symmetry and the equivalence between Schoenflies and Hermann–Mauguin symbols*

In the condensed phase, only the rotation axes of orders 1, 2, 3, 4 and 6 can be used for a periodic paving of the space compatible with crystal lattices. Crystallography counts 32 symmetry groups (see Table 1.3) based on symmetry operations. The various symmetry elements are distributed into seven crystal systems: triclinic, monoclinic, orthorhombic, tetragonal, rhombohedral, hexagonal and cubic.

In addition to these groups, there are special point groups. The continuous point groups have an infinite number of symmetry elements, the axial and spherical symmetry groups that are used to list the groups characterized by the presence of several axes of rotational symmetry of high order *n*. These axes are associated with regular polyhedrons, Plato solids, namely the tetrahedron with four triangular faces, the cube with six square faces, the octahedron with eight triangular faces, the dodecahedron with 12 regular pentagonal faces with three pentagonal faces joining in one point and the icosahedron with 20 equilateral triangular faces with five triangular faces joining in one point.

Chapter 1 of Volume 2 [DAH 19] dealt with the case of N_2O and CO_2, linear molecules characterized by the presence of a rotation axis of infinite order and planes containing the rotation axis in infinite number (group $C_{\infty v}$ for N_2O) possibly with a symmetry plane perpendicular to the rotation axis, as for CO_2, the system being symmetric with respect to its center ($D_{\infty h} = C_{\infty v} \otimes C_i$ for CO_2). Any rotation by a given angle φ about the C_∞ axis (or D_∞) constitutes a rotational group in two dimensions ($C_\infty(\varphi)$ and $C_\infty(-\varphi)$; degeneracy of modes ν_2).

The complete spherical symmetric group comprises rotations by an arbitrary angle about an axis passing through the center of the molecular system and symmetry planes passing also through the center. This group is denoted by K_h for an atom, for example. It contains the subgroup K (spatial) of spatial rotations (the three-dimensional rotation group or the symmetry group SO(3)) that serves to treat the rotational invariance of the Hamiltonian resulting from the space isotropy for a free molecule in the gas phase. Moreover, the existence of a symmetry center requires the use of $K_h = K \otimes C_i$ or symmetry group O(3). It is worth recalling that the irreducible representations of group SO(3) are described by Wigner D-matrices, which are matrix representations in the spherical harmonics basis $Y_m^J(\theta,\varphi)$.

A Plato solid is characterized by the presence of several rotation symmetry axes of high n-th order that belong to groups $T(T_d)$, $O(O_h)$ and $I(I_h)$ such as:

– The group of tetrahedron T (group of rotations that leave the tetrahedron invariant, $h = 12$, symmetry elements: E, $4C_3$, $3C_2$, and symmetry operations: E, $4C_3$, $4(C_3)^2$, $3C_2$) and $T_d = T \otimes i$ (e.g. CCl_4, group of symmetry transformations of the tetrahedron, $h = 24$, symmetry elements: E, $4C_3$, $3C_2$, i, $4S_6$, $3\sigma_h$ and symmetry operations: E, $4C_3$, $4(C_3)^2$, $3C_2$, i, $4S_6$, $4(S_6)^5$, $3\sigma_h$).

– The group of the octahedron O (group of rotations that leave the cube invariant, $h = 24$, symmetry elements: E, $3C_4$, $4C_3$, $6C_2$, and symmetry operations: E, $6C_4$, $8C_3$, $3C_2$, $6C'_2$), $O_h = O \otimes i$ (e.g. SF_6, group of symmetry transformations of the cube, $h = 48$, symmetry elements: E, $3C_4$, $4C_3$, $6C_2$, i, $3S_4$, $4S_6$, $3\sigma_h$, $6\sigma_d$ and symmetry operations: E, $8C_3$, $6C_4$, $3C_2$, $6C'_2$, i, $6S_4$, $8S_6$, $3\sigma_h$, $6\sigma_d$).

– The group of the icosahedron I (group of rotations leaving the icosahedron invariant, Plato solid with 20 triangular faces or the dodecahedron with 12 pentagonal faces with rotation axes of order 5, $h = 60$, symmetry elements: E, $6C_5$, $10C_3$, $15C_2$ and symmetry operations: E, $12C_5$, $12(C_5)^2$, $20C_3$, $15C_2$), I_h (e.g. C_{60}, group of transformations in the presence of symmetry planes that are parallel to the axes of order 5, $h = 120$, symmetry elements: E, $6C_5$, $10C_3$, $15C_2$, i, 15σ, $6S_{10}$, $10S_6$ and symmetry operations: E, $12C_5$, $12(C_5)^2$, $20C_3$, $15C_2$, i, 15σ, $12S_{10}$, $12(S_{10})^3$, $20S_6$).

1.3. Representations by square matrices (general linear group of order *n* on *R* or *C*: GLn(R) or GLn(C))

Symmetry operations can be represented by unitary square matrices in a basis built on the equilibrium configuration of the molecule. The set of matrices associated with each symmetry operation constitutes a group representation. The matrices have the same multiplication table as the group. In mathematical terms, there is an isomorphism between the overlapping group of the molecule (point symmetry group) and the matrix group, a set of invertible matrices that forms a group under matrix multiplication. The law of composition of operations is associated with the law of multiplication of matrices. It is easier to study the effects of symmetry operations on the properties of a molecule based on the set of matrices that represent these operations. The matrix representation makes it possible to replace the symmetry operations by the algebra of matrices.

1.3.1. *Irreducible representations*

A matrix can be associated with each element of group G. Given $\Gamma_{ij}(O_k)$ the matrix associated with the action of operator $\Gamma(O_k)$ identified in the representation Γ, O_k, an element of group G, then $\Gamma(O_kO_l) = \Gamma(O_k)\Gamma(O_l)$, which is written as $\Gamma_{\alpha\beta}(O_kO_l) = \Sigma_\gamma \Gamma_{\alpha\gamma}(O_k)\Gamma_{\gamma\beta}(O_l)$. The set of matrices $\Gamma_{ij}(O_k)$ forms an n-th order representation of the group G. The dimension of the representation is given by the dimension of matrices $\Gamma_{ij}(O_k)$. The linear space $L_n(R)$ (R for real) or $L_n(C)$ (C for complex) in which the matrices are defined is known as the space of representations, and a basis in this space is a basis of the representation. Among all the representations of a group, there is one trivial representation associated with a neutral element corresponding to the identity, and the dimension of this representation is equal to 1.

Given a representation of dimension n built on the basis $\{\Psi_1, \Psi_2, \Psi_3, \ldots, \Psi_n\}$, composed of the set of n x n square matrices associated each with an operation of the group, the three cases can be presented in the form of matrices:

– if all the matrices of the representation involve two blocks, with, respectively, p and n-p lines and columns aligned on the main diagonal such that

$$\begin{pmatrix} \overbrace{\begin{pmatrix} * & * \\ * & * \end{pmatrix}}^{p} & \begin{vmatrix} 0 & 0 & 0 \\ 0 & 0 & 0 \end{vmatrix} \\ \begin{vmatrix} 0 & 0 \\ 0 & 0 \\ 0 & 0 \end{vmatrix} & \underbrace{\begin{pmatrix} \bullet & \bullet & \bullet \\ \bullet & \bullet & \bullet \\ \bullet & \bullet & \bullet \end{pmatrix}}_{n-p} \end{pmatrix} \qquad [1.1]$$

then the representation appears in reduced form. The first p components of the basis are transformed into themselves and constitute the basis of a representation of dimension p; the other n-p components of the basis are also transformed into themselves and constitute the basis of a representation with dimension n-p;

– a representation that does not appear in a reduced form is known as "reducible" if it is possible to find a new basis whose components are n linearly independent combinations of $\{\Psi_1, \Psi_2, \Psi_3, ..., \Psi_n\}$ by a similarity operation and such that the representation built on the new basis $\{\Psi'_1, \Psi'_2, \Psi'_3, ..., \Psi'_n\}$ appears in a reduced form. The two new representations with dimensions p and n-p may be themselves reducible;

– if through a similarity operation, the order of the set of matrices can be reduced, the representation is known as irreducible.

1.3.1.1. *Case of the NH_3 molecule*

For the NH_3 molecule, a representation can be built by determining the matrices associated with each symmetry operation applied to the linear vector space of basis $\vec{e_1}, \vec{e_2}, \vec{e_3}$ or $\vec{e_x}, \vec{e_y}, \vec{e_z}$ (see Figure 1.5). It is worth noting that while the rotations preserve the directions of the reference frame, this is not the case for symmetries which correspond to reflection planes.

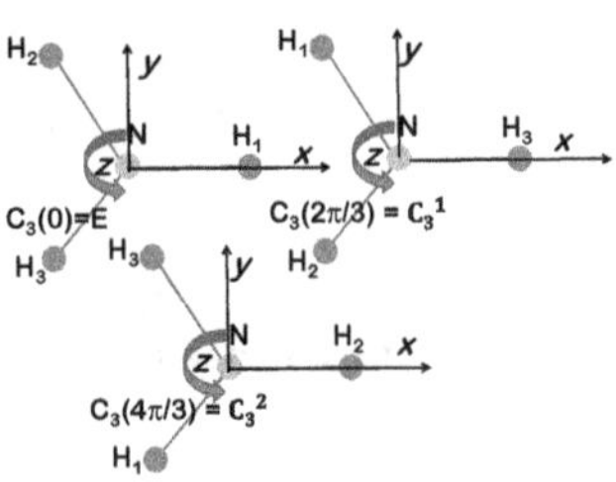

Figure 1.5. *The operations equivalent to rotational symmetry (cyclic group of order 3) overlapping the equilibrium configurations of the NH_3 molecule. For a color version of this figure, see www.iste.co.uk/dahoo/infrared.zip*

For an angle of rotation φ about the Oz axis, the rotation matrix is given by:

$$C_{\varphi}^{z} = \begin{pmatrix} \cos\varphi & -\sin\varphi & 0 \\ \sin\varphi & \cos\varphi & 0 \\ 0 & 0 & 1 \end{pmatrix} \quad [1.2]$$

The character of this matrix is $1 + 2\cos\varphi$.

These basis vectors form a certain representation Γ of the group C_{3v}. The matrices associated with each operation of the group are given by:

$$E = \begin{pmatrix} 1 & 0 & 0 \\ 0 & 1 & 0 \\ 0 & 0 & 1 \end{pmatrix},\ C_3^1 = \begin{pmatrix} -1/2 & -\sqrt{3}/2 & 0 \\ \sqrt{3}/2 & -1/2 & 0 \\ 0 & 0 & 1 \end{pmatrix},\ C_3^2 = \begin{pmatrix} -1/2 & \sqrt{3}/2 & 0 \\ -\sqrt{3}/2 & -1/2 & 0 \\ 0 & 0 & 1 \end{pmatrix}$$

$$\sigma_v = \begin{pmatrix} 1 & 0 & 0 \\ 0 & -1 & 0 \\ 0 & 0 & 1 \end{pmatrix},\ \sigma_v' = \begin{pmatrix} -1/2 & -\sqrt{3}/2 & 0 \\ -\sqrt{3}/2 & 1/2 & 0 \\ 0 & 0 & 1 \end{pmatrix},\ \sigma_v'' = \begin{pmatrix} -1/2 & \sqrt{3}/2 & 0 \\ \sqrt{3}/2 & 1/2 & 0 \\ 0 & 0 & 1 \end{pmatrix} \quad [1.3]$$

All these matrices are diagonal by blocks, and can be decomposed into one 2×2 (two-dimensional) block and one 1×1 (one-dimensional) block; the spaces are disjoint so that the representation Γ has a reduced form. The disjoint spaces of basis vectors $\vec{e_1}, \vec{e_2}$ or $\vec{e_x}, \vec{e_y}$ and $\vec{e_3}$ or $\vec{e_z}$ constitute two different disjoint irreducible representations that can be denoted by Γ_1 and Γ_2, and whose matrices are, respectively, as follows:

– for the Γ_1 representation:

$$E = \begin{pmatrix} 1 & 0 \\ 0 & 1 \end{pmatrix},\ C_3^1 = \begin{pmatrix} -1/2 & -\sqrt{3}/2 \\ \sqrt{3}/2 & -1/2 \end{pmatrix},\ C_3^2 = \begin{pmatrix} -1/2 & \sqrt{3}/2 \\ -\sqrt{3}/2 & -1/2 \end{pmatrix}$$

$$\sigma_v = \begin{pmatrix} 1 & 0 \\ 0 & -1 \end{pmatrix},\ \sigma_v' = \begin{pmatrix} -1/2 & -\sqrt{3}/2 \\ -\sqrt{3}/2 & 1/2 \end{pmatrix},\ \sigma_v'' = \begin{pmatrix} -1/2 & \sqrt{3}/2 \\ \sqrt{3}/2 & 1/2 \end{pmatrix} \quad [1.4]$$

– for the Γ_2 representation:

$$E = 1,\ C_3^1 = 1,\ C_3^2 = 1,\ \sigma_v = 1,\ \sigma_v' = 1,\ \sigma_v'' = 1 \quad [1.5]$$

Therefore, the Γ representation can be expressed as a direct sum of the Γ_1 and Γ_2 representations as follows:

$$\Gamma = \Gamma_1 \oplus \Gamma_2 \quad [1.6]$$

1.3.2. *Equivalent representations*

It must first of all be noted that the elements of a matrix depend on the system of axes that is chosen. For another system of axes in the same vector space, the elements of the matrix are different and we say that another "equivalent" representation is considered. Two representations Γ' and Γ'' are equivalent if there is a non-singular (invertible, of non-zero determinant) matrix P such that:

$$\Gamma'' = P^{-1}\Gamma' P \quad [1.7]$$

A representation is known as irreducible if it is not possible to find a matrix P that transforms all matrices $\Gamma'(O)$ or $\Gamma''(O)$ in the form of blocks or diagonal sub-matrices. A reducible representation is therefore composed of matrices that can all (for each symmetry operation) be diagonalized into blocks and expressed in the form of diagonal sub-matrices (see equation [1.1]).

The matrices of an irreducible representation depend on the chosen basis, but since the trace of matrices (sum of diagonal elements) is an invariant that does not depend on the basis, it is possible to characterize an irreducible representation by the trace, also known as the "character of the representation". It is well known that the eigenvalues of a square matrix associated with an operator O' are the same as those of the conjugate matrix of the equivalent operator $O'' = P^{-1}O'P$ and that in the basis of eigenvectors they are located on the main diagonal of the matrix. The trace being the sum of the diagonal elements, the character of a matrix associated with an operator is therefore an invariant (the same is true for the determinant which is given by the product of the eigenvalues).

An irreducible representation can therefore be represented by the character of the matrix associated with each of the symmetry operations for the considered irreducible representation. The table of characters of all the non-equivalent irreducible representations of the group can thus be built. The dimension of one representation is that of the square matrix (1, 2 (E), 3 (T or F), 4 (G), 5 (H)). The identity representation E is the one that is the easiest to find and for which all the elements of the matrix are on the main diagonal with their value equal to 1 (character = 1 if 1D), (character = 2 if 2D) or (character = 3 if 3D).

The number of irreducible representations of a group is equal to the number of classes of the group. Within the same class, the characters are identical, since the trace of a matrix is an invariant. For an abelian group, when the symmetry operations commute, each operation constitutes a class.

For the group C_{3v}, this number must be equal to 3 since there are three classes. For the above chosen representation (equation [1.6]), this means that the Γ

representation must be completed by the direct sum with another representation, so that it is written as:

$$\Gamma = \Gamma_1 \oplus \Gamma_2 \oplus \Gamma_3 \qquad [1.8]$$

$$\Gamma = A_1 \oplus A_2 \oplus E \qquad [1.9]$$

where Γ_i, $i = 1, 2, 3$, are irreducible representations and are denoted by A_1, A_2 and E (Table 1.5) according to Mulliken notation, where A (symmetric with respect to axis C_n, $\chi(C_n) = 1$) or B (antisymmetric with respect to axis C_n, $\chi(C_n) = -1$) are one-dimensional representations, E is a two-dimensional representation (double degeneracy) for the molecules with rotation axes whose order is greater than 2 and T (in general in electronic spectroscopy) or F (in general in molecular spectroscopy) when there is triple degeneracy, a three-dimensional representation for molecules with more than one rotation axis of order 3. Moreover, the group C_{3v} is isomorphic to groups D_3 and S_3.

1.4. Table of characters and fundamental theorems

The theory of representations involves the substitution of the symmetry elements of a symmetry group by the square matrices representing each symmetry operation in a given basis.

These matrices can be used to build tables of characters that are the trace of the square matrices and to establish theorems that can be used for the construction of tools that facilitate the use of group theory in spectroscopy.

1.4.1. *Tables of characters, classes and irreducible representations*

The considerations developed in sections 1.3.1 and 1.3.2 on the application of the theory of representations of the point-group symmetry of a molecule can be summarized as follows:

– given $\{O_1, O_2, O_3, \ldots, O_p\}$, the operations associated with p symmetry elements of the group G of a molecule;

– given $\{\Psi_1, \Psi_2, \Psi_3, \ldots, \Psi_n\}$ (functions or states) the basis of a representation Γ that generates a vector space L;

– given $\Gamma(O_k)$ the operator associated with the operation O_k and which acts in the vector space L as a linear transformation operator;

– given $\Gamma_{ij}(O_k)$ the matrix associated with the action of operator $\Gamma(O_k)$ in the vector space L_n such that:

$$\Gamma(O_k)\Psi_j = \Sigma_{i=1}^{n} \Psi_i \Gamma_{ij}(O_k) \qquad [1.10]$$

The character of the representation for $\Gamma(O_k)$, the operator associated with the operation O_k, is given by the trace of matrix $\Gamma_{ij}(O_k)$ and its expression is given by:

$$\chi(O_k) = \Sigma_{i=1}^{n} \Gamma_{ii}(O_k) \qquad [1.11]$$

The simultaneous diagonalization of matrices $\Gamma_{ij}(O_k)$ $(k = 1,\ldots, p)$ makes it possible to split each matrix into blocks of sub-matrices and to determine the irreducible representations of the group G. The dimension of each irreducible representation is equal to the dimension (rank) of the sub-matrix.

The following relations can then be written for each O_k:

$$\Gamma(O_k) = \Sigma_{\mu=0}^{d} a_\mu \Gamma^\mu(O_k) \qquad [1.12]$$

$$\chi(O_k) = \Sigma_{\mu=0}^{d} a_\mu \chi^\mu(O_k) \qquad [1.13]$$

where $d + 1$ is the number of irreducible representations (the index 0 corresponds to a totally symmetric representation) and a_μ is the number of times the irreducible representation Γ^μ appears in the representation.

For non-abelian groups, certain operators $\Gamma(O_k)$ can be regrouped into classes, so that a group can be represented by a two-dimensional table with the classes of symmetry operations in the abscissa (for each column) and the irreducible representations in the ordinate (for each line) (see Table 1.4).

Under these conditions, there is a smaller number of symmetry operators and the p operations associated with p symmetry elements of the group G can be re-indexed depending on the number of classes equal to the number of irreducible representations, or $d + 1$, $\{E, O_1, O_2, O_3, \ldots. O_d\}$, where E is the identity operator.

Given that the number of classes and irreducible representations are identical, the table is a square one. The elements of the table are the characters of the matrices associated with symmetry operations.

$^{RI}/_{G}$	E	$h1O_1$	$h2O_2$	...	hdO_d	Scalar vector tensor
Γ_0	$\chi^{\Gamma 0}(E)$	$\chi^{\Gamma 0}(O_1)$	$\chi^{\Gamma 0}(O_2)$	...	$\chi^{\Gamma 0}(O_d)$	α
Γ_1	$\chi^{\Gamma 1}(E)$	$\chi^{\Gamma 1}(O_1)$	$\chi^{\Gamma 1}(O_2)$	...	$\chi^{\Gamma 1}(O_d)$	β
Γ_2	$\chi^{\Gamma 2}(E)$	$\chi^{\Gamma 2}(O_1)$	$\chi^{\Gamma 2}(O_2)$	...	$\chi^{\Gamma 2}(O_d)$	γ
...	...	...	...	...	...	...
Γ_d	$\chi^{\Gamma d}(E)$	$\chi^{\Gamma d}(O_1)$	$\chi^{\Gamma d}(O_2)$	...	$\chi^{\Gamma d}(O_d)$	δ

Table 1.4. *Table of characters giving the classes (number of elements h_i) of the symmetry operations (O_i) and the irreducible representations (IR) of group G*

These tables constitute the table of characters associated with each type of group and to which a column is added. This column contains the coordinates *X, Y* or *Z* (components of the polar vector: T_X, T_Y or T_Z, *T* for translation) and the rotations, J_X, J_Y or J_Z (components of the axial vector: R_X, R_Y or R_Z, *R* for rotation) and the products that form a basis for the irreducible representation of the line. References [BAR 88, BRA 72, BUN 98, COT 63, CRO 63, DEC 77, HAM 62, HER 45, LAN 75, MAR 83, MIX 72, TIN 03, WIG 59, WIL 55, WIL 80] contain more detailed explanations on the application of group theory to atoms and molecules.

1.4.2. *Irreducible representation of group* C_{3v}

In the case of group C_{3v}, the table of characters is presented in Table 1.5(a) and (b); the irreducible representations are identified by Mulliken symbols (*A, E (2D), T* or *F (3D),...*) in the first column. There are three classes comprising the one-dimensional identity operator *E*, the two-dimensional rotational operators C_3 and the three-dimensional plane symmetry operators σ_v. The order *h* of the group that corresponds to the total number of symmetry elements (operators) is equal to 6; this number is obtained by adding the dimensions of each class. Lines 1, 2 and 3 give the characters of the irreducible representations for each operator (Table 1.5(b)). The last three columns recall the symmetries in terms of irreducible representations of the operators involved in the transitions between the energy levels or those of normal coordinates.

C_{3v}	E	C_3^1	C_3^2	σ_v	σ_v'	σ_v''
A_1	1	1	1	1	1	1
A_2	1	1	1	-1	-1	-1
E	2	-1	-1	0	0	0

a)

C_{3v}	E	$2C_3^z$	$3\sigma_v$	Polar vector	Axial vector	Second-order tensor	Polarizability
A_1	1	1	1	z, T_Z		x^2+y^2, z^2	$\alpha_{xx}+\alpha_{yy}+\alpha_{zz}$
A_2	1	1	-1		R_z		
E	2	-1	0	(x,y), (T_X, T_Y)	(R_x, R_y)	(x^2-y^2, xy) (xz,yz)	($\alpha_{xx}-\alpha_{yy}$, α_{xy}); (α_{yz}, α_{zx})

b)

Table 1.5. *Tables of characters of symmetry operations (O_k), classes of operations and irreducible representations (IR) of the group C_{3v}*

Similarly, the table of characters of the five types of symmetry operations for each nucleus of a molecule that is invariant under the symmetry operation can be built as given in Table 1.6.

O_k	E	C_φ^z	σ	S_φ^z	$i=S_\pi^z$	Vector
χ_{Ok}	3	1+2cosφ	1	-1+2cosφ	-3	Polar (T)
χ_{Ok}	3	1+2cosφ	-1	1-2cosφ	3	Axial (R)

Table 1.6. *Table of characters of the symmetry operations (O_k) for each nucleus that is invariant under operation O_k*

The tables of characters given in Tables 1.5 and 1.6 can be used to determine the number of vibrations of the NH_3 molecule as well as their types of symmetry for a given system of Cartesian axes.

A fundamental theorem in group theory applications to the study of molecules is the orthogonality and normalization theorem, which refers to the orthogonality of lines and columns of a table of characters. The proof of this theorem involves Schur's lemma and its corollaries.

1.4.3. *Schur's lemma*

Given $\Gamma^\alpha(O_k)$ and $\Gamma^\beta(O_k)$, two irreducible representations of one group *G*, of dimensions l_α and l_β; if there is a matrix *P*, of dimension l_α x l_β such that for all the O_k, the following can be written:

$$P\Gamma^\alpha(O_k)=\Gamma^\beta(O_k)P \quad [1.14]$$

Then:

– if $l_\alpha \neq l_\beta$, P is a null matrix of dimension l_α x l_β;

– if $l_\alpha = l_\beta$, P is a square non-singular matrix (invertible, non-null determinant) and Γ^α and Γ^β are two equivalent representations.

COROLLARY 1 TO SCHUR'S LEMMA.– A matrix that commutes with each matrix of an irreducible representation of a group G, is a multiple of the identity matrix.

COROLLARY 2 TO SCHUR'S LEMMA.– An irreducible representation of an abelian group G is necessarily one-dimensional.

1.4.4. *Orthogonality and normalization theorem*

Given G a finite group of order h, Γ^α an irreducible representation of G of dimension d_α, $\Gamma^\alpha{}_{ij}(O_k)$ the matrix of the operator $\Gamma(O_k)$ associated with the element O_k in the representation Γ^α (considered as unitary), the following orthogonality relation is verified by the matrices, according to the orthogonality theorem:

$$\Sigma_{O_k \epsilon G} \Gamma^\alpha_{ij}(O_k)\left(\Gamma^\beta_{kl}(O_k)\right)^* = \frac{h}{\sqrt{d_\alpha d_\beta}} \delta_{\alpha\beta}\delta_{jk}\delta_{il} = \frac{h}{d_\alpha}\delta_{\alpha\beta}\delta_{jk}\delta_{il} \quad [1.15]$$

where h is the order of group G and d_α is the dimension of the Γ^α representation. It is worth noting that indices i and j take their values in the set $\{1,2,3, ..., d_\alpha\}$, and indices k and l take their values in the set $\{1,2,3, ..., d_\beta\}$, and that the dimensions of the corresponding matrices are d_α and d_β, respectively. The equivalent representations are considered as being the same representations. This theorem is not commonly used in practice because the tables of the symmetry groups are given in terms of characters. A form based on the characters is used instead (see equation [1.16]).

1.4.5. *Orthogonality of lines*

The characters of the non-equivalent unitary irreducible representations verify the orthogonality relations. The characters of the matrices verify in this case the orthogonality relation such that:

$$\Sigma_{O_k \epsilon G} \chi^\alpha(O_k)\left(\chi^\beta(O_k)\right)^* = h\delta_{\alpha\beta} \quad [1.16]$$

which corresponds to the orthogonality of lines in the table of characters. Referring to Table 1.5(a) of group C_{3v}, for example, it can be verified that for lines 1 and 2 corresponding to IR, A_1 and A_2 ($\alpha = A_1$ and $\beta = A_2$), the relation gives: $1^2 + 2 \times 1^2 + 3 \times 1 \times (-1) = 0$; similarly, for lines 1 and 3 corresponding to IR, A_1 and E ($\alpha = A_1$ and $\beta = E$), the relation gives: $1 \times 2 + 2 \times 1 \times (-1) + 3 \times 1 \times 0 = 0$; finally, for lines 2 and 3 corresponding to IR, A_2 and E ($\alpha = A_2$ and $\beta = E$), the relation gives: $1 \times 2 + 2 \times 1 \times (-1) + 3 \times (-1) \times 0 = 0$.

For the same representation, the sum of squares of characters is equal to the order of the group such that:

$$\Sigma_{O_k \in G} \chi^{\alpha}(O_k)\left(\chi^{\alpha}(O_k)\right)^* = h \qquad [1.17]$$

Referring to Table 1.5(a) of group C_{3v}, for example, it can be verified that for line 1 corresponding to IR, A_1, the relation gives: $1^2 + 2 \times 1^2 + 3 \times 1^2 = 6$; similarly, for line 2 corresponding to IR, A_2, the relation gives: $1^2 + 2 \times 1^2 + 3 \times 1^2 = 6$; finally, for line 3 corresponding to IR, E, the relation gives: $2^2 + 2 \times 1^2 + 3 \times 0 = 6$.

Two equivalent representations have the same system of characters. The operations O_i belonging to the same class and to the same irreducible representation have therefore the same character as a result of the equivalence of irreducible representations (similarity transformation) and the invariance of the trace of a matrix. The presentation of the table of characters of a group G can thus be simplified, grouping the symmetry operations O_i by classes i (new indices where k is replaced by i) of number of elements h_i (see Table 1.5(b)), so that equation [1.17] can be reformulated as follows:

$$\Sigma_{O_i \in G} h_i \chi^{\alpha}(O_i)\left(\chi^{\beta}(O_i)\right)^* = h\delta_{\alpha\beta} \qquad [1.18]$$

where h_i is the number of elements in the class to which the symmetry operation O_i belongs.

Similarly, the relation for the calculation of the order of a group on a line of the table of characters is given by:

$$\Sigma_{O_i \in G} h_i \chi^{\alpha}(O_i)\left(\chi^{\alpha}(O_i)\right)^* = h \qquad [1.19]$$

These relations can be readily verified with Table 1.5(b).

1.4.6. *Orthogonality of columns*

Using Schur's lemma, it is possible to prove that:

$$\Sigma_\alpha \chi^\alpha(O_i)\left(\chi^\alpha(O_j)\right)^* = \frac{h}{h_i}\delta_{ij} \qquad [1.20]$$

which corresponds to the orthogonality of columns in the table of characters.

Referring to Table 1.5(b) of group C_{3v}, for example, it can be verified that for columns 1 and 2 ($i = 1$ and $j = 2$) corresponding to identity operations E and rotations about the axis Oz, C_3^z, it can be written that: $1 \times 1 + 1 \times 1 + 2 \times (-1) = 0$; similarly, for columns 1 and 3 ($i = 1$ and $j = 3$) corresponding to identity operations E and plane symmetry σ_v, it can be written that: $1 \times 1 + 1 \times (-1) + 2 \times 0 = 0$; finally, for columns 2 and 3 ($i = 2$ and $j = 3$) corresponding to rotational operations about the axis Oz, C_3^z and plane symmetry σ_v, it can be written that: $1 \times 1 + 1 \times (-1) + (-1) \times 0 = 0$.

If O_i is assumed to be the identity operation E, then $h_i = 1$ (one class) and the sum of equation [1.20] leads to the order of group G

$$\Sigma_\alpha \chi^\alpha(E)\left(\chi^\alpha(E)\right)^* = h \qquad [1.21]$$

It can be readily verified from Table 1.5(b) of group C_{3v} for which $h = 6$, that indeed: $1 \times 1 + 1 \times 1 + 2 \times 2 = 6$.

This result can be obtained for each operation O_k of the group from equation [1.20], which leads to the calculation of the order of group G. Considering column 2 of Table 1.5(b), for rotations about the Oz axis, or the operation C_3^z it can be written that: $1^2 + 1^2 + (-1)^2 = h/2$; this leads to $h = 3 \times 2 = 6$. Similarly, for column 3, which corresponds to the plane symmetry operation σ_v, it can be written that: $1^2 + (-1)^2 + 0 = h/3$; this leads to $h = 2 \times 3 = 6$.

1.4.7. *Decomposition of a reducible representation on an irreducible basis*

Given a reducible representation Γ of a group G with h elements (order of the group) and K classes, each having a number of elements h_i. Given (l_α, l_β, ...) the dimensions of its irreducible representations(Γ^α, Γ^β...).

The reducible representation Γ of group G is the direct sum of irreducible representations of the group, which is written as follows:

$$\Gamma = a_1\Gamma^1 \oplus a_2\Gamma^2 \oplus \ldots \ldots a_d\Gamma^d = \Sigma_{\oplus\alpha}\, a_\alpha \Gamma^\alpha \qquad [1.22]$$

where:

$$a_\alpha = \frac{1}{h}\Sigma_{O_i \in G}\, h_i \chi(O_i)\left(\chi^\alpha(O_i)\right)^* \qquad [1.23]$$

and

$$\chi(O_k) = \Sigma_\alpha\, a_\alpha \chi^\alpha(O_k) \qquad [1.24]$$

This relation proves to be very useful for the decomposition of a reducible representation into irreducible representations and for finding the number of vibrations belonging to a given irreducible representation from symmetries of the various types of vibrations of a molecule, stretching, bending, torsion, etc.

For the finite groups, the number of irreducible representations is equal to the number of classes of the group. The following relation (sum of squares along a column) is also verified:

$$\Sigma_\alpha\, l_\alpha^2 = h \qquad [1.25]$$

Moreover, the column of operator E (identity) makes it possible to directly determine the dimension of irreducible representations.

1.4.8. *Projection operators for irreducible representations*

Projection operators are used to build wave functions that are the bases of an irreducible representation of the symmetry group of a molecule.

Given $\{\Psi_i^\alpha\}$, the basis of an irreducible representation Γ^α of dimension l_α of the group G of order h, and given $\chi^\alpha(O_k)$ the characters of each operation O_k of the group. Applying the symmetry operation O_k to Ψ_i^α, a linear combination of $\{\Psi_i^\alpha\}$ (restriction of relation 1.10 to the irreducible representation Γ^α) is obtained for the irreducible representation $\Gamma\alpha$ of the group. This combination is transformed depending on the symmetry properties of this representation. It can be shown that these combinations can be determined by applying the projection operator given by:

$$P_{ij}^{\alpha} = \frac{l_{\alpha}}{h} \sum_{O_k \in G} \left(\Gamma_{ij}^{\alpha}(O_k) \right)^* O_k \qquad [1.26]$$

This method requires knowledge of matrices Γ_{ij}^{α} and the definition of the operation $O_k \Psi_i^{\alpha}$.

Since the tables of characters do not give the matrices Γ_{ij}^{α}, but only the characters of these matrices, a simplified projection operator is used, which is given depending on the characters such as:

$$P^{\alpha} = \frac{l_{\alpha}}{h} \sum_{O_k \in G} \left(\chi^{\alpha}(O_k) \right) O_k \qquad [1.27]$$

The linear combinations are not necessarily orthogonal, and this property must be built case by case.

1.4.9. *Characters of irreducible representations of the direct product of two groups*

Using this relation, it can be shown that:

– given $\{\Psi_i^{\alpha}\}$, the basis of an irreducible representation Γ_1^{α} of the group G_1 of characters $\chi_1^{\alpha}(O_k^1)$ for each operation O_k^1 of the group G_1;

– given $\left\{\phi_j^{\beta}\right\}$, the basis of an irreducible representation Γ_2^{β} of the group G_2 of characters $\chi_2^{\beta}(O_k^2)$ for each operation O_k^2 of the group G_2;

– given $\left\{\Psi_i^{\alpha}\phi_j^{\beta}\right\}$, the basis of an irreducible representation $\Gamma^{\alpha\otimes\beta}$ of the group formed by the direct product of G_1 and G_2, $G_1 \otimes G_2$, the characters of the operators that are the products of operators of G_1 by those of G_2 are given by $\left\{O_i^1 O_j^2\right\}$:

$$\chi^{\alpha\otimes\beta}\left(O_i^1 O_j^2\right) = \chi_1^{\alpha}(O_i^1)\chi_2^{\beta}(O_j^2) \qquad [1.28]$$

For example, the direct product of two representations $E \times E$ leads to the sum of three representations: $1 + A_2 + E$.

The rules and formulas given in this section can be used for building the vibration or rotation symmetry coordinates, and thus characterize each state of vibration and rotation by their type of symmetry. The tables of characters are useful for establishing the selection rules in absorption or emission spectroscopy and in Raman spectroscopy and for determining the shape of the observable spectra.

1.5. Overall rotation group symmetry of a molecule

The point-group symmetry of a molecule is defined by the geometry of the molecule and does not involve the overall rotation of this molecule. The symmetry operations of the point group do not commute with the full molecular Hamiltonian, but only with the vibration and electronic Hamiltonian. A more detailed presentation of the considerations given below can be found in the following (non-exhaustive) references, such as: [BUN 98, EDM 74, GRI 64, HER 45, LAN 75, MES 64, SAK 94, WAT 66].

The symmetry group of the Hamiltonian of the rigid rotator is the infinite group of rotations K (J integer; or SO(3)) about an axis passing through the center of mass. The group of spherical symmetry contains the rotations by an arbitrary angle about any axis passing through the center and the reflections in any plane passing through the same point. It is denoted by K_h and contains the group K (group of rotations in the three-dimensional space SO(3)) as subgroup with $K_h = K \otimes C_i$ (group O (3)). All the axes being equivalent in group K, the classes are the rotations by an angle of given absolute value about an arbitrary axis. The group K comprises an infinite number of finite rotations achieved by continuous variation of infinitesimal angles.

The natural approach for the study of overall rotations of a molecule relies on the identification of the rotation axis defined by a unit vector $\hat{n}$ and the angle of rotation ϕ. The two angles (polar angle, θ and azimuthal angle, φ), which in the system of spherical coordinates can be used to identify the unit vector and the angle of rotation, ϕ, show that three parameters are needed for the definition of a rotation in the three-dimensional space. The same rotation can also be defined by means of three Cartesian components of vector $\hat{n}\phi$. Applying this method to study the properties of rotations is not convenient because it is not possible to add vectors of type $\hat{n}\phi$ when characterizing successive rotations.

It is more convenient to use orthogonal 3×3 matrices that form the group SO(3) under the operation of matrix multiplication (S for special with the determinant equal to 1, O for orthogonal and 3 for dimension 3). Hence, it is customary to define the rotation of the reference system attached to the molecule in terms of three Euler angles that can be used for positioning the mobile reference frame with respect to the fixed reference frame by means of three rotations about three axes. These rotations and the axes were defined in section 2.7.3 of Volume 2 [DAH 19] as the rotation by an angle $\theta\,(0 \leq \theta \leq 2\pi)$ about the Z axis of the fixed reference frame, the rotation by an angle φ $(0 \leq \varphi \leq \pi)$ about the line of nodes (new position of the OY axis of the initial fixed reference frame after the first rotation) and the rotation by an angle $\chi\,(0 \leq \chi \leq 2\pi)$ about the axis $z = z'$ of the mobile reference frame (final position of the OZ axis of the initial fixed reference frame).

Given $R(\theta, \varphi, \chi)$ – the operator corresponding to the three rotations defined by: $R(\theta,\varphi,\chi) = exp(\frac{-i\chi}{\hbar}J_z)exp(\frac{-i\varphi}{\hbar}J_{y'})exp(\frac{-i\theta}{\hbar}J_Z)$, it can be proved that the rotation operation can be defined with respect to the fixed axes such that: $R(\theta,\varphi,\chi) = exp(\frac{i\chi}{\hbar}J_Z)exp(\frac{i\varphi}{\hbar}J_Y)exp(\frac{i\theta}{\hbar}J_Z)$, using the transformations: $R_{y'}(\varphi) = R_Z(\theta)\,R_Y(\varphi)\,R_Z^{-1}(\theta)$ and $R_{z'}(\chi) = R_{y'}(\varphi)\,R_Z(\chi)\,R_{y'}^{-1}(\varphi)$, where the three matrices refer to rotations about fixed axes. It is in fact appropriate to consider the rotations with respect to fixed axes rather than to mobile axes, since spin variables can be dealt with by rotation operators following the same properties and for which the notion of mobile axes is not relevant. In the space of spins, R is a 2×2 matrix defined in a two-dimensional space and the matrix is associated with an operator $D(R)$ acting on the spin variables $|\alpha\rangle$ and $|\beta\rangle$ or $|+1/2\rangle$ and $|1/2\rangle$.

The angular momenta operators $\mathcal{J}_X$, $\mathcal{J}_Y$ and $\mathcal{J}_Z$ of a molecule are proportional to the operators associated with rotations by an infinitesimal angle about an axis passing through a center. Consequently, the common eigenfunctions $|JM\rangle$ of the operators $\mathcal{J}^2$ and $\mathcal{J}_Z$ $(\left[H, \mathcal{J}^2\right] = 0, \left[H, \mathcal{J}_Z\right] = 0)$ identified by the quantum numbers J and M constitute a basis for the irreducible representation of the rotation group. Each irreducible representation contains a continuous set of matrices, but the dimension of the representation is finite, meaning that the rank of the matrix remains finite. The number of various irreducible representations is infinite, but countable. At a fixed j, the $2j + 1$ eigenfunctions common to operators J^2 and J_Z, form the basis of the irreducible representation of angular momenta operators. Finite rotations are achieved by iteration of infinitesimal rotations $D(\hat{Z}, d\phi) = 1 - i\left(\frac{J_Z}{\hbar}\right)d\phi$ so that:

$$D_Z(\phi) = \lim_{n \to \infty}\left(1 - i\left(\frac{J_Z}{\hbar}\right)\left(\frac{\phi}{n}\right)\right)^n = exp\left(\frac{-iJ_Z\phi}{\hbar}\right)$$

For a fixed j, the $2j + 1$ eigenfunctions $|JM\rangle$ transform into themselves under the action of operator $D(\theta, \varphi, \chi)$.

Designating by $D(\theta, \varphi, \chi)$ the operator corresponding to three rotations and defined by: $D(\theta,\varphi,\chi) = exp(\frac{-i\chi}{\hbar}J_Z)exp(\frac{-i\varphi}{\hbar}J_Y)exp(\frac{-i\theta}{\hbar}J_Z)$, the following expression can be written:

$$D(\theta,\varphi,\chi)|Jm\rangle = \sum_{m'=-J}^{m'=+J} |Jm'\rangle D^J_{m'm}(\theta,\varphi,\chi) \qquad [1.29]$$

where

$$D^J_{m'm}(\theta,\varphi,\chi) = exp\left(\frac{-im'\chi}{\hbar}\right)\left\langle Jm'\right| exp\left(\frac{-i\varphi}{\hbar}J_{y'}\right)|Jm\rangle\, exp\left(\frac{-im\theta}{\hbar}\right)$$

and

$$D^J_{m'm}(0,\varphi,0) = d^J_{m'm}(\theta,\varphi,\chi) = \left\langle Jm'\right| exp\left(\frac{-i\varphi}{\hbar}J_{y'}\right)|Jm\rangle$$

The irreducible representations of group K are $D^{(0)}$ (totally symmetric representation), $D^{(1)}$, $D^{(2)}$, ...$D^{(j)}$, ..., where j is an integer (for rotations). The dimension of each matrix of the representation j is $(2j + 1)$. The lines and columns can be identified from the indices m' and m that take values in the set $\{-j, -j + 1, ..., j\}$. The trace of a matrix $D^J_{m'm}(\theta,\varphi,\chi)$ for j constant can be evaluated using the equivalence between the operator $R(\theta,\varphi,\chi)$, product of three rotation operators and the operator $R(\hat{n},\varphi)$ (using the fact that the characters of the irreducible elements of a class are identical) to determine the character $\chi^J(\varphi) = \frac{sin(j+1/2)\varphi}{sin(1/2)\varphi}$. For $j = 1$, for example: $\chi^J(\varphi) = 1 + 2cos\varphi$. When j is a half-integer, the character $\chi^J(\varphi)$ can take two distinct values for φ and $\varphi + \pi$, and it can be written that: $\chi^J(\varphi + \pi) = -(\varphi)$ since $\chi^J(\varphi)$ is a periodic function in φ of period 4π, as $\chi^J(\varphi) = sinj\varphi cot\ 1/2\,\varphi + cos\varphi$. Given that the two rotations φ and $\varphi + \pi$ represent the same rotation in the three-dimensional space, there is a 2:1 correspondence between the characters or the matrices, when j is a half-integer and the rotations in the three-dimensional space. A 1:1 correspondence can be established by applying Bethe method [BET 29], which involves the introduction of a new operation Q that commutes with the rotation operators ($QR = RQ$) such that $Q^2 = E$. The new rotation group comprises twice more elements as the initial group and constitutes the double rotation group.

It is worth noting that the analytical expression of the elements of $D^J_{m'm}(\theta,\varphi,\chi)$ (see section 2.7.5, Volume 2 [DAH 19]); (θ,φ,χ) depends on how the rotations are defined. Indeed, the molecule and therefore the reference frame attached to it can be rotated: this point of view is known as "active", or the fixed axes attached to the laboratory system can be given an inverse rotation motion: this point of view is known as "passive". In both cases, the effect of rotations on the rotational Hamiltonian of the molecule is identical. In the first case, the spatial positions $\vec{R}(\vec{r})$ become $\vec{R}(\vec{r}) + \vec{A}$, and in the second case the fixed axes shift by $-\vec{A}$, which leads to the same spatial coordinates of the transformed Hamiltonian. It should be noted that in Volume 2, we use (φ, θ, χ) instead of (θ, φ, χ); it suffices to invert θ and φ to find the correct expressions of the matrices.

For irreducible representations $j = 1/2$ (spin variables) and $j = 1$, the matrix elements of $d^{(j)}_{m'm}(\theta,\varphi,\chi)$ are given by:

$\mathbf{d}^{(\frac{1}{2})}_{\mathbf{m'm}}$		
m'/m	**-1/2**	**1/2**
-1/2	cos(φ/2)	-sin(φ/2)
1/2	sin(φ/2)	cos(φ/2)

a)

$\mathbf{d}^{(1)}_{\mathbf{m'm}}$			
m'/m	$\mathbf{-1}$	**0**	**1**
-1	(1/2)(1+ cos φ)	$-(1/\sqrt{2})$ sin φ	(1/2)(1- cos φ)
0	$(1/\sqrt{2})$ sin φ	cos φ	$-(1/\sqrt{2})$ sin φ
1	(1/2)(1- cos φ)	$(1/\sqrt{2})$ sin φ	(1/2)(1+ cos φ)

b)

Table 1.7. *Elements of the rotation matrix $d^{(j)}_{m'm}$ for irreducible representations: (a) for j = 1/2 and (b) for j = 1*

1.6. Full symmetry group of the Hamiltonian of a molecule

The theory of permutation-inversion (PI) groups is a concept that uses molecular symmetry making it possible to describe both rigid and non-rigid molecular systems. This concept differs from that of Schönflies point-group symmetry based on equilibrium configuration geometry of nucleï, which is limited to rigid species. Non-rigid systems easily modify their conformation, and therefore their symmetry properties, by high-amplitude motions [BUN 98, HOU 62, HOU 63, LON 63]. The effects of high-amplitude motions can be observed in the form of degeneracy lifting by tunneling effect in high resolution spectra. The theory of PI groups can be used to adequately describe the high-amplitude motions that modify the molecular symmetry.

1.6.1. *Permutation operations*

Given a set denoted by $\{abc \bullet \bullet \bullet yz\}$. Given the symbol $(abc \bullet \bullet \bullet yz)$ that represents a permutation such that a is replaced by b, b by c, c by d, …, y by z and z by a. For example, for the set $\{123\}$, a possible permutation is to swap 1 and 2, while 3 maintains its place. In mathematics, this operation is represented by:

$$\binom{123}{213} = (12)$$

where the abbreviated notation *(12)* is used for the permutation operation, which corresponds to a transposition.

A further example is the cyclic permutation of three numbers such that:

$$\binom{123}{231} = (231) = (123)$$

Such a permutation is known as a cycle. A group can be built from all the permutations of 3 numbers by adding the identity element E (no change).

For example, if the permutation operators *(12)* and *(123)* are applied to the ordered set *321*, the following results are obtained:

$$(12)321 = 312$$

$$(123)321 = 132$$

where in the first case, 1 is replaced by 2 and 2 by 1 and in the second case, 1 is replaced by 2, 2 is replaced by 3 and 3 is replaced by 1.

A permutation involving only two elements, for example *(12)*, is known as a transposition. Based on this definition, the following relations can be written:

$$(12) = (21),\ (23) = (32),\ (13) = (31),$$

and

$$(123) = (312) = (231),\ (132) = (321) = (213).$$

1.6.2. *Permutation group* S_n

The set of all the possible permutation operations of n objects meets the axioms of the group. This is known as the n-th degree permutation group and is denoted by S_n. There are $n!$ elements in S_n. It is an $n!$-th order group. This group is composed of transpositions with two elements and cyclic subgroups. The k-cycles (of dimension k) are thus counted by searching the k elements that form a cycle after permutation.

A further important aspect to be considered is that any permutation can be expressed as a product of transpositions. A given permutation is known as even or odd when it has an even or odd number of transpositions in its product of transposition. For example, *(abcdef)(gh)* = *(ab)(bc)(cd)(de)(ef)(gh)* is an even permutation. For example:

$S_2 = \{E, (12)\}$ forms an abelian group;

$S_3 = \{E, (12), (23), (31), (123), (132)\}$.

The transpositions are *(12)*, *(23)* and *(31)*. The cycles are *(123)* and *(132)*. A cycle is built from an even number of transpositions. The set of operations *{E, (123), (132)}* forms a subgroup of S_n. For $n > 2$, the groups S_n are not commutative.

The multiplication or Cayley table of the symmetric group S_3 is given in Table 1.8. An element of line i and of column j is the product of the operator of the first column j and the operator of line i.

S_3	**E**	**(12)**	**(23)**	**(13)**	**(123)**	**(132)**
E	E	(12)	(23)	(13)	(123)	(132)
(12)	(12)	*E*	(123)	(132)	(23)	(13)
(23)	(23)	(132)	*E*	(123)	(13)	(12)
(13)	(13)	(123)	(132)	E	(12)	(23)
(123)	(123)	(13)	(12)	(23)	(132)	*E*
(132)	(132)	(23)	(13)	(12)	*E*	(123)

Table 1.8. *Multiplication (Cayley) table of the symmetric group S_3*

Table 1.9(a) and (b) gives the table of characters of the symmetric groups S_2 and S_3 and the irreducible representations of the groups and the classes.

S_2	E	(12)
	1	**1**
A	1	1
B	1	1

a)

S_3	E	(12)	(123)
	1	**3**	**2**
A_1	1	1	1
A_2	1	-1	1
E	2	0	-1

b)

Table 1.9. *Table of characters of symmetric groups: a) for* S_2 *and b) for* S_3

Two important results are that symmetric groups are finite (of finite dimension) and that each finite group (particularly finite point group) is isomorphic to a subgroup of a given symmetric group (Cayley or Jordan theorem) [HAM 62].

The symmetry group of the Hamiltonian operator of a system of particles should not be confused with the symmetric group. The symmetry group is composed of operators that commute all with the Hamiltonian operator. The total wave function of the Hamiltonian of a molecular system must be invariant in a symmetry operation that leads to a simultaneous inversion of the coordinates of all the particles (electrons and nucleons of the nuclei) and that belongs to the symmetry group C_i isomorphic to the symmetry group S_2. The irreducible representations of group S_2 are one-dimensional and denoted by $\Gamma(2)$ and $\Gamma(1^2)$ for the symmetric and antisymmetric representations, respectively. The wave functions belong to one or the other of these representations and are designated as positive or negative states.

Generally speaking, each symmetric group S_n (of order $n!$) has two one-dimensional irreducible representations, denoted by $\Gamma(n)$ and $\Gamma(1^n)$, for the symmetric and antisymmetric representations, respectively. $\Gamma(n)$ is homomorphic to S_n for $n > 2$ and $\Gamma(1^n)$ is homomorphic to S_n for $n > 3$. The other irreducible representations have dimensions above 1 and are isomorphic to S_n. The latter are known as faithful representations.

Furthermore, given that the identical elementary particles are not discernible, this involves that the Hamiltonian of a system of n indiscernible particles (electrons and nucleons of nuclei) commutes with each operator of the symmetric group S_n so that S_n is a symmetry group of the Hamiltonian. According to Pauli's principle, electrons, nucleons and all the half spin elementary particles belong to the fully antisymmetric irreducible representation $\Gamma(1^n)$. Indeed, if the irreducible representation was fully symmetric $\Gamma(n)$, Bose–Einstein condensate could be possible, which would lead to the formation of aggregates generating a single atom structure and the absence of molecules with a structure.

It is worth noting that $\Gamma(n)$ and $\Gamma(1^n)$ are the only stable irreducible representations to which electrons or nucleons can belong.

1.6.3. *Complete nuclear permutation group (G^{CNP}) of a molecule*

The CNP group of a molecule is a group that contains all the possible permutations of identical nuclei in the molecule. If the chemical formula of the molecule is $A_tB_mC_n$ • • •, the CNP group is the direct product group, $G^{CNP} = S_t \otimes S_m \otimes S_n \otimes \ldots$, and the order of the group is $t!$ x $m!$ x $n!$ x, etc.

A permutation operation (*123*) on the three H nuclei of the CH_3D molecule:

$$
\begin{aligned}
&(123)\,[X_1,Y_1,Z_1,X_2,Y_2,Z_2,X_3,Y_3,Z_3]\\
&=[\tilde{X}_1,\tilde{Y}_1,\tilde{Z}_1,\tilde{X}_2,\tilde{Y}_2,\tilde{Z}_2,\tilde{X}_3,\tilde{Y}_3,\tilde{Z}_3]\\
&=[X_2,Y_2,Z_2,X_3,Y_3,Z_3,X_1,Y_1,Z_1]
\end{aligned}
$$

where (X_i, Y_i, Z_i) are the initial fixed coordinates of H_i nucleus in the space and $(\tilde{X}_i,\tilde{Y}_i,\tilde{Z}_i)$ are its coordinates after permutation. For example:

$$G^{CNP} = S_3 \text{ for } CH_3D \text{ and } G^{CNP} = S_2 \otimes S_2 \text{ for } H_2O_2$$

1.6.4. *Inversion group ε and inversion operations* **E*** *and permutation-inversion operations* **P***

The inversion operation E^* is defined as the inversion operation of the spatial coordinates of all the particles (nuclei and electrons) of a molecule through its center of mass (Figure 1.6.). Using the system of axes (X, Y, Z) fixed in space with the origin in the center of mass, the following can be written: the inversion group $\varepsilon = \{E, E^*\}$.

A permutation-inversion (PI) operation is denoted by P^*. When the inversion operation E^* commutes with an arbitrary permutation operation P, the (PI) operation can be written in the form $P^* = PE^* = E^*P$.

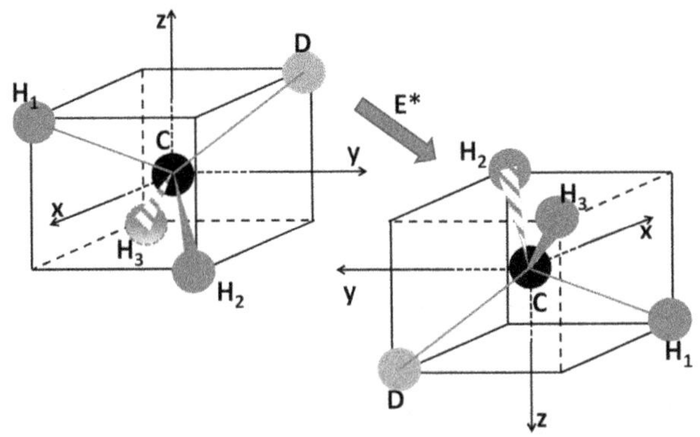

Figure 1.6. *The effect of operation E* on a CH_3D molecule. For a color version of this figure, see www.iste.co.uk/dahoo/infrared.zip*

1.6.5. *Permutation-inversion group G^{CNPI}*

The CNPI (complete nuclear permutation-inversion) group G^{CNPI} is built by the direct tensor product of the G^{CNP} group and of the inversion group ε. $G^{CNPI} = G^{CNP} \otimes \varepsilon$. The CNPI group of a molecule is a group that contains all the possible permutations of identical nuclei in the molecule with and without inversion. The order of the CNPI group is twice that of the CNP group. For example:

$$C_2H_4 \quad G^{CNPI} = G^{CNP} \otimes \varepsilon = S_2 \otimes S_4 \otimes \varepsilon$$

$$H_2O \quad G^{CNPI} = G^{CNP} \otimes \varepsilon = S_2 \otimes \varepsilon = \{E, (12), E^*, (12)^*\}$$

The theory of PI groups makes it possible to determine the spectroscopic selection rules and to calculate the statistical weights of nuclear spin as well as the number and the degeneracy lifting of sub-levels due to tunneling effect. It can therefore be used to predict and analyze the rotation spectra. The complete permutation-inversion (PI) group of a molecule comprises the identity operation E, all the permutations P of identical nuclei, the inversion E^* and the product of E with all the permutations P of identical nuclei. Nevertheless, for many molecules, it is not necessary to take all the elements into account. Limiting the group to achievable operations leads to the “molecular symmetry” (MS) group. The feasibility of an operation with respect to the tunneling effect is limited by the experimental time scale, hence by the resolution of the experiment. The MS group is therefore a subgroup of the complete (PI) group and can be adapted to the problem under study. For nonlinear rigid molecules, the group is isomorphic to the point group of the molecule. The operations of the point-group symmetry have an effect on the vibration and electronic coordinates of a molecule and are denoted by: C_n for rotation, σ for reflection, i for inversion and S_n for improper rotation. These operations are described by permutation, inversion and inversion-permutation in the theory of PI groups, which act on the rotation and spin coordinates. The group SO(3) termed as "molecular" [BUN 98] must also be considered. It contains the rotation group isomorphic to the group of molecular symmetry.

1.6.6. *Group SO(3) isomorphic to permutation-inversion group G^{CNPI}*

In the set of subgroups of SO(3), the cyclic groups as well as the dihedral groups are specific in the sense that their elements can be reduced to two representative operators, either a rotation by β about an axis z attached to the molecule in the mobile reference frame, or a rotation by π about an axis in the plane xy, which forms an angle α with the x axis. For the molecules belonging to groups of symmetry isomorphic to one cyclic group or one dihedral group, it is generally assumed that the rotation motion is approximately described by a symmetric rigid rotator. The eigenfunctions $|JKM\rangle\rangle$ of the rigid rotator can therefore be used as basis states for the solution of Schrödinger's molecular equation. The action of operators representative for the cyclic and dihedral groups is given by:

$$R_z^{\beta}|JKM\rangle = e^{iK\beta}|JKM\rangle \quad [1.30]$$

$$R_{\alpha}^{\pi}|JKM\rangle = (-1)^J e^{-2iK\alpha}|J{-}KM\rangle \quad [1.31]$$

The transformations of Euler angles by these rotations are given in Table 1.10.

R_{α}^{π} is a (direct) rotation of the reference frame attached to the molecule (mobile reference frame: $Oxyz$) by an angle π about the axis making an angle α with the Ox axis. R_z^{β} is a (direct) rotation of the reference frame attached to the molecule (mobile reference frame: $Oxyz$) by an angle β about the Oz axis. The effect of these rotation operators on the Euler angles can be determined by multiplying the rotation matrix corresponding to these operators by the transformation matrix depending on Euler angles (see section 2.7.2 of Chapter 2, Volume 2 [DAH 19]), allowing the switch from the fixed reference frame (OXYZ) to the mobile reference frame ($Oxyz$). Based on the matrix product, the elements are expressed as a function of new Euler angles, allowing the move from the fixed reference frame (OXYZ) to the mobile reference frame ($Oxyz$), taking into account the rotation R_{α}^{π} or R_z^{β}.

The rotation matrices corresponding to the two operators are given by:

$$R_z^\beta = \begin{pmatrix} cos\beta & sin\beta & 0 \\ -sin\beta & cos\beta & 0 \\ 0 & 0 & 1 \end{pmatrix} \qquad [1.32]$$

$$R_\alpha^\pi = \begin{pmatrix} cos2\alpha & sin2\alpha & 0 \\ sin2\alpha & -cos2\alpha & 0 \\ 0 & 0 & -1 \end{pmatrix} \qquad [1.33]$$

	$\mathbf{R}_\alpha^\pi$	$\mathbf{R}_z^\beta$
θ	π-θ	θ
φ	π+φ	φ
χ	2π-2α-χ	χ+β
J_x	$J_x\cos2\alpha+J_y\sin2\alpha$	$J_x\cos\beta+J_y\sin\beta$
J_y	$J_x\sin2\alpha-J_y\cos2\alpha$	$-J_x\sin\beta+J_y\cos\beta$
J_z	$-J_z$	J_z

Table 1.10. *Transformations of Euler angles and components of rovibronic angular momenta during rotations R_α^π and R_z^β*

Consequently, the representations $D^{(J)}$ such that $D^{(J_1)} \otimes D^{(J_2)} = D^{(J_1+J_2)} \oplus D^{(J_1+J_2-1)} \oplus \ldots \oplus D^{(|J_1-J_2|)}$ (vector addition) and generated by functions $|JKM\rangle$ for a fixed J and $K = -J, -J+1, \ldots, J$, can be subdivided into a one-dimensional representation for $K = 0$ (type A) and two-dimensional representation $|J{-}|K|M\rangle$ and $|J|K|M\rangle$ when $K \neq 0$ (type E). The characters are given by equations [1.30] and [1.31].

The molecular symmetry under rotation is thus double. There is a spatial group SO(3), in which the application of an external field may remove the degeneracy in M, and there is a molecular group SO(3), which contains the rotation, a group that is isomorphic to the molecular symmetry group. In Bunker and Jensen [BUN 98], these groups are referred to as K (spatial) and K (mol), respectively. Consequently, the set of functions$(2J+1)^2$ $|JKM\rangle$, here K and M, span each over the set $\{-J, -J+1, \ldots, J\}$, cover the irreducible representations $D^{(J)}$ in spatial SO(3) and the restricted $D^{(J)}$ covers the molecular symmetry group. The isomorphisms of the molecular symmetry groups of A_4, A_5 or S_4 cannot be found by the equivalent rotation technique of Longuet-Higgins [LON 63]. It is nevertheless possible to find the respective isomorphisms with an appropriate approach [HOU 63].

1.7. Correlation between the rotation group and a point-group symmetry of a molecule

The Hamiltonian of a molecule must be invariant under a rotation operation (SO(3) group) and an inversion operation with respect to the origin (I) so that the Hamiltonian must be invariant in a rotation-inversion symmetry operation (O(3) group). Each representation of a rotation group SO(3) generates two representations of the group O(3). The representations of type $D^{(J)}$ (of dimension $2J+1$) in SO(3) for which the character of the operation I is equal to 1, respectively – 1, lead to representations of type $D^{(J+)}$ (positive levels), respectively $D^{(J-)}$ (negative levels), in O(3). When considering the terms that should be added to the zeroth-order Hamiltonian in order to describe the full Hamiltonian of the molecule, it can be considered that these terms constitute a perturbation susceptible to removing the $2J+1$ degeneracy of an irreducible representation of type $D^{(J+)}$.

If it is possible to determine the lower-order symmetry group to which the new Hamiltonian including the higher orders belongs, the degenerate levels will be labeled with the terms identifying the irreducible representation of the lower-order symmetry group. The decomposition of the type of irreducible representation of the higher-order group (here the one associated with the $2J+1$-dimensional space of $D^{(J+)}$ representation) must be determined in terms of those of the lower-order group.

The same procedure is applicable to the decomposition of the type of irreducible representation of the rotation symmetry group of the representation $D^{(J+)}$ in the $2J+1$-dimensional space, in terms of those of the point symmetry group of the molecule (C_{3v}, D_{3h}, T_d, C_{2v}) or of the symmetry group of the environment (point group of a nanocage or of a trapping surface) in which the molecule is placed. In this case the reduction formula [1.23] is used:

$$a_\alpha = \frac{1}{h}\sum_{O_i \in G} h_i \chi(O_i)\left(\chi^\alpha(O_i)\right)^* .$$

Consequently, this requires the identification of the rotational symmetry operations contained in the molecular symmetry group or that imposed by the environment that limits the degrees of freedom of the molecule.

In a three-dimensional crystal structure, the only rotation operations that allow the paving of the three-dimensional space are the rotations of type C_2, C_3, C_4 and C_6. In the case of the symmetry group C_{3v}, the symmetry axes C_3 must therefore be considered, which amounts to considering the subgroup C_3 of the group C_{3v}. Additionally, the second-order rotation symmetry axes of type C_2 perpendicular to the symmetry planes σ_v can also be considered, because a symmetry operation of type σ_v can always be written in the form of a product $i \otimes C_3$.

The characters of the rotation symmetry operations C_2, C_3, C_4 and C_6 are summarized in Table 1.11 for j varying from 0 to 5. They are calculated with the formula: $\chi^J(\varphi) = \frac{sin(j+1/2)\varphi}{sin(1/2)\varphi}$, where φ takes the values π, $2\pi/3$, $\pi/2$ and $\pi/3$ for the operations C_2, C_3, C_4 and C_6, respectively.

J =	**0**	**1**	**2**	**3**	**4**	**5**
$\chi^j(C_2)$	1	-1				
$\chi^j(C_3)$	1	0	1			
$\chi^j(C_4)$	1	1	-1	-1		
$\chi^j(C_6)$	1	2	1	-1	-2	-1

Table 1.11. *Characters of the rotation symmetry operations of type C_2, C_3, C_4 and C_6 in the irreducible representation $D^{(j)}$ of dimension 2j+1*

Considering a nanocage characterized by the symmetry group of type $O_h = I \otimes O$, for example, a substitution site of a face-centered cubic (fcc) structured noble gas matrix, the type of symmetry of the rotation levels of a molecule trapped in the site can be determined as follows. Just as for the irreducible representations of type $D^{(J+)}$ and $D^{(J-)}$ of the group K_h (O(3) group) that can be built from those of group K (SO(3) group), it is sufficient to consider group O in order to establish the symmetries of the rotation levels in the substitutional site and then add the suffix + or – to the types of symmetry when the character of the inversion operation I is equal to 1 or – 1.

Characters of operations of O in the representation $D^{(J)}$ of the rotation group K or SO(3)						**Resolution of $D^{(J)}$ into irreducible representations of the group O**
J	**E**	**C_2**	**C_4^2**	**C_3**	**C_4**	
0 (S)	1	1	1	1	1	A_1
1 (P)	3	-1	-1	0	1	F_1
2 (D)	5	1	1	-1	-1	$E + F_2$
3 (F)	7	-1	-1	1	-1	$A_2 + F_1 + F_2$
4 (G)	9	1	1	0	1	$A_1 + E + F_1 + F_2$
5 (H)	11	-1	-1	-1	1	$E + 2F_1 + F_2$
6 (I)	13	1	1	1	-1	$A_1 + A_2 + E + F_1 + 2F_2$
12 (O)	25	1	1	1	1	$2A_1 + A_2 + 2E + 3F_1 + 3F_2$

Table 1.12. *Table of correlation and characters of symmetry operations of group O in the irreducible representation $D^{(j)}$ of the rotation group SO(3) of dimension 2j+1*

The group O, of order 24, comprises the symmetry operations E, $C_2(6)$, $C_3(8)$, $C_4^2(3)$ and $C_4(6)$. In the symmetry group SO(3), the two operations C_4^2 and C_2 are equivalent. The characters summarized in Table 1.11 are used to determine those of operations of group SO(3) in the representation $D^{(J)}$. Table 1.12 gives the characters of the classes of group O in the representation $D^{(J)}$ of dimension $2J + 1$ of the rotation group as well as the correlations of irreducible representations of O in $D^{(J)}$.

Since 12 is the least common multiple of 2, 3 and 4 associated with operators C_2, C_3 and C_4, it is possible to establish the table of correlation between the group O and the group $D^{(J)}$ up to 13. When examining the decomposition of characters of classes of group O in the representation $D^{(J)}$ of dimension $2J + 1$, it can be noted that the character is given by the number 25 for the identity representation E and that it corresponds to the dimension of the representation $D^{(J)}$, for $J = 12$ since $2J + 1 = 25$. For $J = 12$, the character of each rotation is equal to 1. This for $J = 12$, the characters are the sum of two sets, that of the identity representation for which the character of each operation is 1, and the character of E in O which must be equal to 24 (order of the group), the characters of other symmetry operations being all equal to zero. For the decomposition of the representation $D^{(J)}$, for $J = 12$, $D^{(12)} = A_1 \oplus reg$ (where reg is the regular representation [HAM 62] for which each irreducible representation appears once multiplied by the dimension of the space such that $reg = A_1 \oplus A_2 \oplus 2E \oplus 3F_1 \oplus 3F_2$). Moreover, if the following relation is written as: $D^{(J)} \oplus D^{(11-J)} = reg$ then $D^{(6)} \oplus D^{(5)} = reg$. The table can thus be filled up to $J = 5$ from which the decompositions for $J = 6, 7, 8, 9, 10$ and 11 can be deduced according to the formula. For $J = 12m$, it can be written that:

$$D^{(12m+p)} = m(reg) \oplus D^{(p)} \text{ where } p < 13$$

The table of correspondence between the symmetry group T_d of the spherical top CH_4 and the representation $D^{(J)}$ can similarly be built. The symmetry operations are E, $C_3(8)$, $S_4^2(3)$, $\sigma_d(6)$ and $S_4(6)$ (order 24). The correspondence between the group O and the group T_d makes it possible to easily establish Table 1.13.

The table of correspondence between the symmetry group C_{3v} of the symmetric top NH_3 and the representation $D^{(J)}$ can be established by the same method. The symmetry operations are E, $C_3(2)$ and $\sigma_v(3)$. Either the subgroup C_3 of C_{3v} is used, or the rotation operations equivalent to the symmetry operations $C_3(2)$ and $\sigma_v(3)$, i.e. $R_z^{2\pi/3}$ and $R_{\pi/2}^{\pi}$, are used to build the table of correlation between the rotation group and the group C_{3v}. Using a similar approach to that for group O, Table 1.14 can be established.

Characters of operations of T_d in the representation $D^{(J)}$ of the rotation group K or SO(3)						**Resolution of $D^{(J)}$ into irreducible representations of the group *Td***
Equivalent Rotations	E	C_2	C_4^2	C_3	C_4	
J	E	σ_d	S_4^2	C_3	S_4	
0 (S)	1	1	1	1	1	A_1
1 (P)	3	-1	-1	0	1	F_1
2 (D)	5	1	1	-1	-1	$E + F_2$
3 (F)	7	-1	-1	1	-1	$A_2 + F_1 + F_2$
4 (G)	9	1	1	0	1	$A_1 + E + F_1 + F_2$
5 (H)	11	-1	-1	-1	1	$E + 2F_1 + F_2$
6 (I)	13	1	1	1	-1	$A_1 + A_2 + E + F_1 + 2F_2$
7 (J)	15	-1	-1	0	-1	$A_2 + E + 2F_1 + 2F_2$
8 (K)	17	1	1	-1	1	$A_1 + 2E + 2F_1 + 2F_2$
9 (L)	19	-1	-1	1	1	$A_1 + A_2 + E + 3F_1 + 2F_2$
10 (M)	21	1	1	0	-1	$A_1 + A_2 + 2E + 2F_1 + 3F_2$
11 (N)	23	-1	-1	-1	-1	$A_2 + 2E + 3F_1 + 3F_2$
12 (O)	**25**	**1**	**1**	**1**	**1**	$\mathbf{2A_1 + A_2 + 2E + 3F_1 + 3F_2}$

Table 1.13. *Table of correlation and characters of the symmetry operations of group T_d in the irreducible representation of $D^{(j)}$ of the rotation group SO(3) of dimension 2j+1*

Characters of operations of C_{3v} in the representation $D^{(J)}$ of the rotation group K or SO(3)				**Resolution of $D^{(J)}$ into irreducible representations of group C_{3v}**
Equivalent Rotations	R^0	$R_z^{2\pi/3}$	$R_{\pi/2}\pi$	
J	E	$2C_3$	$3\sigma_v$	
0 (S)	1	1	1	A_1
1 (P)	3	0	-1	$A_2 + E$
2 (D)	5	-1	1	$A_1 + 2E$
3 (F)	7	1	-1	$A_1 + 2A_2 + 2E$
4 (G)	9	0	1	$2A_1 + A_2 + 3E$
5 (H)	11	-1	-1	$A_1 + 2A_2 + 4E$
6 (I)	13	1	1	$3A_1 + 2A_2 + 4E$

Table 1.14. *Table of correlation and characters of symmetry operations of group C_{3v} in the irreducible representation $D^{(j)}$ of the rotation group SO(3) of dimension 2j+1*

The representations $D^{(J)}$ are generated by rotational functions of symmetric tops $|JKM\rangle$. These representations can be divided into a one-dimensional representation when $K = 0$ and a two-dimensional representation for each pair $|J,|K|,M\rangle$ and $|J,-|K|,M\rangle$. The characters of the symmetry operations of group T in this basis are summarized in Table 1.15.

T	$\chi^j(E)$	$\chi^j(C_2)$	$\chi^j(C_3)$	$\chi^j(C_3^2)$
$J = 6p + k$	12p+2k+1	$(-1)^k$	$(2/\sqrt{3})\sin(2(1-k)/3)$	$(2/\sqrt{3})\sin(2(1-k)/3)$

Table 1.15. *Table of characters of representation of group T generated by the basis /JKM⟩. p,k=,1,2,3 … in the irreducible representation $D^{(j)}$ of dimension 2j+1*

When determining the statistical weight of the vibration–rotation levels for a given electronic and vibration state, tensor products of the irreducible representations of the rotation functions and spin functions must be evaluated. Tables X–12 and X–13 of reference [WIL 55] provide the rules of multiplication of direct products of irreducible representations.

By applying these rules, it is possible to build the following tables for the groups C_{3v}, D_{3h} and T_d applicable to symmetric and spherical tops.

C_{3v}	A_1	A_2	E
A_1	A_1	A_2	E
A_2	A_2	A_1	E
E	E	E	$A_1 \oplus A_2 \oplus E$

Table 1.16. *Direct product of irreducible representations of group C_{3v}*

D_{3h}	A'$_1$	A'$_2$	E'	A''$_1$	A''$_2$	E''
A'$_1$	A'$_1$	A'$_2$	E'	A''$_1$	A''$_2$	E''
A'$_2$	A'$_2$	A'$_1$	E'	A''$_2$	A''$_1$	E''
E'	E'	E'	A'$_1$ ⊕ A'$_2$⊕ E'	E''	E''	A''$_1$ ⊕ A''$_2$⊕ E''
A''$_1$	A''$_1$	A''$_2$	E''	A'$_1$	A'$_2$	E'
A''$_2$	A''$_2$	A''$_1$	E''	A_2	A_1	E
E''	E''	E''	A''$_1$ ⊕ A''$_2$⊕ E''	E'	E'	A'$_1$ ⊕ A'$_2$⊕ E'

Table 1.17. *Direct product of irreducible representations of group D_{3h}*

T_d	A_1	A_2	E	T_1	T_2
A_1	A_1	A_2	E	T_1	T_2
A_2	A_2	A_1	E	T_2	T_1
E	E	E	$A_1 \oplus A_2 \oplus E$	$T_1 \oplus T_2$	$T_1 \oplus T_2$
T_1	T_1	T_2	$T_1 \oplus T_2$	$A_1 \oplus E \oplus T_1 \oplus T_2$	$A_2 \oplus E \oplus T_1 \oplus T_2$
T_2	T_2	T_1	$T_1 \oplus T_2$	$A_2 \oplus E \oplus T_1 \oplus T_2$	$A_1 \oplus E \oplus T_1 \oplus T_2$

Table 1.18. *Direct product of irreducible representations of group T_d (for group T, remove the indices 1 and 2)*

1.8. Example of group theory applications

An application of symmetry operations involves the determination of the existence of electric dipole moments in a molecule taking into account the symmetry of its structure. Given that a symmetry operation leaves a molecule in a configuration that is indiscernible from the one it had before the transformation, the direction of the dipole moment vector (polar vector) must be invariant with respect to the operation.

Consequently:

– If the molecule has an *n*-th order symmetry axis, C_n, the permanent dipole moment must be along this axis.

– If the molecule has a symmetry plane, its dipole moment must be in this plane.

– If there are several symmetry planes, the moment must be at the intersection of the planes of symmetry (in NH_3, it is along the axis C_3).

– If the molecule has two or more axes of symmetry, it has no permanent dipole moment (CH_4).

– If the molecule has a center of symmetry, it has no dipole moment, as the inversion changes the direction of a vector.

The molecules whose structures cannot be superimposed to their images in a mirror (enantiomer) can rotate the plane of polarization of light, as the two possible forms (isomer) have opposite optical properties. A molecule that has a symmetry axis S_n is optically inactive ($S_1 = i$ and $S_2 = \sigma$). In general, the absence of axis S_n involves the existence of an optical activity.

1.9. Conclusion

The groups generated by geometric transformations are not the only ones that can be applied in physics, as is revealed by the use of permutation groups that are better adapted for the study of molecules composed of identical nuclei. According to Bose and Fermi statistics, the permutations of N identical particles are associated with two different irreducible representations, but they are very easy to deal with, so the group theory is unnecessary. Point-group symmetry is adapted for dealing with electron, vibrational and rotational degrees of freedom. To deal with the exchange of identical nuclei in a molecule or tunneling effect inversion that are observable in experiments, it is more convenient to use the permutation-inversion group. The elements of the theory of group representation and the methods developed in group theory were presented for the study of molecules from their point-group symmetry. The theory of permutation-inversion groups was also introduced. These methods will be applied in Chapters 2 and 3 in order to determine the symmetries of the wave functions of vibration, rotation and spin degrees of freedom of the various types of molecules considered as well as the selection rules leading to IR spectra of these molecules.

1.10. Appendices: Groups and Lie algebra of SU(2) and SO(3)

1.10.1. *Appendix A: Groups SU(2) and SO(3)*

The study of angular momenta involves operators that can act in a quantum space (spin space with $S = 1/2$) or in a physical space (rotation space with $J = 1$). There is a correspondence between the rotation group SO(3) defined in the physical space of rotations and the group SU(2) defined in the quantum space of spins. The following elements can be found in the references [HAM 62] (see Chapter 8) or [GIL 74] (see Chapter 5) or [LIP 65] (see Chapters 1 and 2).

1.10.1.1. *The relation between the group SU(2) and the group SO(3)*

SU(2) is the set of complex 2 × 2 matrices, M, of determinants equal to + 1 and which have the property that: $M^\dagger M = I$

$$M = \begin{pmatrix} a & b \\ -b^* & a^* \end{pmatrix} \text{ with } |a|^2 + |b|^2 = 1 \qquad [1.34]$$

Only three independent parameters are needed to determine the elements of the matrix M of SU(2). A commonly used form is given by:

$$M = \begin{pmatrix} \cos\frac{\theta}{2} e^{\frac{i}{2}(\psi+\phi)} & \sin\frac{\theta}{2} e^{\frac{i}{2}(\psi-\phi)} \\ -\sin\frac{\theta}{2} e^{\frac{-i}{2}(\psi-\phi)} & \cos\frac{\theta}{2} e^{-\frac{i}{2}(\psi+\phi)} \end{pmatrix} \quad [1.35]$$

with $0 \leq \theta \leq \pi$, $0 \leq \psi \leq 4\pi$ and $0 \leq \phi \leq 2\pi$.

The 2 x 2 matrix that is a representation of the rotation group O(3) is known as the irreducible representation $D_{1/2}$.

The vectors transformed in SU(2) are spinors that can form a basis for the spin operator $\widehat{S}$. The spinors are used to represent the half-spin fermions [LAN 75] used in Dirac's theory (Dirac used Pauli matrices to build 4 x 4 matrices defined from Dirac's[1] equation and referred to this four-dimensional space as the bispinor space, which comprises the particles and the anti-particles; the study is limited to the space of particles) which can be denoted as Φ_α where $\alpha = 1, 2$. An analogy can be drawn with the *3D* vectors in the case of the rotation group SO(3). It can be written as $\Phi_\alpha^{'} = M_{\alpha\beta}\Phi_\beta$.

There is a homomorphism between group SU(2) and group SO(3) (where each element of SO(3) has two corresponding elements in SU(2)). If the image of $M \in$ SU(2) is $R(M) \in$ SO(3), then $R(M) = R(-M)$ The application of SU(2) in SO(3) can be chosen in the following form:

$$R_{ik}(M) = \frac{1}{2} Tr\left(\sigma_i M \sigma_k M^{-1}\right) \text{ with } i,k = 1,2,3 \quad [1.36]$$

where σ_i are Pauli matrices defined by:

$$\sigma_1 = \begin{pmatrix} 0 & 1 \\ 1 & 0 \end{pmatrix}, \sigma_2 = \begin{pmatrix} 0 & -i \\ i & 0 \end{pmatrix} \text{ and } \sigma_3 = \begin{pmatrix} 1 & 0 \\ 0 & -1 \end{pmatrix} \quad [1.37]$$

Therefore, the representations of SO(3) are representations of SU(2). However, the spinor representations of SU(2) have no analogue in SO(3). The representation of SU(2) by 2 x 2 matrices corresponds to the lowest dimension of spinor representations of SU(2).

1.10.1.2. *Representations of SU(2) and SO(3)*

The representation of SU(2) is the one used in the theory of angular momenta in quantum mechanics.

1 Dirac used Pauli matrices to construct the defined 4 x 4 matrices of the Dirac equation, and called this four-dimensional space the bispinor space, which includes particles and antiparticles; we therefore restrict ourselves to particle space.

Indeed, kets are defined from eigenfunctions common to operators l^2 and l_z denoted by $|lm\rangle$ with $l = 0, \frac{1}{2}, 1, \frac{3}{2}, 2, \ldots$ and $m = -l, -l+1, \ldots, l-1, l$. For each l, there is a corresponding distinct irreducible representation of SU(2) and the number of possible values of m, which is $2l + 1$, gives the dimension of the representation.

The representations for which l is an integer number are equivalent to the representations of SO(3) and correspond to the representations of bosons, and those for which l is a half-integer number are equivalent to the representations of SU(2) and correspond to the representations of fermions.

1.10.2. *Appendix B: Lie algebra and SO(3)*

1.10.2.1. *Definition of an algebra*

An algebra is a vector space E over the field K with a law of composition or vector product $E \otimes E \rightarrow E : (a,b) \rightarrow ab \in E, \ (A,b \in E)$ that obeys the following distributive laws:

$$a(\lambda b + \mu c) = \lambda ab + \mu ac \text{ and } (\lambda a + \mu b)c = \lambda ac + \mu bc \qquad [1.38]$$

where $(\lambda,\mu) \in K$ and $(a,b,c) \in E$.

1.10.2.2. *Generators of elements of group SO(3)*

The following 3 x 3 matrix is the starting point:

$$J_x = \begin{pmatrix} 0 & 0 & 0 \\ 0 & 0 & -i \\ 0 & i & 0 \end{pmatrix} \qquad [1.39]$$

The following expansion can be written as:

$$e^{i\phi J_x} = 1 + i\phi J_x + \frac{1}{2!}(i\phi J_x)^2 + \ldots = \begin{pmatrix} 1 & 0 & 0 \\ 0 & \cos\phi & \sin\phi \\ 0 & -\sin\phi & \cos\phi \end{pmatrix} \qquad [1.40]$$

which is a rotation about the Ox axis. J_x is said to be a generator of rotations about the Ox axis. The term generator takes its root from the way the operator is built. An infinitesimal rotation by an angle $\delta\phi$ about the Ox axis is proportional to ϕ and the complete rotation is realized (or generated) by composing an infinity of infinitesimal rotations such that:

$$R_x(\phi) = lim_{N\to\infty}\left(1 + i\frac{\phi}{N}J_x\right)^N \to e^{i\phi J_x} \quad [1.41]$$

Similarly, the rotations about Oy and Oz axes are generated by:

$$J_y = \begin{pmatrix} 0 & 0 & -i \\ 0 & 0 & 0 \\ i & 0 & 0 \end{pmatrix} \text{ and } J_z = \begin{pmatrix} 0 & -i & 0 \\ i & 0 & 0 \\ 0 & 0 & 0 \end{pmatrix} \quad [1.42]$$

The set of three generators J_x, J_y and J_z of SO(3) forms a Lie algebra.

1.10.2.3. *Definition of a Lie algebra*

A real Lie algebra denoted by L, of dimension n ($n \geq 1$), is a real vector space with a Lie product or a commutator $[a, b]$ defined for each element a and b of L such that:

$[a,b] \in L$ for any a and $b \in L$ [1.43a]

$[\alpha a + \beta b,c] = \alpha[a,c] + \beta[b,c]$ for any a and $b \in L$ and any real α and β [1.43b]

$[a,b] = -[b,a]$ for any a and $b \in L$ [1.43c]

$[a,[b,c]] + [b,[c,a]] + [c,[a,b]] = 0$ for any a, b and $c \in L$ [1.43d]

The last property is the Jacobi identity.

1.10.2.4. *Lie algebra of SU(2) and SO(3)*

The algebra of SO(3) generator commutators verifies:

$$[J_x, J_y] = iJ_z \quad [1.44a]$$

$$[J_y, J_z] = iJ_x \quad [1.44b]$$

$$[J_z, J_x] = iJ_y \quad [1.44c]$$

$$[J_i, J_j] = i\varepsilon_{ijk}J_k \quad [1.44d]$$

where J_x, J_y and J_z constitute the basis of the vector space and ε_{ijk} is the Levi-Civita tensor, $\varepsilon_{123} = \varepsilon_{312} = \varepsilon_{213} = 1$ and $\varepsilon_{132} = \varepsilon_{321} = \varepsilon_{231} = -1$ (where $x = 1$, $y = 2$ and

$z = 3$). Each rotation operator is obtained by the integration of three independent infinitesimal rotation operators:

$$1 + i\frac{\phi}{N}J_i,\ i = x,y \text{ or } z$$

For example, the π mesons form a triplet of isotopic spin $I = 1$. The infinitesimal generators are 3 x 3 matrices represented by structure constants of the Lie algebra.

Similarly in SU(2), the following relations between Pauli matrices can be written as:

$$\left[\frac{\sigma_1}{2}, \frac{\sigma_2}{2}\right] = i\frac{\sigma_3}{2} \quad [1.45a]$$

$$\left[\frac{\sigma_2}{2}, \frac{\sigma_3}{2}\right] = i\frac{\sigma_1}{2} \quad [1.45b]$$

$$\left[\frac{\sigma_3}{2}, \frac{\sigma_1}{2}\right] = i\frac{\sigma_2}{2} \quad [1.45c]$$

$$\left[\frac{\sigma_i}{2}, \frac{\sigma_j}{2}\right] = i\varepsilon_{ijk}\frac{\sigma_k}{2} \quad [1.45d]$$

where $\frac{\sigma_1}{2}, \frac{\sigma_2}{2}$ and $\frac{\sigma_3}{2}$ constitute the basis of the vector space.

Though the groups SU(2) and SO(3) are not isomorphic, the Lie algebras of SU(2) and SO(3) are isomorphic. The nucleons form a doublet of isotopic spin $I = 1/2$. The infinitesimal generators are Pauli matrices.

2

Symmetry of Symmetric and Spherical Top Molecules

Molecular vibration–rotation spectroscopy relies on the analytical solutions of Schrödinger's eigenvalue equation, which serves to calculate quantum energy levels and the associated eigenfunctions. The molecular spectra are then determined from the possible transitions between these eigenstates, considering the symmetry and population of the levels. The symmetry properties resulting from the molecular structure influence the profile of its absorption, emission or diffusion spectrum. Rotational spectroscopy is used for the determination of the molecular structure, while vibrational spectroscopy is used for establishing the fields of forces responsible for chemical bonds. Group theory can be applied to identify the type of symmetry characterizing the vibration and rotation degrees of freedom and consequently, to predict the transitions that can be observed when the molecule interacts with light. These transitions correspond to selection rules, either in absorption or emission spectroscopy or in Raman spectroscopy. In order to establish the vibrational–rotational Hamiltonian of the molecules in molecular physics, the theoretical methods developed by Nielsen and Amat use Van Vleck's contact transformations. The research conducted by Watson and Longuet-Higgins showed the necessity to consider the symmetries of the molecular structure for the correlation between the results of the calculations of molecular spectral signatures and the observations, both for rigid molecules, whose vibrations have low amplitudes, and for non-rigid molecules, whose vibrations have high amplitudes or are characterized by the tunneling effect. For the latter, according to the research conducted by Hougen and Bunker, it is important to distinguish between geometric symmetries and symmetries resulting from nuclei permutations. The concept of invariance during a transformation reflects the importance of symmetries in classical mechanics, as shown by the work of Curie or Noether, or in quantum mechanics, as shown by the work of Wigner.

2.1. Introduction

As mentioned in the previous two volumes [DAH 17, DAH 19], the rovibrational energy levels of a molecule are calculated from the molecular Hamiltonian established by Wilson and Howard [WIL 36], reformulated by Darling and Dennison [DAR 40] and later simplified by Watson [WAT 68a]. The calculations are made using Van Vleck's contact transformation method [VAN 29], which can be applied to an arbitrary molecule based on a general formalism following the works of Nielsen, Amat *et al.* and Goldsmith [AMA 57a, AMA 57b, AMA 58, GOL 56, GOL 57, NIE 51]. This method employs unitary transformations to group the Hamiltonian into interacting polyads and to obtain a matrix form in blocks that is easier to diagonalize for energy calculation. The transformation of the initial Hamiltonian into interacting blocks makes it possible to link the experimental observations to the interacting vibrational levels. The reader is referred to various books or chapters that describe the formalism of the contact transformation for the study of IR vibrational–rotational spectra of polyatomic molecules in molecular physics [ALI 85, AMA 71, CAM 85, PAP 83, WAT 67, WAT 68b, WAT 68c].

While diatomic molecules are characterized by having only one vibrational coordinate, this is not the case for polyatomic molecules. In the general case, a molecule with N atoms has $3N$ degrees of freedom. Three of these degrees of freedom are connected to the overall translation of the molecule and therefore to its center of mass motion and three (or two if the molecule is linear) to the overall rotation of the molecule, these degrees of freedom being generally associated with low frequencies in the far infrared or microwave range. The vibrational degrees of freedom that correspond to the deformation of a molecule can therefore be written as the combination of $3N - 6$ (or $3N - 5$) harmonic vibrations, each of which is described by a normal coordinate Q_s. For example, for $N = 4$, the number of normal coordinates is equal to 6 if the molecule is plane (BF_3) but nonlinear, and is still equal to 6 if the atoms are no longer co-planar (NH_3, PF_3), the molecule being nonlinear. The coordinates are characterized by their symmetries, considering the symmetry properties of molecules which depend on the (planar or not) structure and composition in terms of identical or non-identical nuclei (XY_3, XY_2Z, XYZT) and can be fully symmetric, partially symmetric or non-symmetric. The construction of normal coordinates relies on these symmetries.

In order to take these symmetries into account, a common practice is to apply the methods developed in group theory, in order to determine the consequences of symmetries in the spectroscopic study of molecules and to simplify the molecular Hamiltonian. Chapter 1 reviews the group theory methods used for the study of molecules, with a focus on the theory of representations. The various point symmetry groups of a molecule are reviewed in this chapter, which introduces the molecular symmetry group based on the permutation-inversion group for dealing with the exchange of equivalent nuclei in the molecules presented in this volume, or the molecules known as "non-rigid", which are characterized by the inversion phenomenon by the tunneling effect through the potential barrier (XY_3 (NH_3, ND_3), XYZ_2 (NHD_2, NH_2D), XY_4 (CH_4, CD_4), XY_2Z_2 (CH_2D_2), XY_3Z (CH_3D, CHD_3)). The molecular symmetry group can be used for the classification of the wave functions of a molecule relying on the irreducible representations of the group of operators that commute with the Hamiltonian of the molecular system. The symmetry operations associated with these operators do not change the Hamiltonian and consequently, do not modify the energy of the molecule. A more detailed presentation of the group theory and its applications can be found in the literature [BAR 88, BRA 72, BUN 98, COT 63, CRO 63, DEC 77, HAM 62, HER 45, LAN 75, MAR 83, MIZ 72, TIN 03, WIG 59, WIL 55, WIL 80].

This chapter uses group theory to study the symmetry of ammonia and methane molecules, as well as their isotopes. These molecules are characterized by the presence of the hydrogen atom, the lightest in the periodic table, having high amplitude vibration motions. The focus is on stable isotopes, namely H, D, ^{14}N, ^{15}N, ^{12}C and ^{13}C. The approach that is used here differs from the one used in Chapter 1 of Volume 1 [DAH 17], starting with the structure that is the most symmetric and ending with the one that is the least symmetric, thanks to the inversion phenomenon. For ammonia, the molecules of XY_3 (NH_3, ND_3) type are considered first, followed by those of XYZ_2 (NHD_2, NH_2D) type, going from the most elaborate group to the most simple. A similar approach is used for methane, where the molecules of XY_4 (CH_4, CD_4) and XY_2Z_2 (CH_2D_2) types are considered first, followed by the XY_3Z (CH_3D, CHD_3) type.

The various symmetries resulting from the presence of isotopes are reflected by differences in the spectra, particularly in the infrared range, which corresponds to the electromagnetic frequency range of light–matter interaction inducing transitions between the vibrational energy levels of the nuclei, the vibrational–rotational energy levels of the molecule as a whole, or the vibrational energy levels of the nuclei, taking into account the overall rotation of the molecule.

2.2. Symmetry group of molecular Hamiltonian

For the sake of consistency, this chapter summarizes the framework, already presented in Volumes 1 and 2 of the set *Infrared Spectroscopy* [DAH 17, DAH 19] used for the application of the theoretical model, in order to build the energy operator associated with the vibration–rotation motions of a molecule. The eigenvalues of this operator determine the various vibration–rotation levels between which the transitions induced during the light–matter interaction take place. More details on this method can be found in section 2.2 of Chapter 2 of Volume 2 [DAH 19]. It is important to recall that the general motion of a molecule can be expressed as the sum of a translation motion of its center of mass and of the motion of nuclei and electrons with respect to this center of mass. The Hamiltonian associated with the energy operator (see equation [2.6], Volume 1; equation [2.1], Volume 2) is expressed in a reference frame based on Eckart conditions [ECK 35].

According to the first Eckart relation [ECK 35], the origin of the mobile reference frame $(\mathrm{O}, \vec{x}, \vec{y}, \vec{z})$ associated with the equilibrium configuration of the molecule is at the center of mass G of the nuclei such that:

$$\sum_{i=1}^{N} m_i \vec{\eta}_i = \vec{0} \qquad [2.1]$$

The second Eckart condition [ECK 35] fixes the orientation of the molecule with respect to the instantaneous configuration of the nuclei such that:

$$\sum_{i=1}^{N} m_i \vec{\eta}_i^{\mathrm{e}} \wedge \vec{\eta}_i = \vec{0} \qquad [2.2]$$

In equations [2.1] and [2.2], $\vec{\eta}_i$ and $\vec{\eta}_i^{\mathrm{e}}$ correspond to the instantaneous and equilibrium positions of nucleus i, of mass m_i.

The first condition allows the rigorous separation of the translational degrees of freedom from those related to the vibration–rotation motion, while according to the second Eckart condition, only a partial separation of the vibration and rotation modes is possible.

In order to calculate the energy values E of the molecule, Schrödinger's eigenvalue equation must be solved. This can be expressed as follows:

$$(H - E)|\Psi\rangle = (T_e + T_n + V_{en} - E)|\Psi\rangle = 0 \qquad [2.3]$$

where T_e and T_n are operators associated with the kinetic energies of the electrons and nuclei; V_{en} is the interaction potential energy between electrons, between nuclei and between electrons and nuclei. Since it is impossible to solve this equation, the energy levels are determined by successive approximations, applying the Born-Oppenheimer approximation (see section 2.4.2, Chapter 1, Volume 1) and within the framework of contact transformation, a series expansion of the Hamiltonian operator, depending on the normal coordinates for zeroth-order, first-order, etc. approximations.

According to the Born–Oppenheimer approximation, or the adiabatic approximation, electrons and nuclei motions are studied separately: the motion of electrons is studied for a fixed position of the nuclei (see equation [2.4]) and that of the nuclei is studied in the mean field of electrons (see equation [2.5]).

$$H_e|\Psi_e\rangle = (T_e + V_{en} - E_e)|\Psi_e\rangle = 0 \qquad [2.4]$$

$$H_n|\Psi_n\rangle = (T_n + E_e - E)|\Psi_n\rangle = 0 \qquad [2.5]$$

Equation [2.5] defines the stationary electron states whose energy depends on the coordinates of the nuclei as parameters; E_e is the potential energy function, denoted by V, for the study of vibration–rotation in a given electronic state. These energies are generally calculated in the fundamental electronic state. The Born–Oppenheimer approximation is not valid for electronic states characterized by splitting or resonances caused by couplings between the electronic degrees of freedom (orbital or spin moment) and nuclear degrees of freedom (vibration or rotation moment), which is rarely the case for the fundamental electronic state.

The expression of the zeroth-order Hamiltonian, which does not include the coupling terms, is:

$$H_0 = T_{CM} + H_{eVR} + H_{SS} + H_{II} \qquad [2.6]$$

where T_{CM} is the kinetic energy of the center of mass (CM) of the nuclei, $H_{eVR} = H_e + H_V + H_R$ is the sum of the Hamiltonians relative to the electron motions and to the vibration and rotation motions of the nuclei, H_{SS} (Spin–Spin interaction) is related to the interaction between electron spins and H_{II} (Spin–Spin interaction) is related to the interaction between nuclear spins.

According to Bunker [BUN 98], the symmetry group G of the Hamiltonian of a polyatomic molecule is given in the form of a tensor product of a certain number of symmetry groups (see Chapter 1) associated with various components of the molecule, therefore electrons and nuclei and their degrees of freedom. The expression of this group G is given by:

$$G = G_T \otimes K(spatial) \otimes S_n^{(e)} \otimes G^{CNP} \otimes \varepsilon$$

where G_T is the spatial group associated with the translation of the center of mass; *K(spatial)* is the infinite group of rotations about an axis in a fixed reference frame passing through the center of mass, $S_n^{(e)}$ is the group associated with electron permutations, G^{CNP} is the group associated with nuclei permutations (CNP – Complete Nuclear Permutation) and ε is the group associated with the inversion of coordinates of all the components (electrons and nuclei). This general approach is applied to the total Hamiltonian of the polyatomic molecule, which can be used to find the symmetry operations under which the Hamiltonian is invariant and consequently, does not modify the energy of the molecule. Table 7.3 of [BUN 98] summarizes the characteristics of these symmetry operations. G_T and *K(spatial)* are related to the uniformity and isotropy of the space, $S_n^{(e)}$ and G^{CNP} are connected to the indistinguishable property of identical particles and ε is related to the nature of the electromagnetic force and parity conservation (see Chapter 1). Moreover, the complete group is built from the permutation-inversion group, and the tensor product of the symmetry group ε is given by the permutation group, whose elements are the permutations of identical nuclei of a molecule. The labels of the energy levels of a molecule in quantum mechanics, and therefore the nomenclature of the levels, are assigned in accordance with the symmetry properties of the molecule, using the symbols related to the irreducible representations of the symmetry group of the molecule and can be used for their classification depending on these representations.

With the calculation of the energy levels (eigenvalues) and the wave functions (eigenstates) of a molecule based on the resolution of Schrödinger's equation being impossible, the construction of the approximate Hamiltonian operator by Van Vleck's contact transformation methods, a common practice in molecular physics, leads to the definition of approximate symmetries relying on the symmetry groups, as described in Chapter 1 of Volumes 1 and 2 [DAH 17, DAH 19], for the identification of the symmetries of wave functions and energy levels. Each vibration–rotation state can thus be labelled, when the spin properties are considered, by the indices in correspondence with the symmetry properties of the molecule. The selection rules pertaining to the transitions leading to the infrared or microwave spectra can likewise be determined. These rules are modified when couplings are considered.

From the Hamiltonian in equation [2.6], the wave functions of a polyatomic molecule of XY_3 (NH_3) type or XY_4 (CH_4) type of zeroth order can be expressed in the form of a product and the associated eigenenergies can be written as:

$$|\Psi_0\rangle = |\Psi_{CM}\rangle \times |\Psi_e\rangle \times |\Psi_V\rangle \times |\Psi_R\rangle \times |Sm_s\rangle \times |\mathrm{Im}_I\rangle \quad [2.7]$$

$$E_0 = E_{CM} + E_e + E_V + E_R + E_{SS} + E_{II} \quad [2.8]$$

where $\Psi_{CM} = \langle R_{CM} | \Psi_{CM} \rangle$ is the wave function associated with the translational motion of the center of mass of the molecule represented by a plane wave,

$\Psi_e = \langle ..., r_e, ... | \Psi_e \rangle$ is the wave function associated with the motion of electrons and that depends on the positions of the electrons, $\Psi_V = \langle Q | \Psi_V \rangle$ is the wave function associated with the vibrational motions of the nuclei about their equilibrium position and that depend on the normal coordinates (*3N – 6* for a nonlinear molecule with *N* nuclei) gathered in *Q*, $\Psi_R = \langle \theta, \varphi, \chi | JKM \rangle$ for a nonlinear molecule for the wave function associated with the rotational motion of the whole molecule about its center of mass and $|Sm_s\rangle$ and $|\mathrm{Im}_I\rangle$ are wave functions of the electronic and nuclear spins, respectively.

The wave functions associated with various terms of H_0 (see equation [2.6]) are transformed as irreducible representations (see Chapters 1 and 2, Volumes 1 and 2 [DAH 17, DAH 19]) of groups that act in various spaces of various degrees of freedom (translation, vibration, rotation, spin) and under which the Hamiltonian is invariant. The proper quantum numbers can be chosen in order to identify the various vibration–rotation energy levels.

In order to calculate the vibration–rotation energy levels, Schrödinger's eigenvalue equation to be solved is given by:

$$H_{VR} |\Psi\rangle = (T_{VR} + V) |\Psi\rangle = E |\Psi\rangle \qquad [2.9]$$

where T_{VR} is the vibration–rotation kinetic energy, *V* is the electronic potential, invariant under the translational and rotational motions of the set of nuclei and which does not depend on internuclear distances, and *E* is the vibration–rotation energy of the molecule in a reference frame, denoted by $(\mathrm{G}, \vec{\mathrm{X}}, \vec{\mathrm{Y}}, \vec{\mathrm{Z}})$, in uniform translation with respect to a fixed laboratory reference frame.

In what follows, the symmetry considerations are discussed for the study of the vibration–rotation motions of a symmetric molecular top (NH_3) or spherical molecular top (CH_4) and their isotopes in the fundamental electronic state. The terms of the symmetric or spherical top are related to the rotation energy operator of the rigid rotator, as previously described in section 2.4, Volume 2. For the sake of consistency, they are summarized below.

In molecular spectroscopy, the *Gz* axis of the reference frame associated with the equilibrium configuration of a nonlinear molecule is chosen along the highest rotational symmetry axis. Nevertheless, by convention, the chosen reference frame is composed of three main axes of inertia of the molecule, namely: *Ga*, *Gb* and *Gc*, such that its main momenta of inertia are in ascending order, $I_a^e \leq I_b^e \leq I_c^e$, or in descending order, $A^e \geq B^e \geq C^e$, of its rotational constants defined by:

$$A^e = \hbar^2/2I_a^e, \ B^e = \hbar^2/2I_b^e \text{ and } C^e = \hbar^2/2I_c^e.$$

According to this convention, the axes of the mobile reference frame can be named according to the shape of the considered molecule.

Depending on the values of the main momenta of inertia of a molecule, there are five possible cases:

$$I_a^e = 0, \ I_b^e = I_c^e$$

where the molecule is linear, and the *Gz* axis of the mobile reference frame is the bond axis (in Volume 2: CO_2, N_2O);

$$I_a^e < I_b^e = I_c^e$$

where the molecule is a prolate symmetric top, and the *Gz* axis of the mobile reference frame coincides with the main inertia axis Ga (CH_3F, CH_3D);

$$I_a^e = I_b^e < I_c^e$$

where the molecule is an oblate symmetric top, and the *Gz* axis of the mobile reference frame coincides with the main inertia axis Gc (NH_3, ND_3, CHD_3);

$$I_a^e = I_b^e = I_c^e$$

where the molecule is a spherical top (CH_4, CD_4);

$$I_a^e < I_b^e < I_c^e$$

where the molecule is an asymmetric top, and the *Gz* axis of the mobile reference frame coincides with the main inertia axis Ga (NHD_2, NDH_2, CH_2D_2, in Volume 2: H_2O).

The study of the symmetry properties of the Hamiltonian and the vibration–rotation wave functions can be used to simplify the diagonalization of the matrix associated with the Hamiltonian operator. These symmetry properties result from the existence of identical nuclei in the molecule. The rotation Hamiltonian must be invariant under a rotational symmetry operation (*K(spatial)* group related to space isotropy) with respect to each of the eigen axes C_2^a, C_2^b, C_2^c and therefore $[H_O, C_2^a] = 0$, $[H_O, C_2^b] = 0$ and $[H_O, C_2^c] = 0$ for all the molecules independent of their geometry. This makes it possible to use the quantum numbers *J*, *K* and *M* to identify the rotation levels (see section 2.4, Volume 2). At the zeroth order, the rigid rotor approximation, for fixed *J* and *M*, leads to $[H_0, \mathrm{J}^2] = 0$ and $[H_0, J_Z] = 0$, such that a rotation level is expressed as a linear combination of kets $|J\ K\ M\rangle$.

It is important to note that:

$$\begin{aligned} &\mathrm{J}^2 \left|JKM\right\rangle = \hbar^2 J(J+1)\left|JKM\right\rangle \\ &J_Z \left|JKM\right\rangle = \hbar M \left|JKM\right\rangle \\ &J_z \left|JKM\right\rangle = \hbar K \left|JKM\right\rangle \end{aligned} \qquad [2.10]$$

where *Z* and *z* are relative to the OZ and *Oz* axes of the fixed and mobile reference frames, respectively, and the quantum numbers *M* and *K* are relative to the projections of $\vec{\mathrm{J}}$ operator on the OZ and *Oz* axes.

For a spherical top, the rotational energy levels are given by:

$$E_R^0 = BJ(J+1) \qquad [2.11]$$

where for each *J*, in the absence of the external magnetic field, the *K* states have the same energy (*K* is the projection of the angular moment $\vec{\mathrm{J}}$ on the *Oz* axis of the mobile reference frame fixed to the molecule), with the degeneracy in *K* being $2J + 1$.

For a prolate symmetric top, the rotational energy levels are given by (see equation [2.38a], Volume 2):

$$E_R^0 = BJ(J+1) + (A-B)K^2 \qquad [2.12]$$

where for each *J*, the degeneracy in *K* is 2, except for $K = 0$.

For an oblate symmetric top, the rotational energy levels are given by (see equation [2.38b], Volume 2):

$$E_R^0 = BJ(J+1) + (C-B)K^2 \qquad [2.13]$$

where, for each *J*, the degeneracy in *K* is 2, except for $K = 0$.

For a linear molecule, the rotational energy levels are given by:

$$E_R^0 = B(J(J+1) - l^2) \tag{2.14}$$

where l is the component of the vibrational angular momentum associated with double degenerate bending modes.

The energies of rotational levels can be analytically obtained for all the molecules except for the asymmetric top, for which the matrix of the Hamiltonian is determined on the basis of $|J\ K\ M\rangle$ kets of a symmetric top, depending on J^2 operators and on J_+ and J_- ladder operators, as explained in Volume 2 (see section 2.4.3).

2.3. Symmetry of the NH_3 molecule and its isotopologues ND_3, NHD_2 and NDH_2

The profile of the absorption or emission spectrum of a molecule can be predicted by group theory methods. In Volume 2 [DAH 19], the groups considered for three-atom molecules are built for molecules with a second-order axis of symmetry. These groups have fewer elements than those corresponding to ammonia or methane molecules and their isotopic homologues.

2.3.1. *Symmetry group of the symmetric molecular tops NH_3 and ND_3*

Ammonia, which has four nuclei and is pyramid-shaped, has a threefold axis of rotation (see Figure 2.1). Moreover, the splitting of certain vibrational bands observed in the IR spectra of NH_3 in the gas phase, a consequence of the inversion phenomenon related to the fact that the electronic potential, V, has two minima (see Figure 2.2), requires a higher-order group that has irreducible representations for labeling the energy levels and the corresponding functions.

The expansion of the Hamiltonian H_{VR} (see equation [2.10]) must be carried out in a fixed reference frame, generally denoted by $(\mathrm{O}, \vec{\mathrm{X}}, \vec{\mathrm{Y}}, \vec{\mathrm{Z}})$, related to the laboratory and in a mobile reference frame, denoted by $(\mathrm{G}, \vec{x}, \vec{y}, \vec{z})$, related to the reference equilibrium configuration of the molecule. The symmetry properties of the vibration–rotation Hamiltonian result from the presence of identical nuclei. If the molecule is composed of three identical nuclei, as in NH_3, there are three systems of physically indistinguishable, equivalent mobile axes. The transformations used for switching from one system of axes to another superimpose the equilibrium configuration of the molecule to itself. Each transformation is associated with a change in the numbering of nuclei and constitutes the symmetry operations of the overlapping group of the molecule. These various operations are shown in Figure 1.2(a) and (b).

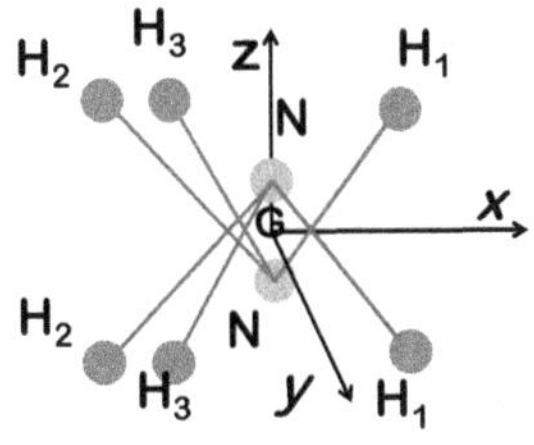

Figure 2.1. *Mobile reference frame and the two equilibrium configurations of the nonlinear molecule NH_3. For a color version of this figure, see www.iste.co.uk/dahoo/infrared.zip*

The space of permutations of indices is a group isomorphic to the symmetry group that is composed of symmetry operations, under which the equilibrium configuration of the molecule is invariant. There are two types of transformations: rotations, with the corresponding transformations of Euler angles in the rotational space (direct operations), and other transformations such as inversion, reflections with respect to a plane and rotations–reflections that reverse the direction of the mobile reference frame (inverse operations). Following the second type of operations, the position of the reference frame fixed to the molecule cannot be identified by Euler angles and requires the application of an additional symmetry operation that reverses the fixed and mobile reference frames. As a result, since the pure rotational Hamiltonian is invariant under this operation, the matrix elements couple the states of similar parity with respect to this operation.

The symmetry elements of NH_3 and ND_3 molecules are a third-order axis of rotation (threefold axis: two rotation operations, denoted by $C_z(2\pi/3)$ and $C_z(4\pi/3)$ or C_3^1 and C_3^2 (see Figures 2.3(a) and Figure 2.4)) about the Oz axis, passing through the nitrogen atom and three planes of reflection symmetry ($\sigma(0)$, $\sigma(2\pi/3)$ and $\sigma(4\pi/3)$), denoted σ_v, σ_v' and σ_v'' (see Figures 1.2(b) and 1.3)). By adding the identity operation E $(C_z(0))$ to these operations, the overlapping group of NH_3 and ND_3 molecules, denoted by C_{3v}, can be built (see Tables 1.2 and 1.5). The Oz axis is usually placed along the highest-order axis of rotation.

Based on the elements on group theory applications reviewed in Chapter 1 (see Figure 1.3), it can be shown (see section 2.4), using irreducible representations of group C_{3v}, that for the molecule with six vibrations ($3n - 6$ degrees of freedom (d.o.f.) of vibration; $12 - 6 = 6$ vibrations), two are non-degenerate of A_1 type and two are degenerate of E type (see Table 1.5).

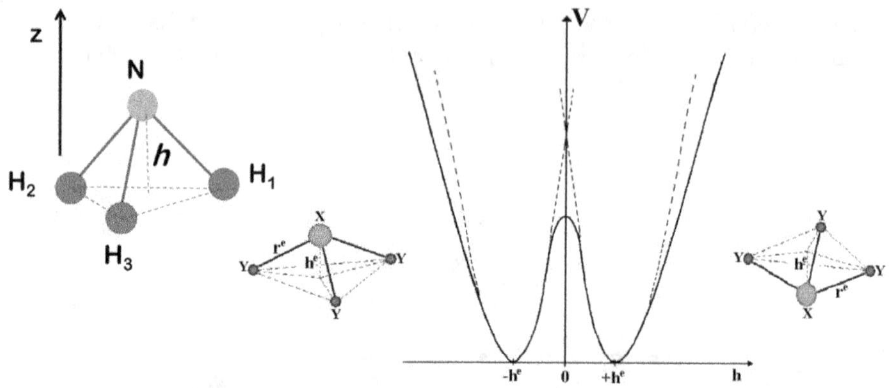

Figure 2.2. *Potential with two minima depending on the distance h between the nitrogen atom and the dotted plane. For a color version of this figure, see www.iste.co.uk/dahoo/infrared.zip*

2.3.2. Symmetry group of asymmetric molecular tops NHD_2 and NDH_2

In the case of the NHD_2 and NDH_2 molecules, which are the isotopic varieties of NH_3 or ND_3, the overlapping group is denoted by C_{1v}. It is composed of the identity operation and the reflection symmetry plane σ_v that exchanges the H nuclei in NDH_2 and the D nuclei in NHD_2 (see Figure 2.3, Tables 2.1 and 2.2). Similar to the NH_3 or ND_3 isotopic variety, the Oz axis is placed along the axis linking the center of mass and the N atom (equivalent to the third-order axis of rotation). If the Oz axis is perpendicular to the previous one, the reflection symmetry plane is denoted by σ_h and the corresponding group by C_{1h}. The NHD_2 and NDH_2 molecules have six vibration eigenmodes, four symmetric modes of A_1 type and two antisymmetric modes of A_2 type.

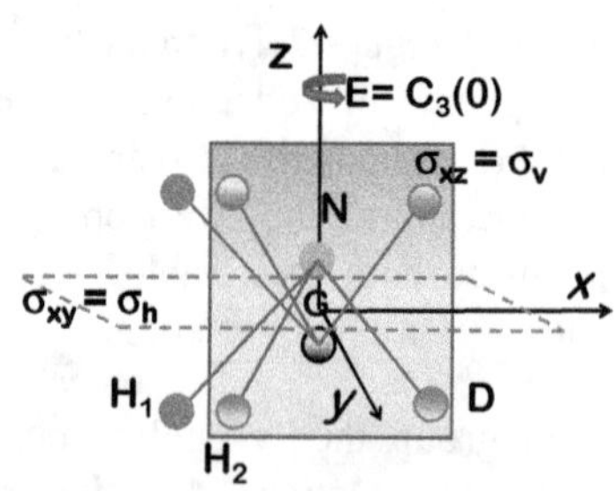

Figure 2.3. *The symmetry elements of the NDH_2 molecule and its reverse equilibrium configuration. For a color version of this figure, see www.iste.co.uk/dahoo/infrared.zip*

C_{1V}	E	σ_v
E	E	σ_v
σ_v	σ_v	E

Table 2.1. *Multiplication table (Cayley) of the C_{1v} group*

C_{1V}	E	$\sigma_{xz}=\sigma_v$	Polar vector	Axial vector	Normal coordinates
A_1	1	1	T_x, T_z	R_y	q_2
A_2	1	-1	T_y	R_x, R_z	

Table 2.2. *Table of characters and symmetry properties of (C_{1v} group) operators*

2.3.3. *Symmetry group of the complete group taking into account the inversion*

To take the inversion phenomenon into account (see Figures 1.3, 2.1 and 2.3), a higher-order group is built by the direct product of the C_{3v} or C_{1v} group, with the group used to move from a configuration, in which *H* and/or *D* atoms point downwards, to a configuration in which they point upwards. It is the group composed of the reflection symmetry plane σ_{xy} (*Oxy* plane) and the identity operation, or the C_s group (see Tables 2.3 and 2.4).

C_S	E	σ_{xy}
E	E	σ_{xy}
σ_{xy}	σ_{xy}	E

Table 2.3. *Multiplication (Cayley) table of the C_S group*

C_s	E	$\sigma_{xy}=\sigma_h$	Polar vector	Axial vector	Second-order tensor	Normal coordinates
A′	1	1	T_x, T_y	R_z	$\alpha_{xx}, \alpha_{yy}, \alpha_{zz}, \alpha_{xy}$	q_2
A″	1	-1	T_z	R_x, R_y	α_{yz}, α_{zx}	

Table 2.4. *Table of characters and symmetry properties of (C_S group) operators*

The exponents ′ and ″ indicate that the symmetry plane is orthogonal to the *Oz* axis and the irreducible representations are denoted by *A*′ and *A*″. In the case of the C_{1v} group, the irreducible representations are denoted by A_1 and A_2.

It is important to recall that it is the direct product group that has the irreducible representations used for labeling the energy levels and the corresponding functions, and that the D_{3h} group (NH_3 and ND_3) is isomorphic to the product group $C_{3v} \otimes C_s$, while the C_{2v} group (NHD_2 and NDH_2) is isomorphic to the product group $C_{1v} \otimes C_s$ for the corresponding isotopic varieties.

Using the rules and formulas presented in Chapter 1, the table of characters of NH_3 (see Table 2.5) and its isotopic varieties (see Table 2.6) can thus be established.

The operations of the C_s and C_{3v} groups commute and make it possible to use the direct product $C_s \otimes C_{3v}$ to generate a new group with 12 elements that is the complete symmetry group of the NH_3 (ND_3) molecule. The table of characters for this group is given in Table 2.5.

D_{3h}			E ⊗			σ(xy) ⊗				
			E	C_3	σ_v	E	C_3	σ_v		
$C_s \otimes C_{3v}$			E	$2C_3$	$3\sigma_v$	σ_h	$2S_3$	$3C_2$	Vectors	Tensors
	A_1	A'_1	1	1	1	1	1	1		
A'⊗	A_2	A'_2	1	1	-1	1	1	-1	R_z	$\alpha_{xx}+\alpha_{yy}+\alpha_{zz}$
	E	E'	2	-1	0	2	-1	0	(T_x, T_y)	$(\alpha_{xx}-\alpha_{yy}, \alpha_{xy})$
	A_1	A''_2	1	1	1	-1	-1	-1	T_z	
A"⊗	A_2	A''_1	1	1	-1	-1	-1	1		
	E	E"	2	-1	0	-2	1	0	(R_x, R_y)	$(\alpha_{yz}, \alpha_{zx})$

Table 2.5. *Table of characters of the direct product group $C_s \otimes C_{3v}$*

The notations used correspond to the irreducible representations of the D_{3h} group, isomorphic to the $C_s \otimes C_{3v}$ group. The C_3 operation defines the symmetries of types A and E, the C_2 operation defines the symmetries of types 1 and 2 and the σ_{xy} operation defines the symmetries of types ′ and ″.

Similarly, from the direct product of C_s by C_{1v}, $C_s \otimes C_{1v}$, which is isomorphic to the group C_{2v}, it is possible to build the complete symmetry group of the isotopic varieties NHD_2 and NDH_2. The group order is 4, where the four elements are the identity operation E and the plane symmetry operations σ_{xz}, σ_{xy} and $\sigma_{xz}\sigma_{xy} = C_2^x$. The table of characters for this group is presented in Table 2.6.

C_{2v}			E ⊗		σ(xy) ⊗			
			E	σ(xz)	E	σ(xz)		
C_s ⊗ C_{1v}			E	σ(xz)	σ(xy)	C_2^x	Vectors	Tensors
A′⊗	A_1	A_1	1	1	1	1	T_z	$\alpha_{xx}, \alpha_{yy}, \alpha_{zz}$
	A_2	B_2	1	-1	1	-1	T_y, R_x	α_{yz}
A″⊗	A_1	B_1	1	1	-1	-1	T_x, R_y	α_{zx}
	A_2	A_2	1	-1	-1	1	R_z	α_{xy}

Table 2.6. *Table of characters of the direct product group $C_s \otimes C_{1v}$*

The notations in the table are those of the irreducible representations of the C_{2v} group. The C_2^x operation defines the symmetries of types A and B, and the σ_{xz} operation defines the symmetries of types 1 and 2.

2.4. Symmetry of CH_4 and its isotopologues CD_4, CHD_3, CDH_3 and CH_2D_2

2.4.1. *Symmetry group of spherical tops CH_4 and CD_4*

The form of the methane molecule, CH_4, or of its isotopic variety CD_4, which involves five nuclei, is inscribed in a cube with the carbon nucleus at its center, as shown in Figure 2.4(a).

The symmetry elements of CH_4 and CD_4 molecules are as follows: eight proper axes of rotation of third order, denoted by $C_3(1\ 1\ 1)$, $C_3^2(1\ 1\ 1)$, $C_3(-1\ 1\ 1)$, $C_3^2(-1\ 1\ 1)$, $C_3(-1\ -1\ 1)$, $C_3^2(-1\ -1\ 1)$, $C_3(1\ -1\ 1)$, $C_3^2(1\ -1\ 1)$ (Figure 2.4(b)); two proper axes of rotation, denoted by $C_2(x)$, $C_2(y)$, $C_2(z)$ (Figure 2.4(b)); six improper axes of rotation, denoted by $S_4(x)$, $S_4^3(x)$, $S_4(y)$, $S_4^3(y)$, $S_4(z)$, $S_4^3(z)$ (Figure 2.4(c)); and six planes of symmetry that contain the ox, oy and oz axes, which are, respectively, denoted by $\sigma_d(z,y)$, $\sigma_d(-y,z)$, $\sigma_d(x,z)$, $\sigma_d(x,-z)$, $\sigma_d(x,y)$, $\sigma_d(-x,y)$ (Figure 2.4(d)).

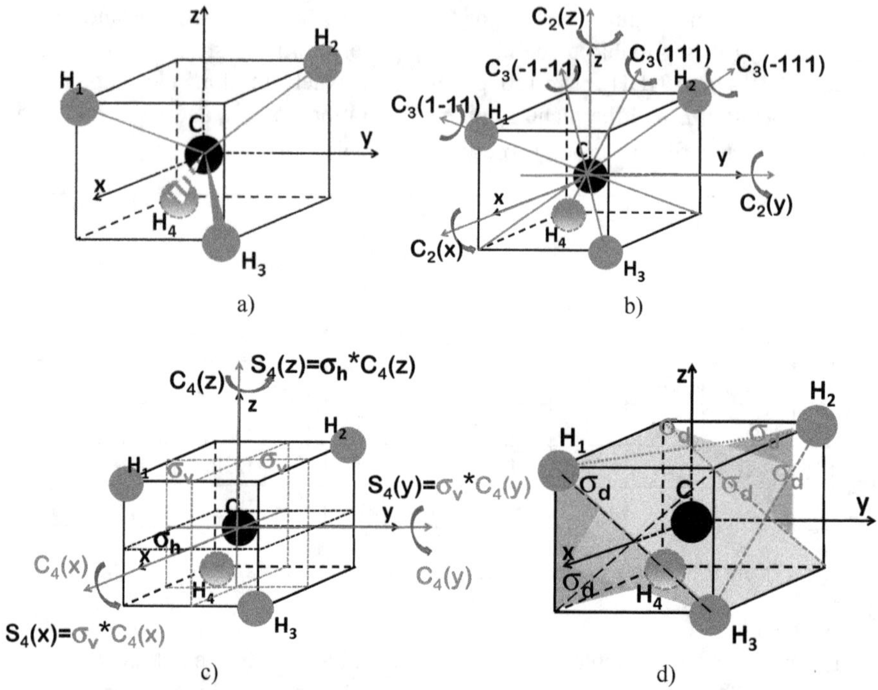

Figure 2.4. *Symmetry elements of the CH_4 molecule. For a color version of this figure, see www.iste.co.uk/dahoo/infrared.zip*

NOTES ON FIGURE 2.4.– *a) CH_4 molecule inscribed in a cube; b) elements of proper rotation symmetry C_3 and C_2 of the CH_4 molecule; c) elements of improper rotation symmetry $S_4 = \sigma_v * C_4$; d) three of six elements of plane symmetry σ_d of the CH_4 molecule.*

The addition of the identity operation E to these operations can be used to build the overlapping group of CH_4 and CD_4 molecules, denoted by T_d (see Table 2.7).

The table of characters given in Table 2.7 presents the one-dimensional irreducible representations, A_1 and A_2, the two-dimensional irreducible representations, E, and the three-dimensional irreducible representations, T_1 and T_2 (or F_1 and F_2), of the T_d group.

The notations 1 and 2 correspond to the symmetric or antisymmetric character with respect to S_4. The group contains 24 symmetry operations.

T_d	E	$8C_3$	$3C_2$	$6S_4$	$6\sigma_d$	Vectors	Tensors
A_1	1	1	1	1	1		$x^2 + y^2 + z^2$ $\alpha_{xx} + \alpha_{yy} + \alpha_{zz}$
A_2	1	1	1	-1	-1		
E	2	-1	2	0	0		$(2z^2 - x^{2-} - y^2, x^{2-} - y^2)$ $(2\alpha_{zz} - \alpha_{xx} - \alpha_{yy}, \alpha_{xx} - \alpha_{yy})$
T_1	3	0	-1	1	-1	R_x, R_y, R_z	
T_2	3	0	-1	-1	1	T_x, T_y, T_z	(xy, xz, yz) $(\alpha_{yz}, \alpha_{zx}, \alpha_{xy})$

Table 2.7. *Table of characters of the T_d group*

2.4.2. *Symmetry group of symmetric tops CHD_3 and CDH_3*

When counting the number of symmetry operations of the isotopic variety CDH_3 or CHD_3 of methane, there is only one three-fold axis left: two rotation operations, denoted by $C_3(2\pi n/3,\ n = 1 \text{ or } 2)$, and three planes of reflection symmetry ($\sigma_v(0)$, $\sigma_v(2\pi/3)$ and $\sigma_v(4\pi/3)$), denoted by σ_v, σ_v' and σ_v'' (see Figure 2.5). The addition of the identity operation E to these operations can be used to build the overlapping group of CDH_3 and CHD_3 molecules, denoted by C_{3v}, similar to that of NH_3.

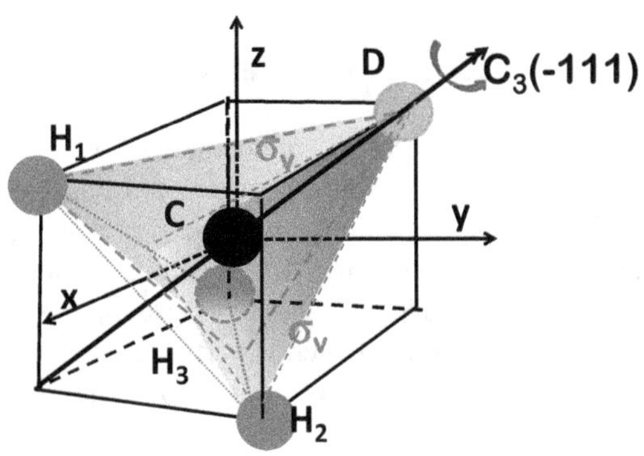

Figure 2.5. *The symmetry element of proper rotation C_3 and two of three planes of symmetry σ_v containing the C_3 axis of the CDH_3 molecule inscribed in a cube. For a color version of this figure, see www.iste.co.uk/dahoo/infrared.zip*

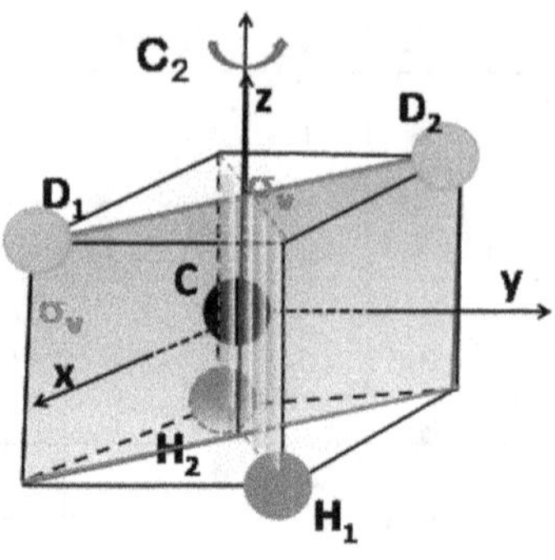

Figure 2.6. *Element of symmetry of proper rotation C_2 and the two planes of symmetry σ_v containing the C_2 axis of the CH_2D_2 molecule inscribed in a cube. For a color version of this figure, see www.iste.co.uk/dahoo/infrared.zip*

2.4.3. *Symmetry group of the asymmetric top CH_2D_2*

When counting the symmetry operations of the isotopic variety CD_2H_2 of methane, there is only one axis of second order left: the rotation operation, denoted by $C_2(2\pi/2)$, and the two planes of reflection symmetry containing the Oz axis, denoted by σ_v' and σ_v'' (see Figure 2.6). The addition of the identity operation E to these operations can be used to build the overlapping group of the CH_2D_2 molecule, denoted by C_{2v} (see Table 2.6), similar to that of H_2O.

2.5. Symmetry group of the complete CNPI group

The point-group theory applies to rigid molecules, whose nuclei motions during a vibration are described by low-amplitude motions and the overall rotation of the molecule is described within the framework of rigid rotators (see section 2.2). The point group is built from the geometric form of the molecule. The operators associated with the elements of a point symmetry group of a molecule generally commute with a reduced Hamiltonian, which is part of the exact or full Hamiltonian (energy operator) of the molecule (zeroth-order approximation). In this case, the focus is on the spatial variables of the position of electrons and nuclei, as well as on the vibration–rotation Schrödinger's equation for a given electronic state.

When considering transformations that involve exchanges of identical nuclei, the point symmetry group is not adapted to the study of symmetries of the wave function that describes a vibration–rotation state for a given electronic state, when the quantum variables related to the spins of identical nuclei are taken into account. The permutation group of identical particles should be considered, i.e. the symmetric group S_n (not to be confused with the point group S_{2n}) of permutations of n identical particles of order $n!$.

In this case, the complete group (CNPI – Complete Nuclear Permutation Inversion) [BUN 98, HOU 62, HOU 63, HOU 65, LON 63] is built from the permutation-inversion group (see Chapter 1). The elements of this group are permutations of the identical nuclei of a molecule (permutation group S_n) and the group built on the operator that reverses all of the coordinates of nuclei and electrons. This group operates on the exchange of nuclei independent of the observable geometric form. The relation between the operators of this group and the Hamiltonian symmetry properties is more direct than between the operators of the point symmetry group. In some studies presenting group theory applications, the construction of symmetry groups is clearly described [BUN 98, HAM 62, LAN 75, WIL 55, WIL 80], building a higher-order group by using the tensor product of this point symmetry group with a second-order group built on the operator that inverts spatial coordinates.

This approach was initially proposed by Bethe to process the lifting of degenerate levels of atomic orbitals of atoms in crystallographic sites by considering the spin variables within the group's theoretical frame. The structures that were not approached by Bethe [BET 29] were dealt with by Opechowski [OPE 40], who showed that the number of classes of a double group G^* built with G, a subgroup of the three-dimensional rotation group SO(3), could be less than twice the number of classes of G.

Moreover, the molecules characterized by a potential function with several minima are not rigid. In order to take inversion into account in the case of semi-rigid molecules, such as NH_3 and its isotopic varieties, a higher-order group should be used. This group should contain the irreducible representations for listing the irreducible representations, allowing the classification of the wave functions of the molecule depending on the symmetry properties of the Hamiltonian. This group is therefore built from the permutation-inversion group that takes the situations in which inversion can be observed by the exchange of identical nuclei in the molecule into account [BUN 98, HOU 62, HOU 63, HOU 65, LON 63].

The molecular symmetry group of NH_3 or ND_3 is given by the G^{CNPI} group, G_{12}.

$$\mathrm{G^{CNPI}} = S_3 \otimes \varepsilon = G_{12}$$
$$\mathrm{G^{CNPI}} = \{E, (12), (23), (31), (123), (132), E^*, (12)^*, (23)^*, (31)^*, (123)^*, (132)^*\} \quad [2.15]$$

The molecular symmetry group of NHD_2 or NDH_2 is given by the G^{CNPI} group, G_{12}.

As highlighted by Longuet-Higgins [LON 63], only the physically feasible or observable symmetry operations should be retained. The point-group symmetry (h order) is generally isomorphic to the subgroup of the complete permutation-inversion group [HOU 62, HOU 63] and a relation should be established between the permutation group and the configuration space. Tables 2.8, 2.9, 2.10 and 2.11 summarize the symmetry groups of G^{CNPI} for NH_3 and CH_4 molecules and their isotopic varieties.

$D_{3h}(M)$	E	(123)	(23)	E*	(123)*	(23)*	Vectors	Tensors
Classes	1	2	3	1	2	3		
D_{3h}	E	$2C_3$	$3C_2$	σ_h	$2S_3$	$3\sigma v$		
Equivalent Rotations	R^0	$R_z^{2\pi/3}$	R_0^{π}	R_z^{π}	$R_z^{5\pi/3}$	$R_{\pi/2}^{\pi}$		
A'_1	1	1	1	1	1	1		$\alpha_{xx}+\alpha_{yy}, \alpha_{zz}$
A''_1	1	1	1	-1	-1	-1		
A'_2	1	1	-1	1	1	-1	R_z	
A''_2	1	1	-1	-1	-1	1	T_z	
E′	2	-1	0	2	-1	0	(T_x, T_y)	$(\alpha_{xx}-\alpha_{yy}, \alpha_{xy})$
E″	2	-1	0	-2	1	0	(R_x, R_y)	$(\alpha_{yz}, \alpha_{zx})$

Table 2.8. *Table of characters of the $D_{3h}(M)$ group for NH_3*

For NH_3 and its isotopic varieties, with the inversion being observable, all the elements of the CNPI group are physically feasible. The complete group corresponds to the group built from point groups. The two groups were established in section 2.3.3. For methane, the complete permutation-inversion group is of order 48, with 10 varieties of symmetry related to the irreducible representations of the group. A classification of the energy levels depending on this group associates a different symmetry to the two components of each doublet. It follows that there are only five types of different doublets, and if the operations converting a right reference frame into a left reference frame are considered non-feasible and removed from the complete permutation-inversion group, the result is a group of 24 elements (corresponding to the point group T_d), with five symmetry species. One classification of the energy levels depending on this latter group associates a different symmetry to each type of doublets, but makes no distinction between the individual components of a doublet.

According to the IUPAC nomenclature, the G^{CNPI} group is denoted by $D_{3h}(M)$, where the letter *M* between the brackets is added to the Schoenflies nomenclature [BUN 97].

The feasible permutation-inversion operations form a group that is isomorphic to T_d. Using the notation of permutation cycles, these are: *E*; *(132), (123), (134), (143), (124), (142), (243), (234); (13) (24), (14) (23), (12) (34); (1432)*, (1234)*, (1342)*, (1243)*, (1324)*, (1423)**; and *(24)*, (13)*, (14)*, (23)*, (34)*, (12)**. According to the Longuet-Higgins [LON 62] notation, it is the G_{24} group, and the * indicates the inversion operation fixed in the laboratory (see section 1.10).

$C_{3v}(M)$	E	(123)	(23)*	Vectors	Tensors
Classes	**1**	**2**	**3**		
C_{3v}	E	$2C_3$	$3\sigma_v$		
Equivalent Rotations	R^0	$R_z^{2\pi/3}$	$R_{\frac{\pi}{2}}^{\pi}$		
A_1	1	1	1	T_z	x^2+y^2, z^2 $\alpha_{xx}+\alpha_{yy}, \alpha_{zz}$
A_2	1	1	-1	R_z	$\Gamma(\mu_A)$
E	2	-1	0	(T_x, T_y) (R_x, R_y)	$(x^{2}-y^2, xy)$ $(\alpha_{xy}, \alpha_{xx}-\alpha_{yy})$

Table 2.9. *Table of characters of the $C_{3v}(M)$ group for CDH_{33}*

$C_{2v}(M)$	E	(12)	E*	(12)*	Vectors	Tensors
Classes	**1**	**1**	**1**	**1**		
C_{2v}	E	C_{2b}	σ_{ab}	σ_{bc}		
Equivalent Rotations	R^0	$R_b^{\pi}=R_0^{\pi}$	$R_c^{\pi}=R_{\frac{\pi}{2}}^{\pi}$	$R_a^{\pi}=R_z^{\pi}$		
A_1	1	1	1	1	T_b	$\alpha_{aa}, \alpha_{bb}, \alpha_{cc}$
A_2	1	1	-1	-1	R_b	$\alpha_{ac}, \Gamma(\mu_A)$ *
B_1	1	-1	-1	1	T_c, R_a	α_{bc}
B_2	1	-1	1	-1	T_a, R_c	α_{ab}

Table 2.10. *Table of characters of the $C_{2v}(M)$ group for CD_2H_2*

$T_d(M)$	E	(123)	(14) (23)	(1423)*	(23)*	Vectors	Tensors
Classes	**1**	**8**	**3**	**6**	**6**		
T_d	E	$8C_3$	$3C_2$	$6S_4$	$6\sigma_d$		
A_1	1	1	1	1	1		$x^2 + y^2 + z^2$ $\alpha_{xx} + \alpha_{yy} + \alpha_{zz}$
A_2	1	1	1	-1	-1		
E	2	-1	2	0	0		$(2z^2 - x^2 - y^2, x^2 - y^2)$ $(2\alpha_{zz} - \alpha_{xx} - \alpha_{yy}, \alpha_{xx} - \alpha_{yy})$
T_1	3	0	-1	1	-1	R_x, R_y, R_z	
T_2	3	0	-1	-1	1	T_x, T_y, T_z	(xy, xz, yz) $(\alpha_{yz}, \alpha_{zx}, \alpha_{xy})$

Table 2.11. *Table of characters of the $T_d(M)$ group for CH_4*

2.6. Conclusion

The point-group symmetry is adapted for electronic, vibration and rotation degrees of freedom. To deal with the exchange of identical nuclei in a molecule or the experimentally observable inversion situations by the tunneling effect, it is more convenient to use the permutation-inversion group. The symmetry properties of NH_3 and CH_4 molecules and their isotopic varieties were presented in this chapter and their geometric symmetry groups were determined. The theory of permutation-inversion groups was also used to determine the corresponding CNPI groups. These various groups will be used in Chapter 3 to determine the symmetries of normal coordinates and of wave functions of various vibration, rotation and spin degrees of freedom in a given electronic state, for various types of molecules considered in this chapter.

3

Line Profiles, Symmetries and Selection Rules According to Group Theory

Chapter 2 presented the establishment of point-group symmetries of symmetric and spherical top molecules based on NH_3 and CH_4 and their isotopic varieties from the symmetry properties resulting from the molecular structure. Using the results given in section 1.10, the methods developed for group theory are applied to the molecules described in Chapter 1, in order to determine the basis of irreducible representations of symmetry groups of each isotopic variety for electronic, vibration, rotation and vibration–rotation degrees of freedom. The objective is to identify the types of symmetry of the energy levels of each type of degree of freedom and to determine the allowed transitions corresponding to the selection rules in absorption, emission or scattering spectroscopy. These transitions are the fingerprints of the molecule when it interacts with light in infrared (IR) absorption or emission spectroscopy or in Raman spectroscopy studies.

The rotation spectra are observed in the microwave and far-IR range (between 25 μm and 1,000 μm or 400 cm^{-1} and 10 cm^{-1}); the vibration–rotation spectra, in the medium-IR range (between 2.5 μm and 25 μm or 4,000 cm^{-1} and 400 cm^{-1}) or near-infrared range (between 0.75 μm and 2.5 μm or 13,333 cm^{-1} and 4,000 cm^{-1}); the electronic spectra in the near-IR, visible or ultraviolet range. The time scales corresponding to these degrees of freedom range between 0.1 fs and 1 fs for the electrons, between 10 fs and 100 fs for vibration and between 0.1 ps and 10 ps for rotation. Permutation-inversion groups are introduced in order to illustrate its use when the molecule is characterized by high-amplitude motions or when the tunneling effect can be observed by the measuring instruments.

3.1. Introduction

As presented in Chapter 1, NH_3 belongs to the point group C_{3v} and CH_4 belongs to group T_d. The substitution of a D atom to an H atom in CH_4 leads to its switching from group T_d to group C_{3v}. Because of lower symmetry, it passes from the spherical tops classification to the symmetric tops classification. The presence of a D isotope in place of one H atom in NH_3, and of two D atoms in place of two H atoms in CH_4, lowers the symmetry of these molecules and, consequently, they pass from C_{3v} or T_d group to the C_{2v} symmetry group. They therefore belong to the class of asymmetric tops, in the same way as the H_2O water molecule, studied in this book. It is important to note that N and C atoms being central in NH_3 and CH_4, respectively, and that the presence of ^{14}N, ^{15}N, ^{12}C and ^{13}C isotopes does not modify the symmetry of the given molecule.

These changes in the new groups of symmetry following the replacement of the H atom by D isotopes lead to modifications in the selection rules. Moreover, by passing from a higher-symmetry group to a lower-symmetry group, the notions of subgroups developed in the group theory recalled in – section 1.10 – can be verified.

The permutation-inversion group was also introduced in order to classify the energy levels as a function of the irreducible representations of the group when considering the high-amplitude motions or the tunneling effect. The question that arises is that of the difference between a geometric symmetry operation and a permutation symmetry operation and the classification of energy levels. It is possible to find permutations of identical nuclei of a molecule for all the geometric symmetry operations that can be used to build the point-group symmetries of the molecule. The nuclear permutation accompanied by an inversion, with respect to the origin of coordinates of all the particles of the molecule, is equivalent to the geometric symmetry operation that converts a right-handed reference frame into a left-handed reference frame.

On the other hand, a permutation of identical nuclei with or without inversion of all the coordinates does not necessarily have its equivalent as a geometric symmetry operation. In order to avoid this difficulty, Longuet-Higgins [BUN 98, HOU 65, LON 63] suggests doing without the complete group of permutations with the inversion, when building the energy levels scheme of a given molecule. He proposes the use of a subgroup of this complete group that contains only feasible operations, i.e. operations corresponding to modifications of the nuclear configuration that can be achieved without passing through an “insurmountable” energy barrier.

The example of methane [BUN 99] can be used to facilitate the understanding of the notion of feasible and non-feasible symmetry operations, and the potential difficulty in classifying the energy levels. Though normally this is not explicitly taken into account, the levels of rotation of methane (similar to those of NH_3) appear in the form of inversion doublets. This phenomenon is associated with the fact that if the hydrogen atoms are numbered, then the equilibrium configuration can be described in two forms: a right-handed form and a left-handed form. Nevertheless, these inversion doublets in the methane molecule cannot be resolved by the available spectroscopic techniques. Therefore, in a practical system for energy-level classification, it is unnecessary to distinguish the two components of a doublet. In fact, a classification system that does not distinguish between the different doublets is sufficient. For methane, the order of the complete permutation-inversion group is equal to 48, containing 10 symmetry operations. A classification of the energy levels depending on this group associates a different species to the two components of each doublet. However, it proves that there are only five types of different doublets, and if the operations converting a right-handed form into a left-handed form are removed from the complete permutation-inversion group as non-feasible, the resulting group has 24 elements (corresponding to the point group T_d), comprising five symmetry operations. A classification of the energy levels by means of the irreducible representations of this latter group associates a different species to each type of doublet but makes no distinction between the individual components of a doublet.

The methane case is in fact an example of a more general physical phenomenon. For many molecules, the energy levels appear in a systematically degenerate set, which corresponds to the fact that the molecule can be considered as existing in one of the forms between which inter-conversion is not feasible. In the case of an inter-conversion, one part of the degeneracy associated with the existence of several equivalent forms is suppressed. From the group theory perspective, as the number of symmetry operations feasible in the molecular symmetry group increases, the previously degenerate levels are separated. The problem of modification of the number of feasible symmetry operations in a molecular symmetry group arises, as there is no absolute criterion for the feasibility of a given element. In order to show the effect of symmetries on the profile of spectra of symmetric and spherical top molecules NH_3 and CH_4 and their isotopic varieties, based on the considerations determined by the application of group theory, and for the sake of consistency across volumes, the following two sections review the results given in Volumes 1 and 2 [DAH 17, DAH 19], given the topics developed in section 1.10 of this book.

Section 3.3 of Volume 1 [DAH 17] presented the profiles of lines most commonly associated with the transitions observed in IR spectroscopy. The profile of the observable spectra depends on the selection rules and the environment in which the molecular system is immersed, the gas phase in the presence or absence of an external (electric or magnetic) field, the dense environment (broadening by pressure) and the stressed environment (molecule in a nanocage or on a surface). In the rest of this chapter, the various expressions used for calculating a line profile, observable in a spectrum, as well as the general method for determining the selection rules, are recalled, taking into account the mathematical form of the complete wave function describing the molecule in zeroth-order approximation, i.e. by neglecting the couplings and its symmetry properties. Group theory methods are then applied to each isotopic variety to determine the selection rules in IR and Raman spectroscopy, based on symmetries, in the sense of the relationship to the irreducible representations of the symmetry group of the molecule, of the various functions describing the electronic, translation, rotation, vibration and vibration–rotation degrees of freedom that depend on the space variables and the quantum spin degrees of freedom.

3.2. Symmetries of the eigenstates of the zeroth-order Hamiltonian

The most common form of the full wave function describing a molecule was presented in the first chapter. The molecular states of a polyatomic molecule of type XY_3 (NH_3, ND_3), XYZ_2 (NHD_2, NH_2D), XY_4 (CH_4, CD_4), XY_2Z_2 (CH_2D_2) and finally XY_3Z (CH_3D, CHD_3) are represented at zeroth order, by the complete wave functions expressed in a first approximation as the product of the translation, electronic, vibration, rotation and spin wave functions (see equation [1.8]): $|\Psi_0\rangle = |\Psi_{CM}\rangle \times |\Psi_e\rangle \times |\Psi_V\rangle \times |\Psi_R\rangle \times |Sm_s\rangle \times |\mathrm{Im}_I\rangle$. This is a general expression that applies to all the molecules and was used in Volumes 1 and 2 in order to characterize diatomic and triatomic molecules [DAH 17, DAH 19]. Eckart's first condition, according to which the center of the mobile reference frame of the equilibrium configuration of the molecule coincides with its center of mass, can be used to rigorously separate the translation degrees of freedom from the other degrees of freedom and consequently allows us to discard the overall translation of the molecule in the description of its degrees of freedom.

The effect of electron and nuclei spins on the symmetry of the complete wave function, $|\Psi_t\rangle = |\Psi_e\rangle \times |\Psi_V\rangle \times |\Psi_R\rangle \times |Sm_s\rangle \times |\mathrm{Im}_I\rangle$, expressed as a product of functions describing a state associated with each degree of freedom, is taken into account by modifying this initial form and regrouping the electron spin variables in the electronic wave function. The electronic degrees of freedom are dealt with separately from those of the nuclei, which facilitates the application of the Born–Oppenheimer approximation. The wave function can be written as: $|m\rangle = |\Psi_t^m\rangle = \left(|\Psi_e^m\rangle \otimes |Sm_s\rangle\right) \times \left(|\Psi_{VR}^m\rangle \otimes |\mathrm{Im}_I\rangle\right) = |m\rangle_e \times |m\rangle_n$, a product of an electronic wave function and a nuclear wave function. It is important to note that the complete wave function of the Hamiltonian of the molecular system must be invariant under a symmetry operation of simultaneous inversion of all the particles (electrons and nuclei (nucleons)). Since the C_i group has two one-dimensional irreducible representations, the wave functions may belong to one or other of these representations. Two possible states are thus obtained, positive states and negative states, depending on whether the function changes its sign or not in this transformation. These states cannot be superimposed by simple rotation (first-order rotation).

$\left|\Psi_{VR}^{m}\right\rangle$ is in a first approximation the product $\left|\Psi_{V}\right\rangle\times\left|\Psi_{R}\right\rangle$. The nuclear Hamiltonian is considered as a perturbation of a given electronic state, and the electronic wave functions depend on the coordinates of electrons and nuclei. In IR spectroscopy, the fundamental electronic state of almost all molecules belongs to the fully symmetric irreducible representation. Consequently, the nuclear wave function $|m\rangle_{n}=\left|\Psi_{n}^{m}\right\rangle=\left|\Psi_{VR}^{m}\right\rangle\otimes\left|\mathrm{Im}_{I}\right\rangle$ is the product of a vibration–rotation function and of a nuclear spin function. Since the spin variables of the electrons of the total wave function are integrated into the full electronic wave function, the symmetry of the complete wave function for an electronic state is determined by the effect of the exchange of identical nuclei. In this case, the mass number M of the nucleus must be known in order to take into account the effect of the exchange of identical nuclei. Given the quantum nature of the nuclear spin wave function, a nucleus of even mass number M has a nuclear spin that is an integer obeying the Bose–Einstein statistics, and when M is odd, the nuclear spin is a half-integer obeying Fermi–Dirac statistics (see Table 1.6 of Volume 1).

Determining the symmetry of an electronic or vibration wave function requires knowledge on how the position coordinates of electrons and nuclei (normal vibration coordinates) are transformed under a symmetry operation [AMA 80a, CAM 88, LAN 75, WIL 55, WIL 80]. The symmetry operation is defined as a change of the system of reference axes *xyz*, accompanied by a permutation of indices assigned to identical nuclei rotating the reference system with respect to the equilibrium or instantaneous configuration of the nuclei. In section 3.6, the irreducible representations of the vibrations of a molecule are built from the irreducible representations of the point-group symmetry of the molecule. This symmetry group is also that of the electronic and vibrational Hamiltonian of the molecule, but it is not that of the rotational Hamiltonian (rigid rotator) and of the spin variables connected to the overall rotation degrees of freedom of the molecule and to the nuclear spin variables. Section 3.8 focuses on the symmetries of the rotation and spin wave function in order to determine the irreducible representations of the products of the rotation and spin wave function. The complete wave function must be symmetric or antisymmetric in a permutation of identical nuclei, i.e. belonging to the irreducible representation of type A of the symmetry group of the molecule and symmetric or antisymmetric, of type A_1 or A_2 conforming to group C_{3v} (symmetric top without inversion, CH_3D) or T_d (spherical top, CH_4), or A'_1 or A'_2 for the representation in a right-handed reference frame or A''_1 or A''_2 for the representation in a left-handed reference frame conforming to group D_{3h} (symmetric top with inversion, NH_3).

3.3. Intensity of the vibration–rotation lines and bar spectrum

This section reviews the main characteristics of the theoretical models developed for calculating the effect of the interaction of IR electromagnetic radiation with the probed molecule, taking into account its environment by considering the theoretical aspects and the measurement or observation under experimental situation, while the effects of the instrumental distortions are considered negligible. The effects of the symmetry of the studied molecules on the shape of the Dirac spectra are then treated.

To discuss the selection rules, some of the elements presented in Chapters 1 and 2 of Volumes 1 and 2 on IR spectroscopy of diatomic and triatomic molecules [DAH 17, DAH 19] are reviewed. The selection rules give the shape of the observable spectrum in the form of lines without structures, i.e. in the form of bars or Dirac spectrum. In order to determine the real spectrum, the interaction between the electromagnetic radiation and the molecule, which interacts with its environment, must be taken into account. The theoretical models developed for the calculation of the effect of the mechanisms responsible for various profiles of a spectral line and its widening were discussed in Chapters 3 and 4 of Volume 1, depending on the type of environment, in gas or condensed medium.

The most common profiles of absorption lines are of Gaussian as a result of the Doppler effect (dependence on temperature of the speed of molecules) or Lorentzian as a result of the broadening under the effect of pressure (effect of collisions between the absorbing molecule and a molecule in its vicinity or environment) or when both effects are present, by a Voigt profile, resulting from the convolution of Gaussian and Lorentzian profiles [DAH 17, DAH 88, GAB 96]. The absorption lines corresponding to transitions between the energy levels are identified experimentally by classical frequency-resolved spectroscopy using a spectrometer or an interferometer, or by laser absorption scanning the absorption spectral range of the line profile. Measurement distortion inherent to spectra observation should be taken into account, as the observed profiles are the products of convolution of the apparatus function of the observation device with the true profile of the absorption line [CLA 81, CLA 84, DAH 88, GHE 79].

This section resumes some of the elements presented in Chapters 3 and 4 of Volume 1 on the profile and form of spectral lines, for weak matter–radiation coupling (weak-field approximation). The expression obtained from the theoretical models described in Volume 1, for a molecule that is not isolated, but interacts with the surrounding environment, is provided. In the gas phase, it can be subjected to collisions under the effect of pressure. In the condensed phase, in a nanocage or on a surface, it can interact with rare gas atoms in inert matrices, carbon atoms in a fullerene structure or water molecules in clathrate structures or graphite surface.

The spectral distribution of the radiation absorbed by the optically active molecules per unit volume, at temperature T, is expressed as the time Fourier transform of the autocorrelation function of the electric dipole moment [BEN 66a, BEN 66b, GOR 68] of the set of molecules, which is also termed spectral density (see equations [3.2], [3.3] and [4.84], Volume 1).

For a $i \rightarrow f$ transition between the initial state i and the final state f, where $E_i, E_f, \ldots.\ and\ |i\rangle \equiv |v_i, J_i, K_i, M_i\rangle, |f\rangle \equiv |v_f, J_f, K_f, M_f\rangle, \ldots,$ are the energies and eigenvectors of the Hamiltonian of the optically active system $\tilde{H}_a$ for the vibration–rotation degrees of freedom, respectively, it was shown in Chapter 4 of Volume 1 (see equation [4.92]) that when the molecule interacts with its environment in the condensed phase, the expression of the absorption coefficient can be written as:

$$K^0(\sigma) = \frac{8\pi^3 N\sigma}{3hc} \sum_{i,f} (\rho_a^i - \rho_a^f) |\langle i|\mu_A|f\rangle|^2$$

$$Re \int_0^\infty dtexp\left(i\left(\omega - \omega_{if} - \Delta\omega_{v_i v_f}\right)t\right) exp(-i\Delta^{(1)}t) exp(-i\Delta^{(2)}t) exp(-\Gamma^{(2)}t) \qquad [3.1]$$

In this expression, ω is the angular frequency of the incident radiation, σ is the wave number, h is the Planck constant, c is the speed of light, k_B is the Boltzmann constant, ρ_a^i and ρ_a^f are, respectively, the populations, at the initial instant, of the initial optical states i and final optical states f, μ_A is the component of the electric dipole moment of an active molecule along the OZ axis of the fixed reference frame (the projection of the electric dipole moment of the set of active molecules along the OZ axis carrying the polarization vector of the electromagnetic field of the incident radiation), N is the number of active molecules per unit volume, $\omega_{if} = \hbar^{-1}(E_f - E_i)$ characterizes frequency of the $i \rightarrow f$ transition between the initial state i and the final state f, associated with the spectral line under consideration, and $\Delta\omega_{v_i v_f}$ is the frequency shift. The interaction with the environment is reflected by the term $\Delta^{(1)}$, which is the frequency shift of the line associated with the $i \rightarrow f$ transition and is given by the first-order term of the overall average expansion of the time evolution operator, and the terms $\Delta^{(2)}$ and $\Gamma^{(2)}$, which are the frequency shift and the width of the same line, respectively, and which are given by the second-order term of the expansion. The contribution $\Delta^{(1)}$ characterizes only a frequency shift of the line, which is due to a thermal coupling with the condensed medium through the phonon bath. Only the two-phonon coupling contributes to this shift, with the average of one-phonon coupling being equal to zero because of the harmonic description of the translation modes of the medium.

The second-order terms are given by:

$$\begin{aligned}\Delta^{(2)} &= \mathrm{Im}\left[\hbar^{-2}\left(\Delta_i^{(2)} + \Delta_f^{(2)} + \Delta_{if}^{(2)}\right)\right]\\ \Gamma^{(2)} &= \mathrm{Re}\left[\hbar^{-2}\left(\Delta_i^{(2)} + \Delta_f^{(2)} + \Delta_{if}^{(2)}\right)\right]\end{aligned} \qquad [3.2]$$

The contributions $\Delta_i^{(2)}$ and $\Delta_f^{(2)}$ (see equations [4.95] and [4.96], Volume 1) characterize inelastic collisions. They depend on the non-diagonal elements of matrix describing the mixing of the initial and final optical states involved by the other optical states through the bath states.

Finally, the contribution $\Delta_{if}^{(2)}$ (see equation [4.97], Volume 1) characterizes only elastic collisions. For the initial and final optical states, it depends on the mean over the bath states of the fluctuations of the coupling Hamiltonian in the vicinity of its average value.

In the construction of the spectrum, in a first approximation, the coupling between the optical system and the thermal bath constituted by the environment is assumed to be negligible, to the extent that the bar spectrum allows the reconstruction of most of the signals present in the experimental spectrum. The working hypothesis is that the coupling is and remains a small perturbation during the evolution of the whole interacting system. The equation obtained from equation [3.1], and which leads to integrated absorption coefficients (lines in the form of bars or Dirac spectrum), can be written as:

$$K^0(\sigma) = \frac{8\pi^3 N\sigma}{3hc}\sum_{i,f}(\rho_a^i - \rho_a^f)|\langle i|\mu_A|f\rangle|^2\delta\left(\omega - \omega_{if} - \Delta\omega_{v_i v_f}\right) \qquad [3.3]$$

In the gas phase, similar expressions are obtained, as shown in section 1.1 of Volume 1. The appendix reviews the theoretical model, leading to these expressions, as well as the parameters that can be measured from observation.

3.4. Transition operator for the selection rules

In equations [3.1] and [3.3], it is the square of the transition moment $\hat{R}_{fi} = \left\langle \Psi_t^f \middle| \hat{R} \middle| \Psi_t^i \right\rangle$ that reflects the interaction between the active molecular system and the electromagnetic radiation.

The type of spectroscopy that can be used to identify a molecule, and characterize the medium in which it is immersed, depends on the nature of operator $\hat{R}$, which depends on the properties of the molecule.

In emission or absorption spectroscopy, it is the electric dipole moment operator, the electric quadrupole moment operator or the magnetic dipole moment operator (vector operator or first-rank tensor operator); in Raman scattering spectroscopy, it is the polarizability tensor operator (second-rank tensor operator) or in collision spectroscopy the collision operator.

Although equation [3.3] shows that the square of the transition moment is but one of the factors (see Chapters 3 and 4, Volume 1 [DAH 17]) playing a role in the setting up of observable spectra, the fact that this moment involves two states between which a transition occurs together with the operator inducing this transition, the transition moment is sufficient by itself for establishing the selection rules in spectroscopy.

To apply the group theory methods, the following general expression can be used as a starting point:

$$\hat{R}_{fi} = \left\langle \Psi_t^f \middle| \hat{R} \middle| \Psi_t^i \right\rangle = \int \Psi_t^{f*} \hat{R} \Psi_t^i d\tau \qquad [3.4]$$

where $\hat{R}$ is the operator responsible for the radiation–matter interaction and $\left|\Psi_t^i\right\rangle = \left|\Psi_{eVR}^i\right\rangle$ is the complete rovibronic wave function corresponding to the initial vibration–rotation state and $\left|\Psi_t^f\right\rangle = \left|\Psi_{eVR}^f\right\rangle$ is the complete rovibronic wave function corresponding to the final vibration–rotation state, for the transition considered, the molecule being in the same given electronic state for both states.

The representation $\Gamma\left(\left|\Psi_t^i\right\rangle\right) \otimes \Gamma\left(\left|\Psi_t^f\right\rangle\right) \otimes \Gamma\left(\left|\hat{R}\right\rangle\right)$ must contain the fully symmetric irreducible representation Γ_{irr}^{sym} so that the integral of equation [3.4] is not zero, such that:

$$\Gamma\left(\left|\Psi_t^i\right\rangle\right) \otimes \Gamma\left(\left|\Psi_t^f\right\rangle\right) \otimes \Gamma\left(\left|\hat{R}\right\rangle\right) \supset \Gamma_{irr}^{sym} \qquad [3.5]$$

This means that the symmetry of the product of the two states between which the transition occurs, in the sense of one of the irreducible representations of the symmetry group of the molecule, must be of the same type as that of operator $\hat{R}R$. The symmetry of states is also important in the calculation of possible transitions between the quantum states that are present in the form of lines in the observed spectra.

According to the Born–Oppenheimer approximation and the zeroth-order approximation, in the absence of an external (electric or magnetic) field, the complete wave function, $\left|\Psi_t\right\rangle = \left|\Psi_e\right\rangle \times \left|\Psi_V\right\rangle \times \left|\Psi_R\right\rangle \times \left|Sm_s\right\rangle \times \left|Im_I\right\rangle$, is expressed as a product of functions describing a state associated with each degree of freedom. The symmetry of such a wave function can be determined from:

$$\Gamma_t = \Gamma(|\Psi_e\rangle) \otimes \Gamma(|\Psi_V\rangle) \otimes \Gamma(|\Psi_R\rangle) \otimes \Gamma(|Sm_s\rangle) \otimes \Gamma(|Im_I\rangle) \quad [3.6]$$

If the electronic wave function is fully symmetric in the fundamental state (for the majority of stable molecules, except for O_2 (Σ_g^-) and NO (π), see section 1.33, Volume 1), then $\Gamma(|\Psi_e\rangle) \otimes \Gamma(|Sm_s\rangle) \supset \Gamma_{irr}^{sym}$.

The symmetry of the complete wave function depends finally only on that of the rotational wave function given the symmetry of the spin state of the respective level $\left(\Gamma_t = \Gamma(|\Psi_R\rangle) \otimes \Gamma(|Im_I\rangle)\right)$ when the molecules are in the fundamental electronic and vibrational states.

A given rotational level cannot be associated with an arbitrary spin function. The nuclear statistical weight of the level is the total number of distinct spin functions that can be combined with a given rotational eigenfunction $|\Psi_R\rangle$. It should be recalled that the eigenfunctions of a rigid rotator are expressed as a function of Euler angles, which can be written as:

$$\langle \theta, \varphi, \chi | \Psi_R^0 \rangle = \Psi_R^0(\theta, \varphi, \chi) = \mathrm{N}_{JKM} \Theta_{JKM}(\theta) e^{iK\chi} e^{iM\varphi} \quad [3.7]$$

with:

$$\Theta_{JKM}(\theta) = \mathrm{t}^{d/2}(1-t)^{s/2} \mathrm{G}_p(1+d+s, 1+d, t)$$

and:

$$\mathrm{N}_{JKM} = |(2p+d+s+1)(p+d+s)!\,(p+d)!/4\pi \mathrm{p}!\,\mathrm{d}^2!\,(p+s)!|^{1/2}$$

with:

$$\mathrm{t} = \frac{1}{2}(1-\cos\theta): \mathrm{d} = |M-K|; \mathrm{s} = |M+K|; \mathrm{p} = J - \frac{1}{2}(\mathrm{d}+\mathrm{s})$$

where G_p is a Jacobi polynomial. The quantum number J takes $(2J+1)$ positive or null integer values, while the values of K and M are at most equal to J. Physically, the quantum numbers M and K represent the projections of the rotational angular moment on the OZ axis of the fixed reference frame and on the Oz axis of the mobile reference frame.

Under a permutation operation, the complete wave function may be invariant or may change sign, which corresponds to the two eigenvalues –1 and +1 of the permutation operator (notation: (12) for the transpositions with two particles of the symmetric group S_2) of two particles. Wave functions can be classified into "*s*" states and "*a*" states depending on whether the wave function is symmetric (+1) or antisymmetric (–1) in a permutation operation of indiscernible particles. In the case of the exchange of two electrons in an atom, a state $|\Psi_e\rangle$ is always antisymmetric according to the Pauli principle, which stipulates that the states of two electrons cannot be characterized by identical values of all the quantum numbers. Two electrons on the same orbital (same space variables) must have different spin orientations ($\pm 1/2$). For a nucleus of mass number *M*, which contains *M* particles of spin $1/2$, exchanging two nuclei amounts to exchanging *M* times the particles of spin $1/2$. The wave function $|\Psi_n\rangle$ changes its sign with each exchange of two particles, so that $|\Psi_n\rangle$ is symmetric or antisymmetric depending on whether *M* is even or odd. Because of its quantum nature, the nuclear spin wave function depends on the mass number *M* of the nucleus (*M* even: integer spin obeying Bose–Einstein statistics; *M* odd: half-integer spin obeying Fermi–Dirac statistics) (see Table 1.6, Volume 1).

If, as shown in Chapter 1 of Volume 1, the spin states of a homonuclear diatomic molecule H_2, or in Chapter 1 of Volume 2 [DAH 19], those of the H_2O molecule, can be easily built, the space dimension being higher for NH_3 and CH_4 or their isotopic varieties, it is more appropriate to use group theory methods for determining the dimension and the spin states of the symmetric and antisymmetric spaces, in order to determine the ortho (symmetric spin states), para (antisymmetric spin states) or meta state of the vibration–rotation levels.

3.5. Dipole moment operator and line profile

Consider, as an example, the electric dipole moment $\vec{\hat{R}}$. If the influence of the nuclear spin is considered negligible, the transition moment can be written as:

$$\vec{\hat{R}}_{fi} = \left\langle \Psi^f_{eVR} \middle| \vec{\hat{R}} \middle| \Psi^i_{eVR} \right\rangle = \int \Psi^{f*}_e \Psi^{f*}_V \Psi^{f*}_R \vec{\hat{R}} \Psi^i_e \Psi^i_V \Psi^i_R d\tau \qquad [3.8]$$

where $d\tau = d\tau_e d\tau_n$ is the space of configuration of space variables connected to electronic coordinates ($d\tau_e$) and nuclear coordinates ($d\tau_n$). This operator has three components, and a transition between two quantum states is possible if at least one of the integrals corresponding to a component differs from zero and is therefore symmetric, with respect to the symmetry operations under which the molecule is invariant.

The expansion of the operator $\vec{\hat{R}}$ in terms of variables of position $\vec{r}_i^e$ of the electrons and $\vec{r}_k^n$ of the nuclei can be written as:

$$\vec{\hat{R}} = \vec{\hat{R}}_e + \vec{\hat{R}}_n = \sum_{i \in electrons} \left(-e\vec{r}_i^{\,e}\right) + \sum_{k \in nuclei} \left(Z_k e\vec{r}_k^{\,n}\right) \quad [3.9]$$

where $Z_k e$ is the nuclear charge of nucleus k. The transition moment associated with the dipole moment can be written as:

$$\vec{\hat{R}}_{fi} = \left\langle \Psi_{VR}^{f} \middle| \vec{\mu} \middle| \Psi_{VR}^{i} \right\rangle \quad [3.10]$$

where:

$$\vec{\mu} = \langle \vec{\hat{R}}_n \rangle_e + \delta_{if} \int \Psi_e^{f*} \vec{\hat{R}}_e \Psi_e^i d\tau_e \quad [3.11]$$

is the dipole moment of the molecule in the given electronic state, which is the same for the initial and final states and is highlighted by the presence of the Kronecker symbol δ_{fi}. It should be noted that the second term of equation [3.11] is in this case $\int \Psi_e^{0*} \vec{\hat{R}}_e \Psi_e^0 d\tau_e$, the average of the electronic positions in the fundamental electronic state.

In a molecule, the positions of electrons are confined to molecular orbitals that are linear combinations of atomic orbitals (LCAO and AO = s, p, d, etc.) and $d\tau_e = dxdydz$ or $d\tau_e = r^2 dr sin\theta d\theta d\varphi$ in Cartesian or spherical coordinates, respectively. These positions are measured with respect to the center of gravity of the masses of nuclei and give the barycenter of the negative charge. If the barycenter of the electronic charges coincides with that of the masses of the nuclei, this integral must be zero, because the positions are measured from the center of mass of the nuclei.

Moreover, the operators resulting from the x, y or z component of $\vec{\hat{R}}_e$ being odd, and the square of the electronic wave function $\left|\Psi_e^0\right|^2$ being symmetric when the electronic state is identical for the vibration–rotation transition, the integral $\int \Psi_e^{0*} \vec{\hat{R}}_e \Psi_e^0 d\tau_e$ is necessarily zero.

Moreover, according to the Born–Oppenheimer approximation, the electronic wave function depends on the electronic coordinates and on the nuclear coordinates in the form of parameters (see equations [1.17] and [1.20a], Volume 1 [DAH 17] for a diatomic molecule). In this case, in equation [3.11], the term $\left\langle \vec{\hat{R}}_n \right\rangle_e$ is given by:

$$\left\langle \vec{\hat{R}}_n \right\rangle_e = \int \Psi_e^{0*} \vec{\hat{R}}_n \Psi_e^0 d\tau_e = \vec{\hat{R}}_n \int \Psi_e^{0*} \Psi_e^0 d\tau_e = \vec{\mu} \qquad [3.12]$$

It is important to note that this term is zero if the transition takes place between different electronic states $\left| \Psi_e^f \right\rangle$ and $\left| \Psi_e^i \right\rangle$ as a result of the orthogonality of electronic states:

$$\int \Psi_e^{f*} \Psi_e^i d\tau_e = 0$$

while the integral $\int \Psi_e^{f*} \vec{\hat{R}}_e \Psi_e^i d\tau_e$ is no longer zero.

Since the dipole moment is expressed in the mobile system attached to the equilibrium configuration of the molecule, and the electric field of the radiation is evaluated with respect to the fixed laboratory reference frame (by space isotropy in the gas phase, the components of the oscillating field E are divided in thirds on each axis), in general, the field is directed along the OZ axis, and for the calculation of the transition operator, the component of the dipole moment on the OZ axis (multiplied by 1/3) is determined, such that:

$$\mu_Z = \sum_{\alpha=x,y,z} \mu_\alpha \Phi_{Z\alpha}$$

depending on the components of the dipole moment μ_α $(\alpha = x, y, z)$ on the mobile axis α $(\alpha = x, y, z)$ and on the direction cosine $\Phi_{Z\alpha}$ of the angles between the OZ axis and the $\mathrm{O}\alpha$ $(\alpha = x, y, z)$ axes.

The expression of $\vec{R}_{fi}$ is then given by:

$$\vec{\hat{R}}_{fi} = \int \Psi_V^{f*} \Psi_R^{f*} \sum\nolimits_{\alpha=x,y,z} \mu_\alpha \Phi_{Z\alpha} \Psi_V^i \Psi_R^i d\tau_n \qquad [3.13]$$

The separation of vibration and rotation leads to:

$$\vec{\hat{R}}_{fi} = \sum_{\alpha=x,y,z} \int \Psi_V^{f*} \mu_\alpha \Psi_V^i d\tau_V \times \int \Psi_R^{f*} \Phi_{Z\alpha} \Psi_R^i d\tau_R \qquad [3.14]$$

and the selection rules can be established from products of elements of vibrational matrices $\langle \Psi_V^f | \mu_\alpha | \Psi_V^i \rangle$ and elements of rotational matrices $\langle \Psi_R^f | \Phi_{Z\alpha} | \Psi_R^i \rangle$.

For low-amplitude motions, the dipole moment can be expanded as a function of internal or normal vibrational coordinates. For the component α (x, y or z), it can be written as:

$$\mu_\alpha = \mu_\alpha^e + \sum_{s,\sigma} \frac{\partial \mu_\alpha}{\partial q_{s\sigma}} q_{s\sigma} + \frac{1}{2} \sum_{s,\sigma,s',\sigma'} \frac{\partial^2 \mu_\alpha}{\partial q_{s\sigma} \partial q_{s'\sigma'}} q_{s\sigma} q_{s'\sigma'} + \ldots \qquad [3.15]$$

where μ_α^e is the component α of the permanent dipole moment of the molecule. Depending on the symmetry of the molecule, this dipole moment is not always present. The same is applicable for the terms of the expansion of the dipole moment as a function of the vibrational internal or normal coordinates (stretching, bending, torsion, etc.).

The bar spectra are calculated as follows:

$$K^0(\sigma) = \frac{8\pi^3 N\sigma}{3hc} \sum_{i,f} \left(\rho_a^i - \rho_a^f\right) \left| \left\langle i \left| \sum_{\alpha=x,y,z} \frac{d\mu_\alpha}{dq_s} q_s \Phi_{Z\alpha} \right| f \right\rangle \right|^2 \times \delta\left(\omega - \omega_{if} - \Delta\omega_{v_i v_f}\right) \qquad [3.16]$$

for the transitions involving two vibration–orientation or vibration–rotation states (different non-degenerate vibration states) $|v_i, J_i, K_i, M_i\rangle \rightarrow |v_f, J_f, K_f, M_f\rangle$ (near-IR spectrum), and:

$$K^0(\sigma) = \frac{8\pi^3 N\sigma}{3hc} \sum_{i,f} \left(\rho_a^i - \rho_a^f\right) \left| \left\langle i \left| \sum_{\alpha=x,y,z} \mu_\alpha^e \Phi_{Z\alpha} \right| f \right\rangle \right|^2 \times \delta\left(\omega - \omega_{if} - \Delta\omega_{v_i v_f}\right) \qquad [3.17]$$

for the transitions involving only orientation or rotation states (same vibration state) $|v_i, J_i, K_i, M_i\rangle \rightarrow |v_i, J_f, K_f, M_f\rangle$ (far-IR spectrum).

In the expressions of equations [3.16] and [3.17], the population ρ_a^i (respectively, ρ_a^f) of the initial (respectively final) state of the optically active system is expressed as a function of its energy E_i (respectively, E_f) and the partition function Z as follows:

$$\rho_a^i = \frac{e^{-E_i/k_BT}}{Z} = \frac{e^{-E_i/k_BT}}{\sum_n e^{-E_n/k_BT}} \qquad [3.18]$$

and the elements of transitions of the dipole moment are expressed in the fixed reference frame.

When the "orientation–thermal bath" coupling is introduced, the shifts of lines $\Delta^{(1)}$ and $\Delta^{(2)}$, and their homogeneous width $\Gamma^{(2)}$ can be calculated. In the hypothesis of the Lorentzian model, the expression of the spectral profile of the molecule in interaction with its environment is:

$$K^0(\sigma) = \frac{8\pi^3 N\sigma}{3hc} \sum_{\substack{v_i \\ v_f}} \left|\langle v_i | q_s | v_f \rangle\right|^2 \sum_{\substack{j_i, m_i, \\ j_f, m_f}} g_i \frac{e^{-E_{v_i j_i m_i}/k_BT} - e^{-E_{v_f j_f m_f}/k_BT^2}}{Z}$$
$$\times \left|\left\langle j_i m_i \left| \frac{d\mu_\alpha}{dq_s} \Phi_{Z\alpha} \right| j_f m_f \right\rangle\right|^2 \frac{\Gamma^{(2)}}{\left(\omega - \omega_{v_i j_i m_i \to v_f j_f m_f} - \Delta\omega_{v_i v_f} - \Delta^{(1)} - \Delta^{(2)}\right)^2 + \left(\Gamma^{(2)}\right)^2} \qquad [3.19]$$

for the vibration–orientation or vibration–rotation transitions for a component α of the dipole moment (cold or fundamental bands) and where $\boldsymbol{g_i}$ is the degeneracy due to the nuclear of the initial level of the transition; this leads to:

$$K^0(\sigma) = \frac{8\pi^3 N\sigma}{3hc} \sum_{\substack{j_i, m_i, \\ j_f, m_f}} g_i \frac{e^{-E_{v_i j_i m_i}/k_BT} - e^{-E_{v_i j_f m_f}/k_BT}}{Z} \left|\langle j_i m_i | \mu_\alpha^e \Phi_{Z\alpha} | j_f m_f \rangle\right|^2$$
$$\times \frac{\Gamma^{(2)}}{\left(\omega - \omega_{v_i j_i m_i \to v_i j_f m_f} - \Delta^{(1)} - \Delta^{(2)}\right)^2 + \left(\Gamma^{(2)}\right)^2} \qquad [3.20]$$

for the orientation or rotation transitions only for a component α of the dipole moment.

The selection rules are established from elements of matrices $\langle v_i | q_s | v_f \rangle$ and $\left\langle j_i m_i \left| \frac{d\mu_\alpha}{dq_s} \Phi_{Z\alpha} \right| j_f m_f \right\rangle$ for the vibration–orientation or vibration–rotation transition and elements of matrices $\langle j_i m_i | \mu_\alpha \Phi_{Z\alpha} | j_f m_f \rangle$ for orientation or rotation transitions.

It should be recalled that in IR spectroscopy, when the operator R is one of the components of the electric dipole moment (μ_x, μ_y, μ_z, see sections 2.3.3 and 2.4.2,

Volume 1 [DAH 17]), then it gives rise to photon absorption or emission spectroscopy and that in elastic scattering spectroscopy (Rayleigh, Mie) or inelastic scattering spectroscopy (Raman, Brillouin), R is one of the operators related to elements $\alpha_{xx}, \alpha_{yy}, \alpha_{zz}, \alpha_{xy}, \alpha_{yz}, \alpha_{xz}$ of the polarizability tensor of the molecule.

3.6. Irreducible representations of the vibrations of the molecules

The irreducible representations labeling the symmetry of the vibrations of a molecule are built from irreducible representations of the point-group symmetry of the molecule. This symmetry group is also that of the electronic and vibrational Hamiltonian of the molecule, but it is not that of the rotational Hamiltonian (rigid rotator) and spin variables connected to the overall rotational degrees of freedom of the molecule, and of the nuclear spin variables as described in section 3.8. The selection rules during a transition between two vibronic states $|\Psi^i_{eV}\rangle = |\Psi^i_e\rangle \times |\Psi^i_V\rangle$ and $|\Psi^f_{eV}\rangle = |\Psi^f_e\rangle \times |\Psi^f_V\rangle$ are determined from the type of symmetry of these states, built from irreducible representations of the point-group symmetry of the molecule.

3.6.1. *Procedure for the decomposition of the reducible representation*

The reducible representation Γ_r of a group G is the direct sum of irreducible representations Γ_i of the group (see equation [1.22]: $\Gamma = a_1\Gamma^1 \oplus a_2\Gamma^2 \oplus \ldots \oplus a_d\Gamma^d = \sum_{\oplus\alpha} a_\alpha \Gamma^\alpha$. This decomposition uses formula [1.23]:

$$a_\alpha = \frac{1}{h}\sum_{O_i \in G} h_i \chi(O_i)\left(\chi^\alpha(O_i)\right)^*$$

used in Chapter 1 to calculate a_α.

This reduction formula provides a procedure for the reduction of the representation generated by a set of basic functions. In formula [1.23]:

– a_α is the number of times the irreducible representation α occurs in representation Γ_r, which is under reduction;

– h is the order of the group, i.e. the number of operations in the group (it should be kept in mind that it is not always equal to the number of classes of the operations of the group);

– $\sum_{O_i \in G} O_i$ denotes an addition over all the operations O_i of the group. If there is more than one operation in a given class, each operation must be included in the sum, multiplied by h_i;

– $\chi(O_i)$ is the character of the reducible representation r under the symmetry operation O_i;

– $\chi^{\alpha}(O_i)$ is the character of the irreducible representation α under the symmetry operation O_i.

In order to determine the reducible representation Γ_{3N}, and the character $\chi(O_i)$ of the reducible representation under the symmetry operation O_i, a function basis is necessary. A system of axes is generally associated with each atom or nucleus (Cartesian reference frame) in the molecule, and a matrix corresponding to each operation of the group is determined in order to calculate the trace, i.e. the character attached to the symmetry operation. It can be shown that in order to calculate this character, it is sufficient to determine the number of invariant nuclei in the symmetry operation and to multiply this number by the character of the symmetry such as that given in Table 1.6 (giving the characters of five basic symmetry operations operating on the basis vectors of a three-dimensional Cartesian space). The character of the reducible representation (Γ_{3N}) is thus obtained for each class of symmetry operation of the group, which makes it possible to determine the reduction of the representation. The characters of the irreducible representations of the group are given in the table of characters of the symmetry group of the molecule. This table also summarizes the type of symmetry attached to the translation degrees of freedom (equivalent to the components of the dipole moment of the molecule), rotation degrees of freedom (equivalent to the components of the rotation operators) and the products of components of types $x^2, y^2, z^2, (x^2 + y^2 + z^2, x^2 - y^2, 2z^2 - x^2 - y^2)$ xy, xz, yz (equivalent to the elements of the polarizability tensor of the molecule) or electronic orbitals (s, p, d, etc.).

In the following, the procedure described is applied to each type of molecule for which the symmetry groups were established in Chapter 1. It should be noted that, if instead of using a basis in a (three-dimensional) Cartesian reference frame, internal coordinates are used, each coordinate represents a basis of a one-dimensional space and the use of Table 1.6 is not required in order to determine, in the reducible representation (Γ_R) considered, the character corresponding to each symmetry operation of the group. The number of coordinates invariant under the symmetry operation gives the corresponding character directly.

3.6.2. *Case of symmetric tops XY_3 and XZY_3 (NH_3, ND_3, CDH_3, CHD_3)*

The symmetry group of symmetric tops XY_3 and XZY_3 (NH_3, ND_3, CDH_3, CHD_3) is denoted by C_{3v}. To build the table of reducible representation that gives the characters of the symmetry operations of the group considering all the nuclei of the molecule, the number of nuclei invariant under the symmetry operation is put on one row, in the column of each operation class; the next row features the character of the operation for each invariant nucleus, and finally the last row features the character associated with each operation by multiplying in each column the elements of the two previous rows.

For NH_3 and ND_3, this leads to Table 3.1.

C_{3v}	E	$2C_3^z$	$3\sigma_v$
Invariant nuclei	4	1	2
Character for a nucleus	3	0	1
Γ_{3N}	12	0	2

Table 3.1. *Table of characters of the reducible representation for NH_3*

Under the identity symmetry operation E, the four nuclei remain in their respective positions and are therefore invariant ($N = 4$). Under the symmetry operation of rotation about the C_3 axis, only the nitrogen nucleus preserves its position, so that only one nucleus is invariant ($N = 1$). Under the symmetry plane operation σ_v, the nitrogen nucleus and one of the hydrogen nuclei that is in the symmetry plane preserve their respective positions and are therefore invariant (N = 2). To establish the characters of operations from Table 1.6, the symmetry operation of rotation about the C_3 axis must be assigned the value $\varphi = 120°$, which leads to $\chi(C_3) = 2(-1/2) + 1 = 0$.

Similarly, for CDH_3 and CHD_3, Table 3.2 is obtained as follows:

C_{3v}	E	$2C_3^z$	$3\sigma_v$
Invariant nuclei	5	2	3
Character for a nucleus	3	0	1
Γ_{3N}	15	0	3

Table 3.2. *Table of characters of the reducible representation for CDH_3*

For CDH_3 and CHD_3, there is one more nucleus on the Oz axis, which is the rotation axis C_3, contained in the symmetry plane σ_v.

The symmetry group of the symmetric tops XY_3 and XZY_3 (NH_3, ND_3, CDH_3, CHD_3) is denoted by C_{3v}. For the sake of clarity, the table with the various elements that can be used to apply the procedure described in section 3.3.1, is given once more below.

C_{3v}	E	$2C_3^z$	$3\sigma_v$	Polar vector	Axial vector	Second-order tensor
A_1	1	1	1	z, T_Z		$x^2 + y^2$, z^2
A_2	1	1	-1		R_z	
E	2	-1	0	(x,y), (T_X, T_Y)	(R_x, R_y)	(x^2 - y^2, xy) (xz,yz)

Table 3.3. *Table of characters and symmetry properties of the operators of group C_{3v}*

Based on the table of characters of the reducible representation (see Tables 3.1 and 3.2), given in Table 3.3, and the reduction formula [1.23], each a_α is calculated for the symmetry group C_{3v} of order 6 ($h = 6$).

For the irreducible representations A_1 and A_2 (Mulliken symbol), Table 3.4a and b are established for NH_3:

A_1	E	$2C_3^z$	$3\sigma_v$	a_{A1}
$\chi^\alpha(O_i)$	1	1	1	
$\chi(O_i)$	12	0	2	
h_i	1	2	3	
Product/6	2	0	1	3

a)

A_2	E	$2C_3^z$	$3\sigma_v$	a_{A2}
$\chi^\alpha(O_i)$	1	1	-1	
$\chi(O_i)$	12	0	2	
h_i	1	2	3	
Product/6	2	0	-1	1

b)

Table 3.4. *Coefficients a_α of the irreducible representations A_1 and A_2 of group C_{3v} for NH_3*

For the two-dimensional irreducible representation, *E* (Mulliken symbol for 2D representation), Table 3.5 is established for NH_3:

E	E	$2C_3^z$	$3\sigma_v$	a_E
$\chi^\alpha(\mathbf{O_i})$	2	-1	0	
$\chi(\mathbf{O_i})$	12	0	2	
h_i	1	2	3	
Product/6	4	0	0	4

Table 3.5. *Coefficients a_α of the irreducible representation E of group C_{3v} for NH_3*

The results summarized in Tables 3.4(a) and (b) and 3.5 lead to:

$$\Gamma_{3N} = 3A_1 \oplus A_2 \oplus 4E \quad [3.21]$$

It can be verified that there are 12 degrees of freedom for four atoms. This decomposition concerns all the translational, overall rotational and vibrational degrees of freedom of the molecule. Referring to the table of characters of group C_{3v} (Table 3.3), it can be written that:

$$\Gamma_{tr} = A_1 \oplus E \quad [3.22]$$

$$\Gamma_{rot} = A_2 \oplus E \quad [3.23]$$

The decomposition for the vibrational degrees of freedom is given by:

$$\Gamma_{vib} = 2A_1 \oplus 2E \quad [3.24]$$

and therefore in terms of two vibrations of type A_1 which are one-dimensional and two vibrations of type *E* which are two-dimensional.

For the irreducible representations A_1 and A_2, the following (Table 3.6a and b) are established for CDH_3:

A_1	E	$2C_3^z$	$3\sigma_v$	a_{A1}
$\chi^\alpha(O_i)$	1	1	1	
$\chi(O_i)$	15	0	3	
h_i	1	2	3	
Product/6	15/6	0	9/6	4

a)

A_2	E	$2C_3^z$	$3\sigma_v$	a_{A2}
$\chi^\alpha(O_i)$	1	1	-1	
$\chi(O_i)$	15	0	3	
h_i	1	2	3	
Product/6	15/6	0	-9/6	1

b)

Table 3.6. *Coefficients a_α of the irreducible representations A_1 and A_2 of group C_{3v} for CDH_3*

For the two-dimensional irreducible representation, *E*, Table 3.7 is established for CDH_3:

E	E	$2C_3^z$	$3\sigma_v$	a_E
$\chi^\alpha(O_i)$	2	-1	0	
$\chi(O_i)$	15	0	3	
h_i	1	2	3	
Product/6	5	0	0	5

Table 3.7. *Coefficients a_α of the irreducible representation E of group C_{3v} for CDH_3*

From the results summarized in Tables 3.6(a) and (b) and Table 3.7, the following can be written:

$$\Gamma_{3N} = 4A_1 \oplus A_2 \oplus 5E \qquad [3.25]$$

It can be verified that there are 15 degrees of freedom for five atoms. This decomposition concerns all the translational, overall rotational and vibrational degrees of freedom of the molecule. Referring to the table of characters of group C_{3v} (Table 3.3), the following still holds: $\Gamma_{tr} = A_1 \oplus$ E and $\Gamma_{rot} = A_2 \oplus E$.

The decomposition for the vibrational degrees of freedom is given by:

$$\Gamma_{vib} = 3A_1 \oplus 3E \qquad [3.26]$$

and therefore in terms of three vibrations of type A_1 which are one-dimensional and three vibrations of type *E* which are two-dimensional.

3.6.3. *Case of spherical top XY_4 (CH_4, CD_4)*

The symmetry group of spherical tops of type XY_4 (CH_4, CD_4) is denoted by T_d. The table of characters of the symmetry operations of the group, considering all the nuclei of the molecule for the considered reducible representation, is given by the elements of Table 3.8.

T_d	E	$8C_3$	$3C_2$	$6S_4$	$6\sigma_d$
Invariant nuclei	5	2	1	1	3
Character for a nucleus	3	0	-1	-1	1
Γ_{3N}	15	0	-1	-1	3

Table 3.8. *Table of characters of the reducible representation for CH_4*

The symmetry group of spherical tops XY_4 is denoted by T_d. Table 2.7 recalls its table of characters.

Using the table of characters of the reducible representation (see Table 3.8) given in Table 2.7, and the reduction formula [1.23], each a_α can be calculated for the symmetry group T_d of order 24 ($h = 24$).

For the irreducible representations A_1 and A_2, Table 3.9a and b can be established for CH_4 or CD_4. For the two-dimensional irreducible representation, *E*, Table 3.9c is established for CH_4 or CD_4. For the three-dimensional irreducible representations, T_1 and T_2, Table 3.9d and e are established for CH_4 or CD_4.

A_1	E	$8C_3$	$3C_2$	$6S_4$	$6\sigma_d$	a_{A1}
$\chi^\alpha(O_i)$	1	1	1	1	1	
$\chi(O_i)$	15	0	-1	-1	3	
h_i	1	8	3	6	6	
Product/24	15/24	0	-3/24	-6/24	18/24	1

a)

A_2	E	$8C_3$	$3C_2$	$6S_4$	$6\sigma_d$	a_{A2}
$\chi^\alpha(O_i)$	1	1	1	-1	-1	
$\chi(O_i)$	15	0	-1	-1	3	
h_i	1	8	3	6	6	
Product/24	15/24	0	-3/24	6/24	-18/24	0

b)

E	E	$8C_3$	$3C_2$	$6S_4$	$6\sigma_d$	a_E
$\chi^{\alpha}(O_i)$	2	-1	2	0	0	
$\chi(O_i)$	15	0	-1	-1	3	
h_i	1	8	3	6	6	
Product/24	30/24	0	-6/24	0/24	0/24	1

c)

T_1	E	$8C_3$	$3C_2$	$6S_4$	$6\sigma_d$	a_{T1}
$\chi^{\alpha}(O_i)$	3	0	-1	1	-1	
$\chi(O_i)$	15	0	-1	-1	3	
h_i	1	8	3	6	6	
Product/24	45/24	0	3/24	-6/24	-18/24	1

d)

T_2	E	$8C_3$	$3C_2$	$6S_4$	$6\sigma_d$	a_{T2}
$\chi^{\alpha}(O_i)$	3	0	-1	-1	1	
$\chi(O_i)$	15	0	-1	-1	3	
h_i	1	8	3	6	6	
Product/24	45/24	0	3/24	6/24	18/24	3

e)

Table 3.9. *Coefficients a_α of the irreducible representations a) A_1 ; b) A_2 ; c) E; d) T_1 and e) T_2 of group T_d for CH_4*

From the results summarized in Table 3.9a–e, it can be written that:

$$\Gamma_{3N} = A_1 \oplus E \oplus T_1 \oplus 3\,T_2 \qquad [3.27]$$

It can be verified that there are 15 degrees of freedom for five atoms. This decomposition concerns all the translational, the overall rotational and vibrational degrees of freedom of the molecule.

Referring to the table of characters of group T_d (Table 2.7), it can be written that:

$$\Gamma_{tr} = T_2 \quad [3.28]$$

$$\Gamma_{rot} = T_1 \quad [3.29]$$

The decomposition for the vibrational degrees of freedom is given by:

$$\Gamma_{vib} = A_1 \oplus E \oplus 2T_2 \quad [3.30]$$

and therefore in terms of one vibration of type A_1, one two-dimensional vibration of type E and two three-dimensional vibrations of type T_2.

3.6.4. *Case of the asymmetric top XY_2Z_2 (CH_2D_2)*

The symmetry group of asymmetric tops of type XY_2Z_2 (CH_2D_2) is denoted by C_{2v}.

The table of characters of the symmetry operations of the group, taking into account all the nuclei of the molecule for the considered reducible representation, is given by the elements in Table 3.10.

C_{2v}	E	C_2	$\sigma_V(xz)$	$\sigma'_V(yz)$
Invariant nuclei	5	1	3	3
Character for a nucleus	3	-1	1	1
Γ_{3N}	15	-1	3	3

Table 3.10. *Table of characters of the reducible representation for CH_2D_2*

The symmetry group of spherical tops XY_2Z_2 is denoted by C_{2v}. Table 3.11 recalls its table of characters.

C_{2v}	E	C_2	$\sigma_V(xz)$	$\sigma'_V(yz)$		
A_1	1	1	1	1	z, T_z	$x^2 + y^2 + z^2$
A_2	1	1	-1	-1	R_z	xy
B_1	1	-1	1	-1	x, T_x, R_y	xz
B_2	1	-1	-1	1	y, T_y, R_z	yz

Table 3.11. *Table of characters and symmetry properties of operators of group C_{2v}*

The table of characters for the reducible representation (see Table 3.10) given in Table 3.11, and the reduction formula [1.23], can be used to calculate each a_α for the symmetry group C_{2v} of order 4 ($h = 4$).

For the irreducible representations A_1 and A_2, Tables 3.12(a) and (b) can be established for CH_2D_2. For the irreducible representations B_1 and B_2, Tables 3.12(c) and (d) can be established for CH_2D_2.

A_1	E	C_2	$\sigma_V(xz)$	$\sigma'_V(yz)$	a_{A1}
$\chi^\alpha(O_i)$	1	1	1	1	
$\chi(O_i)$	15	-1	3	3	
h_i	1	1	1	1	
Product/4	15/4	-1/4	3/4	3/4	5

a)

A_2	E	C_2	$\sigma_V(xz)$	$\sigma'_V(yz)$	a_{A2}
$\chi^\alpha(O_i)$	1	1	-1	-1	
$\chi(O_i)$	15	-1	3	3	
h_i	1	1	1	1	
Product/4	15/4	-1/4	-3/4	-3/4	2

b)

B_1	E	C_2	$\sigma_V(xz)$	$\sigma'_V(yz)$	a_{B1}
$\chi^\alpha(O_i)$	1	-1	1	-1	
$\chi(O_i)$	15	-1	3	3	
h_i	1	1	1	1	
Product/4	15/4	1/4	3/4	-3/4	4

c)

B_2	E	C_2	$\sigma_V(xz)$	$\sigma'_V(yz)$	a_{B2}
$\chi^\alpha(O_i)$	1	-1	-1	1	
$\chi(O_i)$	15	-1	3	3	
h_i	1	1	1	1	
Product/4	15/4	1/4	-3/4	3/4	4

d)

Table 3.12. *Coefficients a_α of the irreducible representations a) A_1 ; b) A_2; c) B_1 and d) B_2 of group C_{2v} for CH_2D_2*

Using the results summarized in Tables 3.12a–d, it can be written that:

$$\Gamma_{3N} = 5A_1 \oplus 2A_2 \oplus 4B_1 \oplus 4B_2 \quad [3.31]$$

It can be verified that there are 15 degrees of freedom for five atoms. This decomposition concerns all the translational, the overall rotational and vibrational degrees of freedom of the molecule. Referring to the table of character for group C_{2v} (Table 3.11), it can be written that:

$$\Gamma_{tr} = A_1 \oplus B_1 \oplus B_2 \quad [3.32]$$

$$\Gamma_{rot} = A_2 \oplus B_1 \oplus B_2 \quad [3.33]$$

The decomposition for the vibration degrees of freedom is given by:

$$\Gamma_{vib} = 4A_1 \oplus A_2 \oplus 2B_1 \oplus 2B_2 \quad [3.34]$$

and therefore in terms of four vibrations of type A_1, one vibration of type A_2, two vibrations of type B_1 and two vibrations of type B_2.

3.6.5. *Case of the asymmetric top XY_2Z (NDH_2 or NHD_2)*

The point-group symmetry of the isotopic varieties NHD_2 and NDH_2 is C_{1v} of order 2, composed of the identity E and the plane σ_{xz}. Applying the methods developed above for the various cases described, it can be easily shown that:

$$\Gamma_{3N} = 7A_1 \oplus 5A_2 \quad [3.35]$$

It can be verified there are 12 degrees of freedom for four atoms. This decomposition concerns all the translational, the overall rotational and vibrational degrees of freedom of the molecule.

Using the table of characters for group C_{1v} (see Table 1.4), it can be written that:

$$\Gamma_{tr} = 2A_1 \oplus A_2 \quad [3.36]$$

$$\Gamma_{rot} = A_1 \oplus 2A_2 \quad [3.37]$$

The decomposition for the vibration degrees of freedom is given by:

$$\Gamma_{vib} = 4A_1 \oplus 2A_2 \quad [3.38]$$

and therefore in terms of four vibrations of type A_1 and two vibrations of type A_2.

3.6.6. *Case of inversion for NH_3, ND_3, NDH_2 and NHD_2*

As highlighted in Chapter 1, the direct product groups, group D_{3h} (NH_3 and ND_3) isomorphic to the product group $C_{3v} \otimes C_s$ and the group C_{2v} (NHD_2 and NDH_2) isomorphic to the product group $C_{1v} \otimes C_s$ for the corresponding isotopic varieties, have irreducible representations for labeling energy levels and the corresponding eigenfunctions.

Bunker and Jensen [BUN 98] describe the method developed for building the molecular symmetry (MS) group when taking inversion into account, when observable ("feasible" configurations, according to Longuet-Higgins) from the permutation-inversion group G^{CNPI}.

The permutation-inversion operations of group G^{CNPI} that are not feasible are excluded from the MS group, as described in Chapter 1. The case of inversion is presented in section 3.8, which is dedicated to rotational and spin degrees of freedom.

3.7. Types of vibrations of irreducible representations

3.7.1. *Case of symmetric tops XY_3 and XZY_3 (NH_3, ND_3, CDH_3, CHD_3)*

It is possible to determine the type of vibration corresponding to each irreducible representation. For this purpose, stretching vibrations and bending vibrations are considered separately, using internal coordinates. For stretching, each bond r_i $(i = 1, 2, 3, ..., n)$ is assigned a displacement along the bond Δr_i $(i = 1, 2, 3, ..., n)$, and each bending characterized by the angle θ_i $(i = 1, 2, 3, ..., n)$ (generally taken in opposition to coordinate r_i) between the bonds is assigned an angle variation $\Delta\theta_i$ $(i = 1, 2, 3, ..., n)$, where n is the number of bonds in the molecule (see Figure 3.1(a) and (b)). Internal coordinates ρ_i $(i = 1, 2, 3, ..., p)$ can also be assigned in the case of torsion, depending on the number of torsion degrees of freedom.

For non-planar molecules, such as NH_3 or ND_3, with a rotation symmetry axis C_3, along which the Oz axis is positioned, there are (see Figure 3.1(a)) three internal stretching coordinates Δr_1, Δr_2, Δr_3, and three internal bending coordinates $\Delta\theta_1$, $\Delta\theta_2$, $\Delta\theta_3$. The correspondences $\Delta\theta_1 \rightarrow \theta_{23}$, $\Delta\theta_2 \rightarrow \theta_{13}$, $\Delta\theta_3 \rightarrow \theta_{12}$ can be noted, where θ_{ij} is the angle between bond r_i and bond r_j.

Similarly to the determination of the type of symmetry corresponding to an irreducible representation, the projection theorem is applied to internal coordinates using the full table of symmetry operations to determine the type of combinations of internal coordinates corresponding to the given irreducible representation.

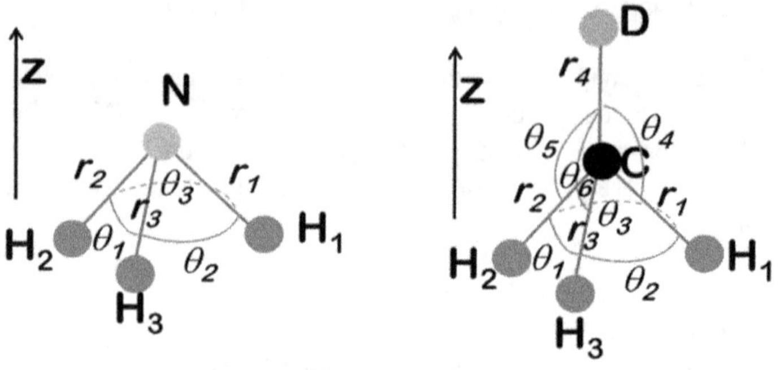

a) b)

Figure 3.1. *Internal coordinates of NH_3 and CDH_3. For a color version of this figure, see www.iste.co.uk/dahoo/infrared.zip*

As a first step, the table giving the result of a group symmetry operation on Δr_i and $\Delta\theta_i$ should be built. Then, for each type of irreducible representation, A_1 and E for NH_3 or ND_3, the type of linear combination that can be used as a basis for the irreducible representation must be determined. Applying the symmetry operations of group C_{3v3}, Table 3.13 is obtained for NH_3 and ND_3 (or a molecule of type XY_3).

$\mathbf{C_{3v}}$	**E**	$\mathbf{C_3^1}$	$\mathbf{C_3^2}$	$\boldsymbol{\sigma_v}$	$\boldsymbol{\sigma_v'}$	$\boldsymbol{\sigma_v''}$
$\mathbf{O_k(r_1)}$	r_1	r_2	r_3	r_1	r_3	r_2
$\mathbf{O_k(\theta_1)}$	θ_1	θ_2	θ_3	θ_1	θ_3	θ_2
$\mathbf{A_1}$	1	1	1	1	1	1
$\mathbf{\chi(A_1)xO_k(r_1)}$	r_1	r_2	r_3	r_1	r_3	r_2
$\mathbf{\chi(A_1)xO_k(\theta_1)}$	θ_1	θ_2	θ_3	θ_1	θ_3	θ_2
$\mathbf{O_k(E)}$	2	-1	−1	0	0	0
$\mathbf{\chi(A_1)xO_k(r_1)}$	$2r_1$	$-r_2$	$-r_3$	0	0	0
$\mathbf{\chi(E)xO_k(\theta_1)}$	$2\theta_1$	$-\theta_2$	$-\theta_3$	0	0	0

Table 3.13. *Table of transformation of r_i and θ_i of group C_{3v} of NH_3*

For the representation of type A_1, up to a normalization constant, the corresponding vibration coordinates of stretching and bending are obtained by applying the projection theorem (see equation [1.27]) using the results in row 4 for $r_1^{A_1}$ and in row 5 for $\theta_1^{A_1}$, which lead to:

$$r_1^{A_1} = \frac{1}{\sqrt{3}}(\Delta r_1 + \Delta r_2 + \Delta r_3) = S_1^{A_1}$$
$$\theta_1^{A_1} = \frac{1}{\sqrt{3}}(\Delta\theta_1 + \Delta\theta_2 + \Delta\theta_3) = S_2^{A_1} \qquad [3.39]$$

For stretching, the length of the three bonds increases or decreases at the same time (symmetric stretching). Similarly, the angles between bonds increase or decrease at the same time (respiration mode). These two symmetric vibrations are reproduced in Figure 4.3.

For the representation of type *E*, from the results in rows 7 and 8, up to a normalization constant, the corresponding stretching and bending vibration coordinates are given by:

$$r_{21}^{E} = \frac{1}{\sqrt{6}}(2\Delta r_1 - \Delta r_2 - \Delta r_3) = S_{3a}^{E}$$
$$\theta_{21}^{E} = \frac{1}{\sqrt{6}}(2\Delta\theta_1 - \Delta\theta_2 - \Delta\theta_3) = S_{4a}^{E} \qquad [3.40]$$

For stretching, a bond increases (respectively decreases) with an amplitude that is twice the one corresponding to the simultaneous decrease (respectively increase) of the other two bonds (antisymmetric stretching). The same is applicable to angles between bonds. An angle increases (respectively decreases) with an amplitude that is twice the one corresponding to the simultaneous decrease (respectively increase) of the other two angles (antisymmetric stretching).

With the irreducible representation of type *E* being two dimensional, the second stretching and bending vibration can be determined by applying to r_{21}^{E} and θ_{21}^{E}, the symmetry operation C_3^1, and by subtracting the result from r_{21}^{E} or θ_{21}^{E}. This leads to:

$$r_{22}^{E} = \frac{1}{\sqrt{2}}(\Delta r_1 - \Delta r_2) = S_{3b}^{E}$$
$$\theta_{22}^{E} = \frac{1}{\sqrt{2}}(\Delta\theta_1 - \Delta\theta_2) = S_{4b}^{E} \qquad [3.41]$$

For stretching, one bond remains fixed and one bond increases (respectively decreases), while the other decreases (respectively increases). The same is applicable to angles between bonds. One angle remains fixed and one angle increases (respectively decreases), while the other decreases (respectively increases). These are likewise antisymmetric stretching and bending vibrations.

The vibration modes are thus split as:

$$\Gamma_{vib}^{stretching} = A_1 \oplus E$$
$$\Gamma_{vib}^{bending} = A_1 \oplus E \qquad [3.42]$$

with the vibrations of type A_1 being symmetric, while those of type *E* are antisymmetric.

For the selection rules, the table of characters in Table 3.3 shows that NH_3 (ND_3) is active in both IR and Raman spectra since Z and α_{xx} are of type A_1, (X,Y) and $(\alpha_{xz}, \alpha_{yz})$ are of type E and vibrations of both types are possible.

The symmetries of the bending and stretching coordinates can be determined without finding the form of these coordinates, which is given by the linear combination of these internal coordinates. It is sufficient to determine the number of internal coordinates that are invariant under a symmetry operation of the group and then apply formula [1.23].

The second line in Table 3.5(a) and (b) gives the number of internal coordinates for stretching and bending, respectively, invariant under a symmetry operation of group C_{3v}, which makes it possible to calculate the coefficients a_α as the sum of products obtained for each type of irreducible representation.

The data in Tables 3.14(a) and (b) lead to the same results as those given in equation [3.43].

Stretching	E	$2C_3^z$	$3\sigma_v$	a_α
h_i	1	2	3	
$\chi(O_i)$	3	0	1	
$\chi^{A1}(O_i)$	1	1	1	
$\pi^{A1}/6$	0.5	0	0.5	1
$\chi^{A2}(O_i)$	1	1	-1	
$\pi^{A2}/6$	0.5	0	-0.5	0
$\chi^{E}(O_i)$	2	-1	0	
$\pi^{E}/6$	1	0	0	1

a)

Bending	E	$2C_3^z$	$3\sigma_v$	a_α
h_i	1	2	3	
$\chi(O_i)$	3	0	1	
$\chi^{A1}(O_i)$	1	1	1	
$\pi^{A1}/6$	0.5	0	0.5	1
$\chi^{A2}(O_i)$	1	1	-1	
$\pi^{A2}/6$	0.5	0	-0.5	0
$\chi^{E}(O_i)$	2	-1	0	
$\pi^{E}/6$	1	0	0	1

b)

Table 3.14. *Coefficients a_α of the irreducible representations of group C_{3v} for the internal stretching and bending coordinates of NH_3*

For the CHD_3 or CDH_3 molecules, there are four stretching internal coordinates Δr_1, Δr_2, Δr_3, Δr_4 and six bending internal coordinates $\Delta\theta_1$, $\Delta\theta_2$, $\Delta\theta_3$, $\Delta\theta_4$, $\Delta\theta_5$, $\Delta\theta_6$. With the same notations used for NH_3 (see Figure 3.1(a) and (b)), the bond CD is identified by r_4 and the additional angles θ_4, θ_5 and θ_6 are chosen between r_1 and r_4, r_2 and r_4, and r_3 and r_4, respectively. The following correspondences should be noted: $\Delta\theta_1 \rightarrow \theta_{23}$, $\Delta\theta_2 \rightarrow \theta_{13}$, $\Delta\theta_3 \rightarrow \theta_{12}$, $\Delta\theta_4 \rightarrow \theta_{14}$, $\Delta\theta_5 \rightarrow \theta_{24}$, $\Delta\theta_6 \rightarrow \theta_{34}$, where θ_{ij} is the angle between bond r_i and bond r_j.

Stretching	E	$2C_3^z$	$3\sigma_v$	a_α
h_i	1	2	3	
$\chi(O_i)$	4	1	2	
$\chi^{A1}(O_i)$	1	1	1	
$\pi^{A1}/6$	4/6	2/6	6/6	2
$\chi^{A2}(O_i)$	1	1	-1	
$\pi^{A2}/6$	4/6	2/6	-1	0
$\chi^{E}(O_i)$	2	-1	0	
$\pi^{E}/6$	8/6	-2/6	0	1

a)

Bending	E	$2C_3^z$	$3\sigma_v$	a_α
h_i	1	2	3	
$\chi(O_i)$	6	0	2	
$\chi^{A1}(O_i)$	1	1	1	
$\pi^{A1}/6$	1	0	1	2
$\chi^{A2}(O_i)$	1	1	-1	
$\pi^{A2}/6$	1	0	-1	0
$\chi^{E}(O_i)$	2	-1	0	
$\pi^{E}/6$	2	0	0	2

b)

Table 3.15. *Coefficients a_α of the irreducible representations of group C_{3v} for the stretching and bending internal coordinates of CDH_3*

Given that the number of vibrations is nine, 10 internal coordinates are depicted because, in contrast to NH_3, there is one constraint on the variation of angles, which cannot all increase simultaneously, which must be considered. This redundancy can be removed by inspection, by comparing the decomposition of the possible modes of vibrations, hence three vibrations of type A_1 and three vibrations of type E with two dimensions with the linear combinations of the internal coordinates that lead to the bases of the irreducible representations.

In the first step, the method involving the determination of the type of stretching and bending symmetry is applied, without trying to determine the form of the respective stretching or bending coordinate. Similarly to the NH_3 case, the number of coordinates that are invariant under the symmetry operations is determined, and then the coefficients a_α are calculated.

As shown in Tables 3.15(a) and (b), there are two stretchings of type A_1 and one of type E, and two bendings of type A_1 and two of type E, for a total of 10 vibrations.

In comparison with equation [3.26], there must be three vibrations of type A_1 and three of type E, which makes it possible to eliminate one bending vibration of type A_1. The following decomposition is obtained for the vibrations of CDH_3 in stretching and bending modes:

$$\Gamma_{vib}^{stretching} = 2A_1 \oplus E$$
$$\Gamma_{vib}^{bending} = A_1 \oplus 2E \qquad [3.43]$$

C_{3v}	E	C_3^1	C_3^2	σ_v	σ_v'	σ_v''
$O_k(r_1)$	r_1	r_2	r_3	r_1	r_3	r_2
$O_k(r_4)$	r_4	r_4	r_4	r_4	r_4	r_4
$O_k(\theta_1)$	θ_1	θ_2	θ_3	θ_1	θ_3	θ_2
$O_k(\theta_4)$	θ_4	θ_5	θ_6	θ_4	θ_6	θ_5
A_1	1	1	1	1	1	1
$\chi(A_1)xO_k(r_1)$	r_1	r_2	r_3	r_1	r_3	r_2
$\chi(A_1)xO_k(r_4)$	r_4	r_4	r_4	r_4	r_4	r_4
$\chi(A_1)xO_k(\theta_1)$	θ_1	θ_2	θ_3	θ_1	θ_3	θ_2
$\chi(A_1)xO_k(\theta_4)$	θ_4	θ_5	θ_6	θ_4	θ_6	θ_5
$O_k(E)$	2	-1	-1	0	0	0
$\chi(E)xO_k(r_1)$	$2r_1$	$-r_2$	$-r_3$	0	0	0
$\chi(E)xO_k(r_4)$	$2r_4$	$-r_4$	$-r_4$	0	0	0
$\chi(E)xO_k(\theta_1)$	$2\theta_1$	$-\theta_2$	$-\theta_3$	0	0	0
$\chi(E)xO_k(\theta_4)$	$2\theta_4$	$-\theta_5$	$-\theta_6$	0	0	0

Table 3.16. *Table of transformations of r_i and θ_i of group C_{3v} of CDH_3*

In order to determine the explicit form of these vibrations, obtained by a linear combination of internal coordinates, the same method used for NH_3 can be applied here.

The table that gives the results of the symmetry operations for group C_{3v} is first built on the internal coordinates, then the projection operator is applied (see equation [1.27] for each irreducible representation of the group.

For the representation of type A_1, from the results in rows 6 to 9 of Table 3.16, the following vibrations are thus determined:

$$\begin{aligned} r_{11}^{A_1} &= \frac{1}{\sqrt{3}}(\Delta r_1 + \Delta r_2 + \Delta r_3) = S_1^{A_1} \\ r_{12}^{A_1} &= \Delta r_4 = S_2^{A_1} \\ \theta_{11}^{A_1} &= \frac{1}{\sqrt{3}}(\Delta\theta_1 + \Delta\theta_2 + \Delta\theta_3) \\ \theta_{12}^{A_1} &= \frac{1}{\sqrt{3}}(\Delta\theta_4 + \Delta\theta_5 + \Delta\theta_6) \end{aligned} \quad [3.44]$$

Referring to the decomposition of vibrations of CDH_3 into stretching and bending modes (equation [3.26]), a redundant bending vibration is obtained. There is no argument allowing the choice of a form of bending $\theta_{11}^{A_1}$ relative to another form $\theta_{12}^{A_1}$. The two forms can be linked by building a Wang basis by symmetric and antisymmetric linear combination of the two forms such that:

$$\begin{aligned} \theta_s^{A_1} &= \frac{1}{\sqrt{2}}\left(\theta_{11}^{A_1} + \theta_{12}^{A_1}\right) \\ \theta_{as}^{A_1} &= \frac{1}{\sqrt{2}}\left(\theta_{11}^{A_1} - \theta_{12}^{A_1}\right) \end{aligned} \qquad [3.45]$$

Nevertheless, if the form $\theta_{11}^{A_1}$ is represented by vectors perpendicular to the bonds, it can be noted that when $\theta_{11}^{A_1}$ increases, the bonds are pushed up and $\theta_{12}^{A_1}$ decreases and vice versa. Consequently, only one bending vibration of type A_I is to be retained, from the antisymmetric linear combination of $\theta_{11}^{A_1}$ and $\theta_{12}^{A_1}$.

$$\theta_{as}^{A_1} = \frac{1}{\sqrt{6}}(\Delta\theta_1 + \Delta\theta_2 + \Delta\theta_3 - \Delta\theta_4 + \Delta\theta_5 + \Delta\theta_6) = S_3^{A_1} \qquad [3.46]$$

For the three symmetric vibrations of type A_I, there is a mode for stretching, characterized by three C–H bonds, whose length increases or decreases simultaneously (symmetric stretching) while the C–D bond remains fixed, and a mode for which the reverse situation is applicable, i.e. the C–D bond increases or decreases while the other three C–H bonds remain fixed. For the bending vibration, the angles between the C–H bonds increase while the angles between these bonds and the C–D bonds decrease and vice versa, simultaneously (respiration mode).

For the representation of type E, the corresponding coordinates of the stretching and bending vibration are determined from Table 3.16, similarly to NH_3, with a bending coordinate more, as:

$$\begin{aligned} r_{21}^{E} &= \frac{1}{\sqrt{6}}(2\Delta r_1 - \Delta r_2 - \Delta r_3) = S_{4a}^{E} \\ \theta_{21}^{E} &= \frac{1}{\sqrt{6}}(2\Delta\theta_1 - \Delta\theta_2 - \Delta\theta_3) = S_{5a}^{E} \\ \theta_{23}^{E} &= \frac{1}{\sqrt{6}}(2\Delta\theta_4 - \Delta\theta_5 - \Delta\theta_6) == S_{6a}^{E} \end{aligned} \qquad [3.47]$$

For stretching, a bond increases (respectively decreases) with an amplitude that is twice the one corresponding to decrease (respectively to increase) of the two other bonds at the same time (antisymmetric stretching). The same is applicable to the angles between the bonds. An angle increases (respectively decreases) with an amplitude that is twice the one corresponding to the decrease (respectively increase) of the two other angles at the same time (antisymmetric stretching).

Similarly to NH_3, the irreducible representation of type *E*, the second stretching and bending vibration is determined by applying the symmetry operation C_3^1 to r_{21}^{E}, θ_{21}^{E} and θ_{23}^{E}, and subtracting the result from r_{21}^{E}, θ_{21}^{E} or θ_{23}^{E}. This yields:

$$\begin{aligned} r_{22}^{E} &= \frac{1}{\sqrt{2}}(\Delta r_1 - \Delta r_2) = S_{4b}^{E} \\ \theta_{22}^{E} &= \frac{1}{\sqrt{2}}(\Delta\theta_1 - \Delta\theta_2) = S_{5b}^{E} \\ \theta_{23}^{E} &= \frac{1}{\sqrt{2}}(\Delta\theta_4 - \Delta\theta_5) = S_{6b}^{E} \end{aligned} \qquad [3.48]$$

For stretching, one bond remains fixed and one bond increases (respectively decreases), while the other one decreases (respectively increases). The same is applicable for the angles between bonds. One angle remains fixed and one angle increases (respectively decreases), while the other decreases (respectively increases). These are also antisymmetric stretching and bending vibrations.

3.7.2. *Case of spherical top XY_4 (CH_4, CD_4)*

The method that was applied involves the determination of the type of stretching and bending symmetry, without trying to determine the form of the respective stretching or bending coordinate. In the case of CH_4 or CD_4 molecules (similar to CDH_3 and CHD_3), there are four stretching internal coordinates Δr_1, Δr_2, Δr_3, Δr_4 and six bending internal coordinates $\Delta\theta_1$, $\Delta\theta_2$, $\Delta\theta_3$, $\Delta\theta_4$, $\Delta\theta_5$, $\Delta\theta_6$. As the number of vibrations must be equal to nine, the combination of the 10 internal coordinates must be reduced by one unit, which is physically related to the necessary limitation on the variation of angles, which cannot all increase simultaneously. This redundancy of the bending coordinates is removed by inspection, comparing the decomposition of the possible vibration modes (see equation [3.30]) to the total decomposition calculated with stretchings and bendings.

The number of coordinates that are invariant under symmetry operations is calculated similarly to CDH_3, and then the coefficients a_α are calculated. Tables 3.19(a) and (b) are established for stretchings and bendings:

Stretching	E	$8C_3$	$3C_2$	$6S_4$	$6\sigma_d$	a_α
h_i	1	8	3	6	6	
$\chi(O_i)$	4	1	0	0	2	
$\chi^{A1}(O_i)$	1	1	1	1	1	
$\pi^{A1}/24$	4/24	8/24	0	0	12/24	1
$\chi^{A2}(O_i)$	1	1	1	-1	-1	
$\pi^{A2}/24$	4/24	8/24	0	0	-12/24	0
$\chi^{E}(O_i)$	2	-1	2	0	0	
$\pi^{E}/24$	8/24	-8/24	0	0	0	0
$\chi^{T1}(O_i)$	3	0	-1	1	-1	
$\pi^{T1}/24$	12/24	0	0	0	-12/24	0
$\chi^{T2}(O_i)$	3	0	-1	-1	1	
$\pi^{T2}/24$	12/24	0	0	0	12/24	1

a)

Bending	E	$8C_3$	$3C_2$	$6S_4$	$6\sigma_d$	a_α
h_i	1	8	3	6	6	
$\chi(O_i)$	6	0	2	0	2	
$\chi^{A1}(O_i)$	1	1	1	1	1	
$\pi^{A1}/24$	6/24	0	6/24	0	12/24	1
$\chi^{A2}(O_i)$	1	1	1	-1	-1	
$\pi^{A2}/24$	6/24	0	6/24	0	-12/24	0
$\chi^{E}(O_i)$	2	-1	2	0	0	
$\pi^{E}/24$	12/24	0	12/24	0	0	1
$\chi^{T1}(O_i)$	3	0	-1	1	-1	
$\pi^{T1}/24$	18/24	0	-6/24	0	-12/24	0
$\chi^{T2}(O_i)$	3	0	-1	-1	1	
$\pi^{T2}/24$	18/24	0	-6/24	0	12/24	1

b)

Table 3.17. *Coefficients a_α of the irreducible representations of group T_d for the internal stretching and bending coordinates of CH_4*

According to Table 3.17(a) and (b), there is one stretching mode of type A_1 and one of type T_2, and one bending mode of type A_1, one of type E and one of type T_2 for a total of 10 vibrations. In comparison with equation [3.30], there must be one vibration of type A_1, one of type E and two of type T_2, which makes it possible to eliminate one bending vibrational mode of type A_1, which is redundant. The following decomposition is thus obtained for CH_4 or CD_4 vibrations, in stretching or bending mode:

$$\Gamma_{vib}^{bending} = \mathrm{A}_1 \oplus \mathrm{T}_1$$
$$\Gamma_{vib}^{bending} = \mathrm{E} \oplus \mathrm{T}_1 \quad [3.49]$$

In order to determine the explicit form of these vibrations, obtained from the linear combination of internal coordinates, the same method used for NH_3 can be applied. A slightly different method developed by Kim [KIM 81a, KIM 81b, KIM 81c] will be applied here to determine these combinations for CH_4. There are four stretching internal coordinates Δr_1, Δr_2, Δr_3, Δr_4 (see Figure 3.2) and instead of considering six bending internal coordinates $\Delta\theta_1$, $\Delta\theta_2$, $\Delta\theta_3$, $\Delta\theta_4$, $\Delta\theta_5$, $\Delta\theta_6$, six angle bisector vectors are used. These are directed towards the faces of the cube in which the molecule can be inscribed, as shown in Figure 3.3.

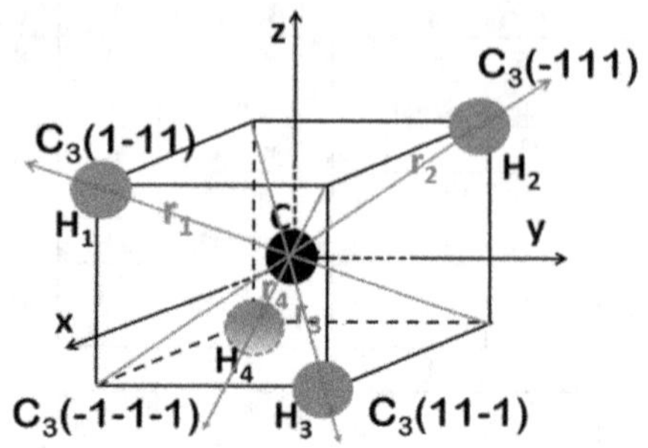

Figure 3.2. *Internal coordinates r_i of the CH_4 molecule inscribed in a cube. For a color version of this figure, see www.iste.co.uk/dahoo/infrared.zip*

The internal coordinates r_i are defined as follows: $r_1(1\ -1\ 1)$, $r_2(-1\ 1\ 1)$, $r_3(1\ 1\ -1)$ and $r_4(-1\ -1\ -1)$. The table of characters of group T_d shows that the cubic function xyz is of the fully symmetric type A_1. If the components x, y and z of each internal coordinate r_i are multiplied, it can be noted that each coordinate is multiplied by -1, ($x_iy_iz_i = -1$), which makes it possible to form the following linear combination, which is fully symmetric to the irreducible representation of type A_1, based on r_i, such that the corresponding stretchings Δr_i lead to the following vibration form expressed as:

$$S^{A_1}_{stretch} = \frac{1}{2}(\Delta r_1 + \Delta r_2 + \Delta r_3 + \Delta r_4) = S_1 \qquad [3.50]$$

The normalization constant is chosen as positive.

Similarly, it can be noted that x, y, and z form a basis for the irreducible representation of type T_1. The combination of r_i can be successively determined by considering each component separately, as shown in Tables 3.18(a) and (b).

Stretching	r_1	r_2	r_3	r_4
x	r_1	$-r_2$	r_3	$-r_4$
y	$-r_1$	r_2	r_3	$-r_4$
z	r_1	r_2	$-r_3$	$-r_4$

a)

Bending	b_1	b_2	b_3	b_4	b_5	b_6
x	b_1	$-b_2$	0	0	0	0
y	0	0	b_3	$-b_4$	0	0
z	0	0	0	0	b_5	$-b_6$

b)

Table 3.18. *Coefficients of the stretching and bending internal coordinates in the irreducible representation T_1 of group T_d for CH_4*

From the results in Table 3.17(a), the form of the following stretching symmetry coordinates is obtained:

$$\begin{aligned} S_x^{T_1} &= \frac{1}{2}(\Delta r_1 - \Delta r_2 + \Delta r_3 - \Delta r_4) = S_{3x} \\ S_y^{T_1} &= \frac{1}{2}(-\Delta r_1 + \Delta r_2 + \Delta r_3 - \Delta r_4) = S_{3y} \\ S_z^{T_1} &= \frac{1}{2}(\Delta r_1 + \Delta r_2 - \Delta r_3 - \Delta r_4) = S_{3z} \end{aligned} \quad [3.51]$$

The internal coordinates b_i are thus defined as: $b_1(1\ 0\ 0)$, $b_2(-1\ 0\ 0)$, $b_3(0\ 1\ 0)$, $b_4(0\ \text{-}1\ 0)$, $b_5(0\ 0\ 1)$ and $b_6(0\ 0\ -1)$.

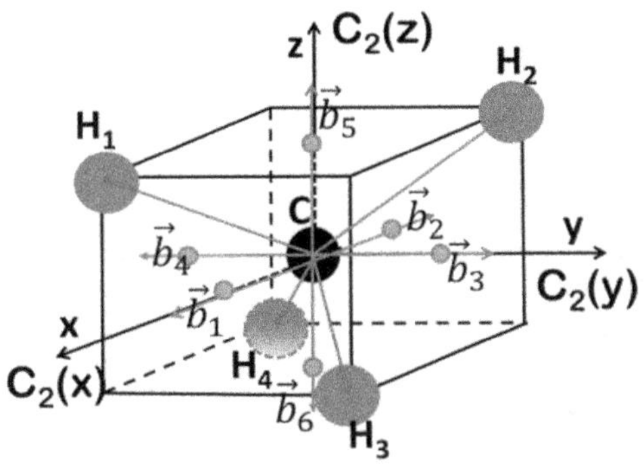

Figure 3.3. *Internal coordinates b_i of the CH_4 $_4$molecule inscribed in a cube. For a color version of this figure, see www.iste.co.uk/dahoo/infrared.zip*

To determine the type of linear combination of internal coordinates for the fully symmetric irreducible representation A_1, the quadratic function $x^2 + y^2 + z^2$ is used. Considering the case of $b_1(1\ 0\ 0)$, $x^2 = 1$, $y^2 = 0$ and $z^2 = 0$, hence $x^2 + y^2 + z^2 = 1$, so that b_1 is multiplied by +1. In this case, the coordinates are $b_1(1\ 0\ 0)$, $b_2(-1\ 0\ 0)$, $b_3(0\ 1\ 0)$, $b_4(0\ -1\ 0)$, $b_5(0\ 0\ 1)$ and $b_6(0\ 0\ -1)$, and the fully symmetric bending coordinate can be written as:

$$S_{bending}^{A_1} = \frac{1}{\sqrt{6}}(b_1 + b_2 + b_3 + b_4 + b_5 + b_6).$$

Given that the bending coordinate cannot be of type A_1, this coordinate is redundant and removed.

Using the results in Table 3.18(a), the following bending symmetry coordinates are obtained:

$$
\begin{aligned}
S_{xbending}^{T_1} &= \frac{1}{\sqrt{2}}(b_1 - b_2) \\
S_{ybending}^{T_1} &= \frac{1}{\sqrt{2}}(b_3 - b_4) \\
S_{zbending}^{T_1} &= \frac{1}{\sqrt{2}}(b_5 - b_6)
\end{aligned} \qquad [3.52]
$$

For the bending coordinate of type E, the basis $(2z^2–x^2–y^2, x^2–y^2)$ is used to build the linear combinations, giving the form of the coordinates from the coefficients summarized in Table 3.19.

Bending	**b_1**	**b_2**	**b_3**	**b_4**	**b_5**	**b_6**
x^2	1	1	0	0	0	0
y^2	0	0	1	1	0	0
z^2	0	0	0	0	1	1
$2z^2$-x^2-y^2	-1	-1	-1	-1	2	2
x^2-y^2	1	1	-1	-1	0	0

Table 3.19. *Coefficients of bending internal coordinates in the irreducible representation E of group T_d of CH_4*

The symmetric bending coordinates of type E can be written as:

$$
\begin{aligned}
S_{1bending}^{E} &= \frac{1}{\sqrt{12}}(2b_5 + 2b_6 - b_1 - b_2 - b_3 - b_4) \\
S_{2bending}^{E} &= \frac{1}{2}(b_1 + b_2 - b_3 - b_4)
\end{aligned} \qquad [3.53]
$$

The bisector vectors can then be replaced by angle variations. As Figures 3.1(a) and (b) show, the following correspondences can be noted: $b_1 \rightarrow \theta_{13}$, $b_2 \rightarrow \theta_{24}$, $b_3 \rightarrow \theta_{23}$, $b_4 \rightarrow \theta_{14}$, $b_5 \rightarrow \theta_{12}$, $b_6 \rightarrow \theta_{34}$, where θ_{ij} is the angle between bond r_i and r_j. After the substitution of b_i, this yields:

$$
\begin{aligned}
S_{xbending}^{T_1} &= \frac{1}{\sqrt{2}}(\Delta\theta_{13} - \Delta\theta_{24}) = S_{4x} \\
S_{ybending}^{T_1} &= \frac{1}{\sqrt{2}}(\Delta\theta_{23} - \Delta\theta_{14}) = S_{4y} \\
S_{zbending}^{T_1} &= \frac{1}{\sqrt{2}}(\Delta\theta_{12} - \Delta\theta_{34}) = S_{4z} \\
S_{1bending}^{E} &= \frac{1}{\sqrt{12}}(2\Delta\theta_{12} + 2\Delta\theta_{34} - \Delta\theta_{13} - \Delta\theta_{24} - \Delta\theta_{23} - \Delta\theta_{14}) = S_{2a} \\
S_{2bending}^{E} &= \frac{1}{2}(\Delta\theta_{13} + \Delta\theta_{24} - \Delta\theta_{23} - \Delta\theta_{14}) = S_{2b}
\end{aligned} \qquad [3.54]
$$

It can be noted that Kim's method leads to the same expressions of stretching and bending vibrations as those given, for example, by references [HOU 76, GRA 79, WIL 55, WIL 80].

3.7.3. *Case of the asymmetric top XY_2Z_2 (CD_2H_2)*

The method applied here involves the determination of the type of stretching and bending symmetry, without trying to determine the form of the respective stretching or bending coordinate.

Similarly to CH_4 or CD_4, and CDH_3 or CHD_3, there are four stretching internal coordinates Δr_1, Δr_2, Δr_3, Δr_4 and six bending internal coordinates $\Delta\theta_1$, $\Delta\theta_2$, $\Delta\theta_3$, $\Delta\theta_4$, $\Delta\theta_5$, $\Delta\theta_6$.

Stretching	E	C_2	σ_v	σ'_v	a_α
h_i	1	1	1	1	
$\chi(O_i)$ Invariant coordinates	4	0	2	2	
$\chi^{A1}(O_i)$	1	1	1	1	
$\pi^{A1}/4$	1	0	0.5	0.5	2
$\chi^{A2}(O_i)$	1	1	-1	-1	
$\pi^{A2}/4$	1	0	-0.5	-0.5	0
$\chi^{B1}(O_i)$	1	-1	-1	1	
$\pi^{B1}/4$	1	0	-0.5	0.5	1
$\chi^{B2}(O_i)$	1	-1	1	-1	
$\pi^{B2}/4$	1	0	0.5	-0.5	1

a)

Bending	E	C_2	σ_v	σ'_v	a_α
h_i	1	1	1	1	
$\chi(O_i)$ Invariant coordinates	6	2	2	2	
$\chi^{A1}(O_i)$	1	1	1	1	
$\pi^{A1}/4$	6/4	2/4	2/4	2/4	3
$\chi^{A2}(O_i)$	1	1	-1	-1	
$\pi^{A2}/4$	6/4	2/4	-2/4	-2/4	1
$\chi^{B1}(O_i)$	1	-1	-1	1	
$\pi^{B1}/4$	6/4	-2/4	-2/4	2/4	1
$\chi^{B2}(O_i)$	1	-1	1	-1	
$\pi^{B2}/4$	6/4	-2/4	2/4	-2/4	1

b)

Table 3.20. *Coefficients a_α of the irreducible representations of group C_{2v} for the internal stretching and bending coordinates of CD_2H_2*

As the number of vibrations is equal to nine, the combination of the 10 internal coordinates must be reduced by one unit, which is physically related to the limitation on the variation of angles, which cannot all increase simultaneously.

This redundancy on the bending coordinates is removed by inspection, comparing the decomposition of the possible vibration modes (see equation [3.34]) to the full decomposition calculated with stretchings and bendings.

The number of coordinates that are invariant under symmetry operations is determined similarly to CD_2H_2, and then the coefficients a_α are calculated. Tables 3.20(a) and (b) are established for stretchings and bendings:

According to Tables 3.20(a) and (b), there are two stretchings of type A_1, one stretching of type B_1 and one stretching of type B_2. For the bending vibrations, there are three of type A_1, one of type A_2, one of type B_1 and one of type B_2 for a total of 10 vibrations with stretchings. In comparison with equation [3.34], there must be four vibrations of type A_1, one vibration of type A_2, two vibrations of type B_1 and two vibrations of type B_2, which makes it possible to eliminate one redundant bending vibration of type A_1. The following decomposition is thus obtained for stretching and bending vibrations of CD_2H_2:

$$\begin{aligned} \Gamma_{vib}^{bending} &= 2\mathrm{A}_1 \oplus \mathrm{B}_1 \oplus \mathrm{B}_2 \\ \Gamma_{vib}^{bending} &= 2\mathrm{A}_1 \oplus \mathrm{A}_2 \oplus \mathrm{B}_1 \oplus \mathrm{B}_2 \end{aligned} \qquad [3.55]$$

The decomposition determined for the vibrational degrees of freedom (equation [3.34]) is thus obtained:

$$\Gamma_{vib} = 4\mathrm{A}_1 \oplus \mathrm{A}_2 \oplus 2\mathrm{B}_1 \oplus 2\mathrm{B}_2,$$

with four vibrations of type A_1, one vibration of type A_2, two vibrations of type B_1 and two vibrations of type B_2.

In order to determine the explicit form of these vibrations, obtained by a linear combination of internal coordinates, Kim's method [KIM 81a, KIM 81b, KIM 81c] can be applied to determine these combinations, as in the case of CH_4. Using the same notations as for CH_4 in Figures 3.1(a) and (b), with D_3 and D_4 instead of H_3 and H_4, the method is applied by noting that z and x^2 are of type A_1, x is of type B_1 and y is of type B_2. The bending vibrational coordinates are built from the internal coordinates of the bonds.

Starting from x^2, each internal coordinate r_i is multiplied by 1 and a fully symmetric combination of type A_1 is obtained as:

$$S_{x^2}^{A_1} = \frac{1}{2}(\Delta r_1 + \Delta r_2 + \Delta r_3 + \Delta r_4) = S_1 \qquad [3.56]$$

Starting from z, the internal coordinates r_1 and r_2 are multiplied by 1 and the internal coordinates r_3 and r_4 are multiplied by –1. The fully symmetric combination of type A_1 is obtained as:

$$S_{z}^{A_1} = \frac{1}{2}(\Delta r_1 + \Delta r_2 - \Delta r_3 - \Delta r_4) = S_2 \qquad [3.57]$$

Starting from x, the internal coordinates r_1 and r_3 are multiplied by 1 and the internal coordinates r_2 and r_4 are multiplied by –1. The fully symmetric combination of type B_1 is obtained as:

$$S_x^{B_1} = \frac{1}{2}(\Delta r_1 - \Delta r_2 + \Delta r_3 - \Delta r_4) = S_3 \qquad [3.58]$$

Starting from y, the internal coordinates r_2 and r_3 are multiplied by 1 and the internal coordinates r_1 and r_4 are multiplied by –1. The fully symmetric combination of type B_2 is obtained as:

$$S_y^{B_2} = \frac{1}{2}(-\Delta r_1 + \Delta r_2 + \dot{\Delta} r_3 - \Delta r_4) = S_4 \qquad [3.59]$$

Moving from the symmetry group T_d to the symmetry group C_{2v}, the decomposition of the irreducible representation T_2 of T_d in three irreducible representations of C_{2v} is verified, such that: $T_2 = A_1 \oplus B_1 \oplus B_2$.

From the internal coordinates defined by: $b_1(1\ 0\ 0)$, $b_2(-1\ 0\ 0)$, $b_3(0\ 1\ 0)$, $b_4(0\ {-1}\ 0)$, $b_5(0\ 0\ 1)$ and $b_6(0\ 0\ {-1})$, the following linear combinations are obtained by considering the coordinate z of type A_1, x of type B_1 and y of type B_2:

$$\begin{aligned} S_x^{B_1} &= \frac{1}{\sqrt{2}}(b_1 - b_2) = S_6 \\ S_y^{B_2} &= \frac{1}{\sqrt{2}}(b_3 - b_4) = S_7 \\ S_z^{A_1} &= \frac{1}{\sqrt{2}}(b_5 - b_6) = S_5 \end{aligned} \qquad [3.60]$$

The decomposition of the irreducible representation T_1 of T_d into three irreducible representations of C_{2v} can be verified, such that: $T_1 = A_1 \oplus B_1 \oplus B_2$.

For the bending coordinate of type E, the Wang basis built from symmetric and antisymmetric linear combinations of the vibration coordinates of CH_4 in the irreducible representation E of group T_d is built.

The Wang basis is given by the following expressions:

$$\begin{aligned} S_{bending}^{S} &= \frac{1}{\sqrt{2}}\left(S_{1bending}^{E} + S_{2bending}^{E}\right) \\ S_{bending}^{AS} &= \frac{1}{\sqrt{2}}\left(S_{1bending}^{E} - S_{2bending}^{E}\right) \end{aligned} \qquad [3.61]$$

which leads to:

$$S^{A_1}_{bending} = \frac{1}{\sqrt{24}}(2b_5 + 2b_6 + C_1b_1 + C_1b_2 - C_2b_3 - C_2b_4) = S_8$$
$$S^{A_2}_{bending} = \frac{1}{\sqrt{24}}(2b_5 + 2b_6 + C_2b_1 + C_2b_2 - C_1b_3 - C_1b_4) = S_9 \qquad [3.62]$$

where $C_1 = (\sqrt{3} - 1)$ and $C_2 = (\sqrt{3} + 1)$. It can be readily verified from the transformations of vectors b_i, by the symmetry operations of group C_{2v} given in Table 3.21, that the symmetric basis $S^{S}_{bending}$ is of type A_1 and the antisymmetric basis $S^{AS}_{bending}$ is of type A_2. The decomposition of the irreducible representation E of T_d into two irreducible representations of C_{2v} can also be verified, such that: $E = A_1 \oplus A_2$.

C_{2v}	E	C_2	$\sigma_v(xz)$	$\sigma'_v(yz)$
b_1	b_1	b_2	b_4	b_3
b_2	b_2	b_1	b_3	b_4
b_3	b_3	b_4	b_2	b_1
b_4	b_4	b_3	b_1	b_2
b_5	b_5	b_5	b_5	b_5
b_6	b_6	b_6	b_6	b_6

C_{2v}	E	C_2	$\sigma_v(xz)$	$\sigma'_v(yz)$	
$\chi(S^{S}{}_{bending})$	1	1	1	1	A_1
$\chi(S^{AS}{}_{bending})$	1	1	-1	-1	A_2

Table 3.21. *Transformation of vectors b_i under the symmetry operations of group C_{2v} and symmetry of the bending internal coordinates of CD_2H_2*

3.7.4. Case of the asymmetric top XY_2Z (NDH_2 or NHD_2)

The method applied in this case involves the determination of the type of stretching and bending symmetry, without trying to determine the form of the given stretching or bending coordinate. Similarly to NH_3, there are three stretching internal coordinates Δr_1, Δr_2, Δr_3 and three bending internal coordinates $\Delta\theta_1$, $\Delta\theta_2$, $\Delta\theta_3$. As the number of vibrations is six, there are three internal coordinates as there is no constraint on the variation of angles such as that they can all increase at the same time. There is no redundancy on the bending coordinates. There are low-amplitude motions. The symmetry group of the isotopic varieties NHD_2 and NDH_2 is the point group, C_{1v} of order 2, composed of the identity E and plane σ_{xz}.

It is easy to show from Tables 3.22(a) and (b) that:

$$\Gamma^{Stretching}_{vib} = 2A_1 \oplus A_2$$
$$\Gamma^{Bending}_{vib} = 2A_1 \oplus A_2 \qquad [3.63]$$

Stretching	E	σ_v	a_α
h_i	1	1	
$\chi(O_i)$ Invariant coordinates	3	1	
$\chi^{A1}(O_i)$	1	1	
$\pi^{A1}/2$	3/2	1/2	2
$\chi^{A2}(O_i)$	1	-1	
$\pi^{A2}/2$	3/2	-1/2	1

a)

Bending	E	σ_v	a_α
h_i	1	1	
$\chi(O_i)$ Invariant coordinates	3	1	
$\chi^{A1}(O_i)$	1	1	
$\pi^{A1}/2$	3/2	1/2	2
$\chi^{A2}(O_i)$	1	-1	
$\pi^{A2}/2$	3/2	-1/2	1

b)

Table 3.22. *Coefficients a_α of the irreducible representations of group C_{1v} for the internal stretching and bending coordinates of NDH_2*

In order to determine the form of the coordinates, those determined for NH_3 can be used. The symmetry group C_{1v} (see Tables 1.4 and 3.25(b)) of NH_2D or of ND_2H is a subgroup of the point group C_{3v} (see Tables 1.2 and 3.25(a)) of NH_3. In this case, the irreducible representations of group C_{3v} are also representations of subgroup C_{1v}, but are necessarily irreducible.

C_{3v}	E	$3\sigma_v$	$2C_3^z$
A_1	1	1	1
A_2	1	-1	1
E	2	0	-1

a)

C_{1V}	E	$\sigma_{xz}=\sigma_v$
A_1	1	1
A_2	1	-1

b)

Table 3.23. *Table of characters of group C_{3v} end of sub-group C_{1v}*

The reduction formula [1.23] can be used in this case, which makes it possible to calculate each a_α for the symmetry group C_{1v} and to decompose the representation of group C_{3v3} into irreducible representations of group C_{1v}. The representations of types A_1 and A_2 are the same in the two groups. Upon inspection, it can be noted that the two-dimensional representation of type E in C_{3v} is decomposed such that: $E = A_1 \oplus A_2$ by adding the characters.

The symmetric stretching and bending vibrations of NH_3 can therefore be used for the description of the two vibrations of type A_1 for NDH_2.

$$r_1^{A_1} = \frac{1}{\sqrt{3}}(\Delta r_1 + \Delta r_2 + \Delta r_3)$$
$$\theta_1^{A_1} = \frac{1}{\sqrt{3}}(\Delta\theta_1 + \Delta\theta_2 + \Delta\theta_3) \quad [3.64]$$

Then, using the two vibrations of type *E*, a linear combination gives for stretching and bending, respectively, the coordinates corresponding to the stretching and bending vibration for NDH_2. The simplified projection operator (see equation [1.27]), which is expressed in terms of characters, is used. For the representation of type A_1, the operator can be written as:

$$P^{A_1} = \frac{1}{2}(\chi^{A_1}(E)E + \chi^{A_1}(\sigma_v)\sigma_v)$$

that can be applied to the functions describing the vibrations of type *E* of group C_{3v}. Denoting by r_1 the internal coordinate between N and D, by r_2 the coordinate between N and H_1, by r_3 the coordinate between N and H_2 (Figure 3.4), and by θ_i the angles on the opposite side of the bonds, and knowing that the identity operator *E* does not change these coordinates, the following transformations are obtained by applying the operator σ_v to the variations of coordinates:

$$\sigma_v(\Delta r_1) = \Delta r_1, \sigma_v(\Delta r_2) = \Delta r_3, \sigma_v(\Delta r_3) = \Delta r_2, \sigma_v(\Delta\theta_1) = \Delta\theta_1, \sigma_v(\Delta\theta_2) = \Delta\theta_3, \sigma_v(\Delta\theta_3) = \Delta\theta_2 .$$

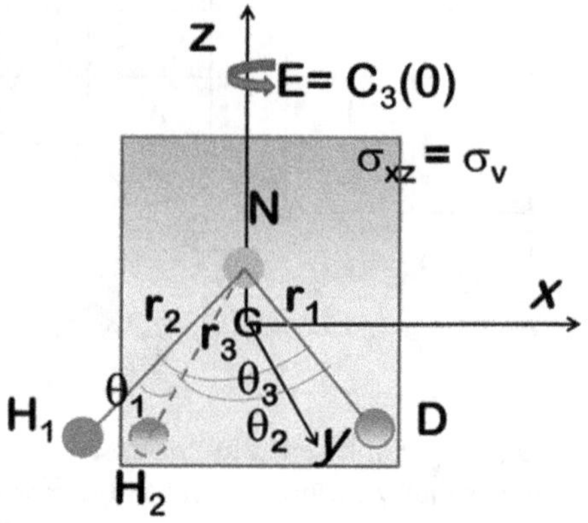

Figure 3.4. *Internal coordinates of the NDH₂ molecule. For a color version of this figure, see www.iste.co.uk/dahoo/infrared.zip*

The application of the projection operators of the representations of types A_1 and A_2 leads to:

$$P^{A_1}(r_{21}^{E}) = \frac{1}{\sqrt{6}}P^{A_1}(2\Delta r_1 - \Delta r_2 - \Delta r_3) = \frac{1}{2}\left\{\frac{1}{\sqrt{6}}(2\Delta r_1 - \Delta r_2 - \Delta r_3) + \frac{1}{\sqrt{6}}(2\Delta r_1 - \Delta r_3 - \Delta r_2)\right\}$$

$$= r_{21}^{E}\ P^{A_2}(r_{21}^{E}) = \frac{1}{2}\left\{\frac{1}{\sqrt{6}}(2\Delta r_1 - \Delta r_2 - \Delta r_3) - \frac{1}{\sqrt{6}}(2\Delta r_1 - \Delta r_3 - \Delta r_2)\right\} = 0 \qquad [3.65]$$

$$P^{A_1}(r_{22}^{E}) = \frac{1}{\sqrt{2}}P^{A_1}((\Delta r_1 - \Delta r_2)) = \frac{1}{2}\left\{\frac{1}{\sqrt{2}}(\Delta r_1 - \Delta r_2) + \frac{1}{\sqrt{2}}(\Delta r_1 - \Delta r_3)\right\} \propto r_{21}^{E}$$

$$P^{A_2}(r_{21}^{E}) = \frac{1}{\sqrt{2}}P^{A_2}((\Delta r_1 - \Delta r_2)) = \frac{1}{2}\left\{\frac{1}{\sqrt{2}}(\Delta r_1 - \Delta r_2) - \frac{1}{\sqrt{2}}(\Delta r_1 - \Delta r_3)\right\} \propto (\Delta r_3 - \Delta r_2) \qquad [3.66]$$

It can be noted that the result is a linear combination of the stretching coordinates $r_2^{A_1} = \frac{1}{\sqrt{6}}(2\Delta r_1 - \Delta r_2 - \Delta r_3)$, similar to that for NH_3 and a coordinate $r_3^{A_2} = \frac{1}{\sqrt{2}}(\Delta r_3 - \Delta r_2)$ that does not involve the coordinate of the ND bond. Given the choice of coordinates θ_i, for the bending vibration coordinate, the result is $\theta_2^{A_1} = \frac{1}{\sqrt{6}}(2\Delta\theta_1 - \Delta\theta_2 - \Delta\theta_3)$ and $\theta_3^{A_2} = \frac{1}{\sqrt{2}}(\Delta\theta_3 - \Delta\theta_2)$.

3.8. Rotation and spin Hamiltonian symmetries

In various space media (planetary atmosphere, interstellar medium, comets, etc.) in which a molecule can be found in the condensed phase, the spectra resulting from the light–matter interaction between an electromagnetic wave of the IR spectral range and the molecule are characterized by the presence of cold bands and/or combination bands, and hot bands depending on the temperature of the environment. The observations most often concern the cold bands originating in the transitions from the fundamental level of the molecule and less frequently, the hot bands issued from excited levels. As highlighted in section 3.4, the fundamental electronic state, similar to the fundamental vibrational state of nearly all the molecules, leads to a product of the vibronic wave function $|\Psi_{eV}\rangle = |\Psi_e\rangle \times |\Psi_V\rangle$ belonging to the fully symmetric irreducible representation of the symmetry group of the molecule.

3.8.1. *Vibronic degrees of freedom of NH_3 and CH_4*

For the electronic state, it is important to consider the symmetry of the wave functions that depend on the space variables and the symmetry of the electronic spin functions. The electronic wave functions are built from atomic orbitals (AO) in the form of linear combinations (LCAO), leading to molecular orbitals (MO). They are determined with group theory methods (see Chapter 1) using the point-group symmetry of the molecule or the molecular symmetry group (CNPI group limited to feasible permutation-inversion operations) and by combining the AO of neighboring energies.

The resulting MO are classified in ascending order of energy and are allotted 2, 1 or 0 electrons per MO, according to the Pauli principle, depending on the total number of electrons of the atoms in the molecule.

The methods described in Chapter 1 (see section 1.10) and used in section 3.6 can be applied to determine the symmetry coordinates of the vibrations by using the projection operators to build the MO from AO.

For the symmetric top NH_3, there are three H atoms for three electrons, each of which is steered by an atomic orbital of type *1s*, and for the central N atom, there are seven electrons localized in 5 *AO* $1s^2 2s^2 2p_x{}^1 2p_y{}^1 2p_z{}^1$. The linear combinations corresponding to three 1s orbitals of H atoms that generate the irreducible representations of group C_{3v} are built first. Then, given that the N atom is central, and therefore invariant under all the symmetry operations of the molecule, the possible linear combinations between the orbitals of the N atom and the three MO built from the three AO of H atoms are composed.

For the H atoms, the irreducible representation given by $\Gamma_{3\otimes 1sH} = A_1 \oplus E$ is generated by the functions:

$$\Psi_{A_1} = 1/\sqrt{3}\,(1s_{H1} + 1s_{H2} + 1s_{H3})$$

$$\Psi_{Ea} = 1/\sqrt{6}\,(2\,1s_{H1} - 1s_{H2} - 1s_{H3})$$

$$\Psi_{Eb} = 1/\sqrt{2}\,(\,1s_{H1} - 1s_{H2}).$$

Taking the energies of the AO of the N atom into account, the following MO are built:

$$\Psi_{A_1} = \mathrm{N}_{A_1}\left(2s_N + 1/\sqrt{3}\,(1s_{H1} + 1s_{H2} + 1s_{H3})\right)$$

$$\Psi_{E_a} = \mathrm{N}_{E_a}\left(2p_{Ny} + 1/\sqrt{6}\,(2\,1s_{H1} - 1s_{H2} - 1s_{H3})\right)$$

$$\Psi_{E_b} = \mathrm{N}_{E_b}\left(2p_{Nx} + 1/\sqrt{2}\,(\,1s_{H1} - 1s_{H2})\right)$$

The AO $2p_{Nz}$ is a free pair of electrons, and the AO $1s_N$ is not used to bind the N atom to the three H atoms. There are 10 electrons to be placed on the MO: two on $1s_N$, two on $2p_{Nz}$ and the remaining six on the three MO: Ψ_{A_1}, Ψ_{Ea} and Ψ_{Eb}.

The MO of the spherical top CH_4 can be built in a similar manner. In this case, there are four H atoms for four electrons, each of which is steered by an atomic orbital of type 1s, and for the central atom C, there are six electrons localized in five AO $1s^2 2s^2 2p_x{}^1 2p_y{}^1 2p_z{}^0$. The linear combinations corresponding to four 1s orbitals of the H atoms that generate the irreducible representations of group T_d are built first. Then, since the C atom is central, and therefore invariant under all of the symmetry operations of the molecule, the possible linear combinations between the orbitals of the C atom and the four MO built from the four AO of the H atoms are composed.

For the H atoms, the irreducible representation is given by: $\Gamma_{4\otimes 1sH} = A_1 \oplus T_2$ generated by the functions:

$$\Psi_{A_1} = 1/2\,(1s_{H1} + 1s_{H2} + 1s_{H3} + 1s_{H4})$$

$$\Psi_{T_{2a}} = 1/2\,(1s_{H1} + 1s_{H2} - 1s_{H3} - 1s_{H4})$$

$$\Psi_{T_{2b}} = 1/2\,(1s_{H1} - 1s_{H2} - 1s_{H3} + 1s_{H4})$$

$$\Psi_{T_{2c}} = 1/2\,(1s_{H1} - 1s_{H2} + 1s_{H3} - 1s_{H4}).$$

Taking the energies of the AO of the C atom into account, the following MO are built:

$$\Psi_{A_1} = \mathrm{N}_{A_1}\left(2s_C + 1/2\,(1s_{H1} + 1s_{H2} + 1s_{H3} + 1s_{H4})\right)$$

$$\Psi_{T_{2a}} = \mathrm{N}_{T_{2a}}\left(2p_{Cz} + 1/2\,(1s_{H1} + 1s_{H2} - 1s_{H3} - 1s_{H4})\right)$$

$$\Psi_{T_{2b}} = \mathrm{N}_{T_{2b}}\left(2p_{Cy} + 1/2\,(1s_{H1} - 1s_{H2} - 1s_{H3} + 1s_{H4})\right)$$

$$\Psi_{T_{2c}} = \mathrm{N}_{T_{ca}}\left(2p_{Cx} + 1/2\,(1s_{H1} - 1s_{H2} + 1s_{H3} - 1s_{H4})\right).$$

The AO $1s_C$ is not used to bind the C atom to the four H atoms. There are 10 electrons to be placed on the MO: two on $1s_N$ and the remaining eight on the four MO: $\Psi_{A_1}, \Psi_{T_{2a}}, \Psi_{T_{2b}}$ and $\Psi_{T_{2c}}$.

The symmetry of the MO is multiplied by the corresponding representation generated by the spin functions. The resulting symmetry depends on the strength of the spin–orbit characterizing the molecule. The spin–orbit interaction results from the magnetic coupling between the electron spin and the orbital motions of the electrons and nuclei.

When this coupling is weak or negligible, the rules termed as Hund's case b) are applied. In this case, the individual spin operators are quantized along the axes of the laboratory reference frame (fixed space axes). The symmetry of the spin wave functions is not affected by the symmetry operations of the symmetry group of the molecule, which is based on the mobile axes attached to the molecule, as the spin functions are invariant under a symmetry operation of the molecule. The symmetry and the classification of electronic wave functions can be established without taking into consideration those of the spin functions.

When this coupling cannot be neglected, the rules termed as Hund's case a) are applied. In this case, the electrons are attached to the mobile axes and the individual spin operators are quantized along the axes of the mobile reference frame (axes attached to the molecule). The symmetry of the spin wave functions is such that the z component of the spin is expressed as a function of Euler angles defining the orientation of the molecule with respect to fixed axes. The spin functions are no longer invariant under the symmetry operations of the symmetry group of the molecule that are based on the mobile axes attached to the latter. The symmetry and the classification of the electronic wave functions are established based on those of the electronic spin functions.

When the electronic wave function is symmetric, the symmetry of the complete wave function is given by its nuclear component $\Gamma_t = \Gamma(|\Psi_V\rangle) \otimes \Gamma(|\Psi_R\rangle) \otimes \Gamma(|\mathrm{Im}_I\rangle)$. The fundamental vibrational state of all the molecules is a symmetric wave function. For the vibration, this result is based on the following rule: if the symmetries of the vibration coordinates of a molecule are denoted by $\Gamma^{(1)},\Gamma^{(2)}, \ldots, \Gamma^{(n)}$ and this molecule is in a state identified by the quantum numbers $v_1, v_2, \ldots, v_n$, given the ket $|v_1, v_2, \ldots, v_n\rangle$, the symmetry of the vibration wave function associated with this ket is determined by the tensor product: $\Gamma_{vib} = \left[\Gamma^{(1)}\right]^{v_1} \otimes \left[\Gamma^{2}\right]^{v_2} \otimes \ldots \otimes \left[\Gamma^{(n)}\right]^{v_n}$ where $\left[\Gamma^{(j)}\right]^{v_j}$ is the v_j^{th} symmetric product for a degenerate state and the v_j^{th} ordinary product for a non-degenerate state. In a fundamental state, $v_1 = v_2 = \cdots = v_n = 0$, so that the fundamental state is always symmetric.

3.8.2. *Rovibronic degrees of freedom*

In order to study the symmetries of rotational states (space variables) and spin states (spin variables), it is assumed, unless otherwise mentioned, that the molecule is in its fundamental vibronic state, which is fully symmetric. The complete wave function (see section 3.2):

$$
\begin{aligned}
|m\rangle &= |\Psi_t^m\rangle = \{(|\Psi_e^m\rangle \otimes |Sm_s\rangle) \times |\Psi_V\rangle\} \times |Im_I\rangle \\
&= |m\rangle_{eV} \otimes |m\rangle_{Rns} = |\Psi_{eV}\rangle \times |\Psi_R\rangle \times |\Psi_{ns}\rangle
\end{aligned}
$$

is the product of a vibronic wave function, a rotational wave function and a nuclear spin wave function if the vibration–rotation interaction is considered negligible. In a first approximation, $|m\rangle_{Rns} = |\Psi_{Rns}^m\rangle$ is given by the product $|\Psi_R\rangle \times |\Psi_{ns}\rangle$.

When the molecules are in their fundamental electronic and vibrational states, the irreducible representation that is appropriate for identifying the type of symmetry of the complete wave function depends only on the type of symmetry of the rotational wave function given the symmetry of the spin state of the level $(\Gamma_t = \Gamma(|\Psi_R\rangle) \otimes \Gamma(|\Psi_{ns}\rangle))$.

The method that is generally applied involves, first, the determination of the irreducible representation of the rovibronic wave function $|\Psi_{eVR}\rangle = |\Psi_e\rangle \times |\Psi_V\rangle \times |\Psi_R\rangle$, the product of the electronic wave function and the vibration–rotation wave function. The type of allowed product $|\Psi_{eVR}\rangle \times |\Psi_{ns}\rangle$ is then determined. According to the Fermi–Dirac or Bose–Einstein statistics that govern the nuclei spin properties in the molecule (see sections 3.2 and 3.4), we proceed differently from in the case of electrons, which have half-integral (1/2) spins, which governs Fermi–Dirac statistics. As for the nuclei, they can have half-integral or integral spins, such that the process of counting the number of nuclear spin states and determining their symmetries is more complex. The nomenclature used to label a level relies on the irreducible representations of the symmetry group *G* of the molecule (called the approximate symmetry group [BUN 98]). At the zeroth-order approximation, each element *g* of group *G* commutes with the zeroth-order Hamiltonian $\left(\left[\hat{H}_0, g\right] = 0\right)$.

The rule followed is, first, to determine the irreducible representation that can be used to classify the rovibronic wave function and then to combine this wave function with the nuclear wave function in order to apply the statistics corresponding to the combinations allowed by symmetry considerations so that the complete wave function is of type *A* symmetric (A_1, A'_1, A''_1) or antisymmetric (A_2, A'_2, A''_2). Therefore, a spin function cannot be arbitrarily associated with a given rotation level. As a result, the total number of distinct spin functions that can be combined with a given rotational wave function $|\Psi_R\rangle$ leads to defining a nuclear statistical weight for a given vibronic level, determined as the number of spin states of each type of symmetry, and which is not necessarily identical for each symmetry. This statistical weight shapes the structure of the vibration–rotation spectrum observable in the IR range.

3.8.3. *Rotational degrees of freedom*

The rotational Hamiltonian must be invariant under a rotational symmetry operation (group *K*(*spatial*), rotation with respect to fixed axes, section 1.2, which relates to space isotropy) with respect to each of the eigen axes C_2^a, C_2^b, C_2^c (group D_2). It is a Hermitian energy operator that is also invariant under a time reversal operation. These axes are the three main axes of inertia of the molecule, denoted as G_a, G_b and G_c, so that the main moments of inertia, arranged in ascending order of magnitude, are termed as $I_a^e \leq I_b^e \leq I_c^e$, or, in descending order of magnitude of rotational constants, designated by $A^e \geq B^e \geq C^e$ and defined as $A^e = \hbar^2/2I_a^e, B^e = \hbar^2/2I_b^e$ and $C^e = \hbar^2/2I_c^e$.

The rotational Hamiltonian belongs to the fully symmetric irreducible representation of the point-group symmetry of the molecule. Not all the symmetry operations of the point-group symmetry of a molecule transform the Euler angles that give the orientation of the mobile reference frame attached to the molecule with respect to a fixed reference frame. The same is true of certain operations on the components of the angular momentum and the nuclear spin coordinates.

The Hamiltonian of the rigid rotator commutes with the elements of the rotation group D_2 for an asymmetric top (see Volume 2 [DAH 19], the nonlinear triatomic H_2O molecule), group D_∞ for a symmetric top or a linear polyatomic molecule and group *K* for a spherical top. Consequently, the labeling of the rotation states of a molecule makes use of the irreducible representations of group *K* (a spherical top, dependence in *J*; spherical symmetry) or of group D_∞ (a symmetric top, dependence in *J* and *K*; cylindrical symmetry) or of group D_{22} (or *V*, see Chapter 1, Volume 2, dependence in *J*, K_a and K_c; conical symmetry) for asymmetric tops. It must be noted that each element of group G^{CNPI} has its equivalent in the rotation group *K*.

At zeroth order, which is the rigid rotator approximation, and as for fixed *J* and *M*, $\left[H_0, \hat{J}^2\right]=0, \left[H_0, \hat{J}_Z\right]=0, \left[H_0, \hat{J}_z\right]=0$, a rotational level is expressed as a linear combination of kets $|JKM\rangle$. The quantum numbers *J*, *K* and *M* are used to identify the rotational levels (see equations [1.10] and [2.33], Volume 2) from the rotational molecular wave functions in representation *J*. Given that the generator of the irreducible representation $D^{(J)}$ is the angular momentum **J** and that the eigenfunctions common to the three operators $\hat{J}^2$, $\hat{J}_Z$ and $\hat{J}_z$ are proportional to the matrices (or rotation matrices) of the irreducible representation of the rotation group K (SO(3)), the rotational states are identified by the quantum numbers J, K and M. The eigenfunctions $|JKM\rangle$ are written in the form (see equation [3.7]):

$$\langle \phi, \theta, \chi | JKM \rangle = N_{JKM} \Theta_{JKM} e^{iM\phi} e^{iK\chi}$$

where N_{JKM} is a normalization constant and $\Theta_{JKM}(\theta)$ is a function of the nutation coordinate θ.

The irreducible representations of group K(SO(3)) being described by Wigner $D^{(J)}$ matrices, which are the matrix representations in the spherical harmonics basis $\left(Y_m^J(\theta,\phi)\right)$, it can be shown that the two irreducible representations of group K lead to the well-known results on the vector addition model of angular momentum operators (see sections 1.5 and 1.6.6), such that:

$$D^{(J_1)} \otimes D^{(J_2)} = D^{(J_1+J_2)} \oplus D^{(J_1+J_2-1)} \oplus \ldots \oplus D^{(|J_1-J_2|)}$$

The representations $D^{(J)}$ generated by the functions $|JKM\rangle$ for J fixed and $K = -J, -J+1, \ldots, J$ can be subdivided (see section 1.6.6) into a one-dimensional representation for $K = 0$ (type A) and a two-dimensional representation $J, -|K|, M\rangle$ and $J, |K|, M\rangle$ when $K \neq 0$ (type E).

For a symmetric top, this corresponds to the rotation energy of the rigid rotator as a function of J and K with a double degeneracy in K when $K \neq 0$ (see equations [2.12] and [2.13]) in the absence of an external field (degeneracy of $2J+1$ in M). The general effect of rotation operations on the eigenfunctions of the rigid rotator $|JKM\rangle$ used as basis states for the solution of Schrödinger's molecular equation is given in Chapter 1. The action of the operators' representative for the cyclic and dihedral groups is given (see equations [1.30] and [1.31]) by:

$$\boldsymbol{R}_Z^{\beta} |JKM\rangle = e^{iK\beta} |JKM\rangle$$

$$\boldsymbol{R}_{\alpha}^{\pi} |JKM\rangle = (-1)^J e^{-2iK\alpha} |J - KM\rangle$$

where $\boldsymbol{R}_{\alpha}^{\pi}$ is a (direct) rotation of the reference frame attached to the molecule (mobile reference frame: $Oxyz$) by an angle π about the axis making an angle α with the Ox axis. $\boldsymbol{R}_Z^{\beta}$ is a (direct) rotation of the reference frame attached to the molecule (mobile reference frame: $Oxyz$) by an angle β about the Oz axis.

Based on these results, it is possible to determine the transformation properties and the symmetries of the wave functions of the symmetric top for integral J in the symmetry group of the molecule by using the equivalent rotations (see section 1.6.6). The characters of the irreducible representations are given by:

$$\chi_J(E) = \chi_J(R^0) = 1 \text{ and } \chi_J(R_{\vec{n}}^{\phi}) = \frac{\sin\left((2J+1)\phi/2\right)}{\sin\left(\phi/2\right)}$$

which depend on the maximum value of J.

Various approaches were proposed to determine the symmetry of the complete wave function from that of the product function $|\Psi_R\rangle \times |\Psi_{ns}\rangle$ for a fully symmetric vibronic wave function.

Herzberg [HER 45] determines, first, the type of symmetry of the rotational wave function from the rotation subgroup of the point-group symmetry of the molecule. For example, for group C_{3v} (symmetric top NH_3, symmetries of types A_1, A_2, E), subgroup C_3 is used (symmetries of types A and E; A_1 and A_2 of C_{3v} are of type A in C_3); for group D_{3h}, group D_3 is used (A'_1 and A''_2 of D_{3h} are of type A_1 in D_3); for group T_d, (spherical top CH_4, symmetries of types A_1 and A, T_1 or F_1 and T_2 or F_2), group T is used (symmetries of types A, E and F; A_1 and A_2 of T_d are of type A in T and T_1 or F_1 and T_2 or F_2 in T_d are of type T or F in the group T), etc. The statistical weight of the levels is then determined by taking into account the symmetry properties of the spin functions.

For the symmetric top NH_3, of the symmetry group C_{3v}, if the rotation wave function is of type A or E, the rotational level is of type A or E if the vibrational state is of type A_1 or A_2. If the vibrational state is of type E, there are two distinct cases: one in which the rotational wave function is of type A ($K \neq 0$ and $K = 3p$, multiple of 3, a one-dimensional rotation space), and therefore the complete wave function is of type $A \otimes E = E$ (two doubly degenerate states of type E, given two vibrational states); one in which the rotational wave function is of type E ($K \neq 0$ and $K \neq 3p$, $K = 3p \pm 1$, a two-dimensional rotation space), and therefore the complete wave function is of type $E \otimes E = A \oplus A \oplus E$ (three states, two of type A and one doubly degenerate of type E). Taking into account the influence of the vibration–rotation coupling, of the Coriolis or centrifugal type, the levels with the same energy for a given J ($K \neq 0$), but with different symmetries, undergo degeneracy lifting in two levels. When $K \neq 0$ and $K = 3p$, the two doubly degenerate states split (+l and -1) into two doublets of type E of the same energy each, and when $K \neq 0$ and $K \neq 3p$, $K = 3p \pm 1$, the four states split into a doubly degenerate doublet ($K = 3p - 1$ and +l) of type E of the same energy and two levels ($K = 3p + 1$ and -l) of type A of neighboring energies (doubly degenerate of type *K*). Then, the statistical weight of the levels is determined, taking into account the symmetry of the spin functions that depend on quantum variables and not on space variables.

For CH_4, of the symmetry group T_d, the symmetry species *A, E* and *F* of the group *T*, a rotation subgroup of T_d, are considered. It is known that for a fully symmetric vibronic function, the symmetry of the rotational wave function determines that of the vibration–rotation function, when the spin component is neglected. The three types of symmetry *A, E* and *F* are possible for a given *J* and for each type of vibration *A, E* or *F* (see Figure 1.38 of [HER 45]). When the vibration is of type *E* or *F*, the type of symmetry of the vibration must be multiplied by that of the rotation (see Table 2.7).

Moreover, Watson [WAT 67] developed a method for determining the type of symmetry of the rotational wave functions from the expression of the rotational Hamiltonian in terms of powers of the rotation operator **J**. It is possible to calculate the wave functions and eigenvectors of the Hamiltonian and to determine the irreducible representations to which the eigenfunctions belong. Watson showed that, when the higher-order terms are included in the zeroth-order approximation of the rigid rotator, to take into account the non-rigidity of the molecule (the Coriolis effect, the centrifugal effect, etc.), the rotational Hamiltonian can be written as:

$$H_{rot} = \sum_{\substack{p+q+r \\ even}} h_{pqr} \left(J_x^p J_y^q J_z^r + J_z^r J_y^q J_x^p \right) \qquad [3.67]$$

where the coefficients h_{pqr} are real numbers, given that the Hamiltonian must be invariant under a time reversal symmetry operation and must be a Hermitian operator. Generally speaking, in the power series expansion, the number of terms of degree $n = p + q + r$ is equal to $1/2\ (n + 1)(n + 2)$. If *e* and *o* (*e* for "even" and *o* for "odd") denote the type of parity of powers *p*, *q* and *r*, there are $1/8(n + 2)(n + 4)$ of type *eee*, and $1/8n(n + 2)$ of type *eoo*, *oeo* and *ooe* each. In general, there are six terms of degree 2, 15 terms of degree 4, and 28 terms of degree 6. The terms of degree 2 are diagonal, taking into account the rotation axes attached to the main axes of inertia of the molecule.

Given that the rotational Hamiltonian must belong to the fully symmetric irreducible representation of the symmetry group of the given molecule, the terms of degree *n* are transformed under the transformations of the symmetry group of the molecule, according to the symmetric *n* powers of the representation of *J*. This representation is generally given in the tables of characters as the type of irreducible representation of each of the three rotations with respect to the axes attached to the molecule. For a given point group, the number of terms of degree *n* can be obtained from the number of times the fully symmetric irreducible representation is contained in the symmetric representation of power n of the representation J. The symmetric powers of n in the representation of operator $\boldsymbol{J}$ in the full rotation–reflection group, K_h, can be determined. Therefore, $\left[\Gamma\left(\hat{J}^2\right)\right]_{sym} = S_g + D_g$ $\left[\Gamma\left(\hat{J}^4\right)\right]_{sym} = S_g + D_g + G_g$, $\left[\Gamma\left(\hat{J}^6\right)\right]_{sym} = S_g + D_g + G_g + I_g$, etc. The table of correlation between the (rotation–reflection) group K_h and the point-group symmetry of the molecule is then used to determine the representations of symmetric powers of n in this point group. In the case of groups C_{3v} and T_d, of interest for the symmetric tops (NH_3 and ND_3 or CH_3D or CHD_3) and for spherical tops (CH_4 or CD_4), the method described by Hamermesh (see Table 9.4) in Chapter 9 of [HAM 62] can be applied, for example (and using the formulas in Chapter 1), to show the following equivalences between the rotation group and the symmetry group of the molecule (see Tables 1.12, 1.13 and 1.14 for the correlation between group K and the point-group symmetry of a molecule).

$\mathbf{K_h}$	$\mathbf{C_{3v}}$			$\mathbf{T_d}$				
S_g	A_1			A_1				
D_g	A_1		$+ 2E$			E	$+ F_2$	
G_g	$2A_1$	$+ A_2$	$+ 3E$	A_1		$+ E$	$+ F_1$	$+ F_2$
I_g	$3A_1$	$+ 2A_2$	$+ 4E$	A_1	$+ A_2$	$+ E$	$+ F_1$	$+ 2F_2$

Table 3.24. *Table of correlation of groups K_h or (SO(3)), C_{3v} and T_d*

Only the terms of the fully symmetric type A_1 must appear in the Hamiltonian, corresponding to the number of independent terms in H_{rot}. As shown in Table 3.24, for a molecule of the symmetry group C_{3v} (NH_3 or CH_3D), the number of terms is equal to two for degree 2 (S_g and D_g), four for the following degree, which is 4 (S_g, D_g and G_g), and seven for degree 6 (S_g, D_g, G_g and I_g). For a molecule of the symmetry group T_d (CH_4, CD_4), the number of terms is equal to 1 for degree 2 (S_g and D_g), 2 for degree 4 (S_g, D_g and G_g) and 3 for degree 6 (S_g, D_g, G_g and I_g).

The terms of the Hamiltonian can be built from powers of operators $J_\pm = J_x \pm iJ_y$ and J_z. All the symmetry operators of a point group that are equivalent to rotations form a rotational subgroup of the molecule. In C_{3v}, the rotations with respect to the C_3 axis belong to a rotation subgroup, but not the reflections with respect to planes σ_v. Therefore, as proposed by Herzberg [HER 45], subgroup C_3 is used. Similarly, for D_{3h}, subgroup D_3 is used, and for T_d, subgroup T is used.

The rotational eigenfunctions can belong to any type of symmetry, in the sense of the irreducible representation, of the rotation subgroup. For NH_3 or CH_3D, the *A* or *E* nomenclature is used, since $C_{3V} \supset C_3$. The tables of correlation can be used to establish the correspondence between the types of symmetry of one group and one subgroup. In the general case, the correlation between the type of symmetry of the irreducible representations of the point-group symmetry of the molecule and the type of symmetry of the irreducible representations of the space group of rotation in the three-dimensional space, limited to odd powers of the operator *J*, can be used to identify the type of symmetry of the rotational wave functions.

If the CNPI symmetry group is used (see section 1.10) [BUN 98], the types of symmetry of the rotational wave functions are determined from symmetry operations equivalent to the permutation-inversion operators that constitute the symmetry group of the molecule (see section 1.10). The method used by Bunker and Jensen is extensively described in [BUN 98].

$D_{3h}(M)$	E	2(123)	3(23)	E*	2(123)*	3(23)*
D_{3h}	E	$2C_3$	$3C_2$	σ_h	$2S_3$	$3\sigma v$
Equivalent Rotations	R^0	$R_z^{2\pi/3}$	R_0^{π}	R_z^{π}	$R_z^{5\pi/3}$	$R_{\pi/2}^{\pi}$
A'_1	1	1	1	1	1	1
A''_1	1	1	1	-1	-1	-1
A'_2	1	1	-1	1	1	-1
A''_2	1	1	-1	-1	-1	1
E'	2	-1	0	2	-1	0
E''	2	-1	0	-2	1	0
J, K=0	1	1	$(-1)^J$	1	1	$(-1)^J$
J, $\lvert K\rvert \neq 0$	2	$2\cos(2K\pi/3)$	0	$2(-1)^K$	$2(-1)^K\cos(2K\pi/3)$	0
J even, K=0 : A'_1 J odd, K=0 : A'_2	1 1	1 1	1 -1	1 1	1 1	1 -1
J, K=±3 : E= $A''_1 \oplus A''_2$	2	2	0	-2	-2	0
J, K=±6 : E= $A'_1 \oplus A'_2$	2	2	0	2	2	0
J, K=±1 : E''	2	-2	0	-2	-2	0
J, K=±2 : E'	2	-2	0	2	2	0

Table 3.25. *Table of characters of group D_{3h} (M) and D_{3h} for NH_{3h}*

The classification, in terms of irreducible representations of the rotational wave functions of the rigid rotator from the wave functions of the symmetric tops $|JKM\rangle$ used as basis states for the solution of Schrödinger's molecular equation, is done for integer values of J and half-integer values of J. The action of operators $\boldsymbol{R}_{\alpha}^{\pi}$ and $\boldsymbol{R}_{Z}^{\beta}$ representative of the cyclic and dihedral groups is determined (see section 1.10) for $K = 0$ on the ket $|J00\rangle$ in a one-dimensional space (type A) and for $K \neq 0$ in a two-dimensional space (type E) on the kets $|J \pm KM\rangle$ by using the transformation properties of ket $|J00\rangle$ and those of ladder operators $J_m^{\pm}$ and $J_s^{\pm}$ defined, respectively, in the mobile reference frame (axes attached to the molecule) and the laboratory reference frame (axes fixed in space).

The results of these operations are summarized in Table 3.25 for the group D_{3h} that can be used for NH_3 when considering the tunneling effect.

It can be noted that for $K = 0$, the representation is fully symmetric of type A_1' if J is even, and of type A_2' if J is odd. For $K > 0$, the situations for $K = 6n$, multiple of 6, and $K = 3n$, multiple of 3, for which the symmetry of type E is reducible, must be distinguished from the cases where $K = 3n \pm 1$ and $K = 3n \pm 2$ for which the symmetry of type E is irreducible.

$C_{3v}(M)$	E	2(123)	3 (23)*
C_{3v}	E	$2C_3$	$3\sigma_v$
Equivalent Rotations	$\mathbf{R}^0$	$\mathbf{R}_Z^{2\pi/3}$	$\mathbf{R}_{\frac{\pi}{2}}^{\pi}$
A_1	1	1	1
A_2	1	1	-1
E	2	-1	0
J, K=0	1	1	$(-1)^J$
J, $\lvert K\rvert \neq 0$	2	$2\cos(2K\pi/3)$	0
K=0 J even: A_1 J odd: A_2	1 1	1 1	1 -1
K > 0 K=3: E= $A_1 \oplus A_2$	2	2	0
$\lvert K\rvert \neq 0$ K=1 and -1 : E	2	-1	0

Table 3.26. *Table of characters of group C_{3v} (M) and C_{3v} for CDH_3*

For the group C_{3v}, the results given in Table 3.26 can be used for CH_3D or CHD_3. It can be noted that for $K = 0$, the representation is fully symmetric of type A_1 if J is even and of type A_2 if J is odd. For $K > 0$, the situations for which $K = 3n$, multiple of 3 for which the symmetry of type E of the two-dimensional space is reducible to symmetry of type A_1 and of type A_2 in one-dimensional spaces, must be distinguished from the cases in which $K = 3n \pm 1$ for which the symmetry is of type E in a two-dimensional space.

Concerning the molecule methane CH_4 or its isotopologue CD_4, of symmetry group T_d, the rotations can be defined from several axes. The reduction of the representation $D^{(J)}$ in the point group or the molecular symmetry group T_d for the classification of the rotation levels can be established from the tables of correlation (see Table 1.13).

Landau [LAN 75] determines, first, the symmetry of the spin function and then applies the rotation subgroup of the extended point-group symmetry to the equivalent rotation operators (e.g. each reflection symmetry plane σ_v of the NH_3 molecule is represented by the product of the operator of rotation by an angle π with respect to an axis perpendicular to the plane and the operator that reverses the axes) to identify the rotational wave functions allowed by symmetry considerations. This method will be discussed in the following section, which focuses on the quantum spin degrees of freedom.

3.8.4. *Spin degrees of freedom*

To establish the selection rules, the nuclear spin function must be taken into account. These functions are explained for NH_3 and ND_3 molecules, as examples of symmetric tops, and for CH_4 and CD_4 molecules, as examples of spherical tops, with spin 1/2 exchangeable H nuclei for NH_3 and CH_4 and spin 1 D nuclei for ND_3 and CD_4. It should be noted that the N nucleus has a spin of 1 and that the C nucleus has a spin equal to 0.

In the case of NH_3, there are three H nuclei that can be exchanged. The spin of the H atom being equal to 1/2, each nucleus H_i can be associated with a three-dimensional space of variables of spin S_i^2. This space has the basis:

$$\left\{\begin{matrix} |\alpha_i\rangle = |s_i = 1/2, m_{s_i} = 1/2\rangle \\ |\beta_i\rangle = |s_i = 1/2, m_{s_i} = -1/2\rangle, i = 1,2,3 \end{matrix}\right\}.$$

The full spin space can be built by the tensor product of each of the subspaces, which leads to a spin space of dimension 8 (2^3), $S^8 = S_1^2 \otimes S_2^2 \otimes S_3^2$. The reduction of this space to irreducible representations makes it possible to determine the dimension of the symmetric and antisymmetric spaces associated with the spin variables. For CH_4, the dimension of the full spin space is 16. In the case of isotopic varieties, ND_3 and CD_4, since deuterium D has a spin equal to 1, each spin evolves in a space of dimension $2s + 1 = 3$; consequently, the dimensions of the full spin space are $3^3 = 27$ and $3^4 = 81$ for ND_3 and CD_4, respectively.

In order to determine the statistical weight of each rotation energy level of a fully symmetric vibronic level, Landau [LAN 75] established a formula, applied by Bunker and Jensen (see section 8.4 [BUN 98]), to calculate the characters $\chi_{sp}(G)$ of the representation of the spin function in the symmetry group G of the molecule. The complete wave function must include the part concerned by the spatial coordinates and the part concerned by the spin quantum coordinates in the form of a product at zeroth order. The spin function is the function of the projection of the spins of all the nuclei on a given direction in space. For a nucleus of spin I, the projection yields $2I + 1$ values that are assigned to m_I. If the molecule is composed of N nuclei, the various values corresponding to the spin factor are given by a number equal to $\prod_a (2I_a + 1) = (2I_1 + 1)(2I_2 + 1)...(2I_N + 1)$, which corresponds to the total number of spin functions. In a symmetry operation, the nuclei exchange their places, hence the spin value of the nuclei is exchanged. The various spin factors are mutually transformed, and a certain reducible representation of the symmetry group G of the molecule is achieved. The types of symmetry of the spin wave function are determined by decomposing the reducible representation on the irreducible representations of group G. The general formula of the characters $\chi_{sp}(G)$ of the representation achieved by the spin factors $(2I_\alpha + 1)$ is established by considering the nuclei with the same spin projections when the nuclei are permuted in the transformation g of group G. Otherwise, the spin factor does not contribute to $\chi_{sp}(G)$. The following expression is determined for $\chi_{sp}(G)$: $\chi_{sp}(G) = \prod_a (2I_a + 1)$ where the product involves the groups of nuclei that exchange their places in a transformation g of group G and which, consequently, contribute to the character of the transformation, so that each index a represents a group of nuclei of spin I_α and contributes to a factor in the product, including the group composed of only one nucleus, if this nucleus is invariant under the operation (exchanges with itself).

Given that the final objective of the calculation is not the symmetry of the spin functions, but that of the wave function of space coordinates, which is that of the rotation for a symmetric vibronic state (when the nuclei are interchanged, the coordinates of the electrons are considered to be invariant, as electrons move faster than nuclei), the calculation must include the symmetry of the complete wave function. The latter must be invariant or change its sign when a pair of nuclei is interchanged depending on whether the permutation concerns, respectively, a couple of nuclei governed by Bose–Einstein statistics or Fermi–Dirac statistics. The contribution ($2I_a + 1$) of the groups of nuclei to the character $\chi_{sp}(G)$ must be multiplied by $(-1)^{2Ia}$, where I_a is the spin of the nuclei of the group that undergoes the exchange. If n_a is the number of nuclei belonging to group a, factor ($2I_a$) must be multiplied by ($n_a - 1$) so that the product (($2I_a$)($n_a - 1$)) is odd only for an even permutation of fermions. Finally, the system of characters $\chi_{sp}(G)$ of the representation that involves all the irreducible representations according to which the coordinate wave functions are transformed is given by:

$$\chi_{sp}(G) = \prod_a (2I_a + 1)(-1)^{(2I_a)(n_a - 1)} \qquad [3.68]$$

where I_a is the spin of the nuclei of group a that undergoes the exchange and n_a is the number of nuclei in the group a.

The decomposition of this representation into an irreducible representation of the symmetry group G of the molecule makes it possible to determine the possible types of symmetry of the space coordinate wave functions of the molecule, weighted by the degeneracy multiplicities, according to the various spin states of the nuclei.

In the case of the ammonia molecule, NH_3, which belongs to the symmetry group C_{3v}, if inversion is considered negligible, the spin of nucleus N is equal to 1 and that of nucleus H is equal to 1/2. The formula of equation [3.68] can be used to determine the reducible character for each symmetry operation of group C_{3v}. The group is composed of the identity operation E, the operation $2C_3$ and the operation $3\sigma_v$.

In an identity operation E, the four nuclei remain in the same position. Each nucleus constitutes one group; therefore, a takes values from 1 to 4. For the nucleus N, when the identity transformation E is applied, 3 is obtained as the element of the product

$$(2I_a + 1) = 3 \text{ and } (-1)^{(2I_a)(n_a - 1)} = (-1)^{(0)} = 1$$

and for each nucleus H, 2 is obtained as the element of the product

$$(2I_a + 1) = 2 \text{ and } (-1)^{(2I_a)(n_a - 1)} = (-1)^{(0)} = 1$$

so that the product (3 x 2 x 2 x 2) yields $\chi_{sp}(E) = 24$.

Similarly, when the transformation C_3 is applied, two groups of atoms are identified: N, which is invariant, and the three H nuclei, which are exchanged. For the N nucleus, when the transformation C_3 is applied, 3 is obtained as the element of the product

$$(2I_a+1)=3 \text{ and } (-1)^{(2I_a)(n_a-1)}=(-1)^{(0)}=1$$

and for the group of three H nuclei, 2 is obtained as the element of the product

$$(2I_a+1)=2 \text{ and } (-1)^{(2I_a)(n_a-1)}=(-1)^{(2)}=1$$

so that $\chi_{sp}(C_3)=6$.

Similarly, when the transformation σ_v is applied, three groups of atoms are identified: N, which is invariant, an H atom, which is also in the symmetry plane and does not change its place in the operation and the two remaining H nuclei, which are exchanged. For the N nucleus, when applying the transformation σ_v, 3 is obtained as the element of the product

$$(2I_a+1)=3 \text{ and } (-1)^{(2I_a)(n_a-1)}=(-1)^{(0)}=1$$

for the H nucleus, 2 is obtained as the element of the product

$$(2I_a+1)=2 \text{ and } (-1)^{(2I_a)(n_a-1)}=(-1)^{(2)}=1$$

and for the group of the two H nuclei, -2 is obtained as the element of the product

$$(2I_a+1)=2 \text{ and } (-1)^{(2I_a)(n_a-1)}=(-1)^{(1)}=-1$$

so that $\chi_{sp}(\sigma v)=-12$.

This representation can be reduced by applying the reduction formula [1.23], given in Chapter 1.

The irreducible representations are given by: $12A_2 \oplus 6E$, which correspond to total spins 3/2 and 1/2, respectively. The statistical weights of the two types of levels A_2 and E are thus 12 and 6, respectively.

The symmetry of the complete wave function being of type A symmetric or antisymmetric, the statistical weights of the rotational levels of type A or E

(Tables 3.25 and 3.26) can be determined for a fully symmetric vibronic state, as reported in Table 3.27.

If this method is applied to the ND_3 molecule, the irreducible representations are given by: $30A_1 \oplus 3A_2 \oplus 24E$.

a)

NH_3 (C_{3v})			
Γ_{rve}	Γ_{ns}	Γ_{int}	g_{ns}
A_1	$12A_1$	A_1	12
A_2	$12A_1$	A_2	12
E	6E	A_1 A_2	12

b)

NH_3 (D_{3h})			
Γ_{rve}	Γ_{ns}	Γ_{int}	g_{ns}
A'_1			0
A''_1			0
A'_2	$12A'_1$	A'_2	12
A''_2	$12A'_1$	A''_2	12
E'	6 E'	A'_2	6
E''	6 E'	A''_2	6

Table 3.27. *Statistical weights of rovibronic states for NH_3 without inversion C_{3v} and with inversion D_{3h}*

In the case of the CDH_3 molecule, which belongs to the symmetry group C_{3v}, the spin of the C nucleus is equal to 0, that of the D nucleus is equal to 1 and that of the H nucleus is equal to 1/2. The group is composed of the identity operation E, the operation $2C_3$ and the operation $3\sigma_v$. The formula of equation [3.68] can be used to determine the reducible character for each symmetry operation of group C_{3v}.

In an identity operation E, the five nuclei remain at the same position. Each nucleus constitutes one group; therefore, the values of a range from 1 to 5. For the C nucleus, when the identity transformation E is applied, 1 is obtained as the element of the product

$$(2I_a+1)=1 \text{ and } (-1)^{(2I_a)(n_a-1)}=(-1)^{(0)}=1$$

for the D nucleus, 3 is obtained as the element of the product

$$(2I_a+1)=3 \text{ and } (-1)^{(2I_a)(n_a-1)}=(-1)^{(0)}=1$$

and for each H nucleus, 2 is obtained as the element of the product

$$(2I_a+1)=2 \text{ and } (-1)^{(2I_a)(n_a-1)}=(-1)^{(0)}=1$$

so that the product (1 x 3 x 2 x 2 x 2) yields $\chi_{sp}(E)=24$.

Similarly, when the transformation C_3 is applied, three groups of atoms are identified: C, which is invariant, the D nucleus, which is invariant and the three other H nuclei, which are exchanged. For the C nuclei, when the transformation C_3 is applied, 1 is obtained as the element of the product

$$(2I_a+1)=1 \text{ and } (-1)^{(2I_a)(n_a-1)}=(-1)^{(0)}=1$$

for the D nuclei, 3 is obtained as the element of the product

$$(2I_a+1)=3 \text{ and } (-1)^{(2I_a)(n_a-1)}=(-1)^{(0)}=1$$

and for the group of three H nuclei, 2 is obtained as the element of the product

$$(2I_a+1)=2 \text{ and } (-1)^{(2I_a)(n_a-1)}=(-1)^{(2)}=1$$

so that the product (1 x 3 x 2) yields $\chi_{sp}(C_3)=6$.

Similarly, when the transformation σ_v is applied, four groups of atoms are identified: C, which is invariant, the D nucleus, also in the plane of symmetry, and an H nucleus, also in the plane of symmetry and which does not change its place in the operation and the two remaining H nuclei, which are exchanged. For the C nucleus, when the σ_v transformation is applied, 1 is obtained as the element of the product

$$(2I_a+1)=1 \text{ and } (-1)^{(2I_a)(n_a-1)}=(-1)^{(0)}=1$$

for the D nucleus, also in the plane of symmetry, 3 is obtained as the element of the product

$$(2I_a+1)=3 \text{ and } (-1)^{(2I_a)(n_a-1)}=(-1)^{(0)}=1$$

for the H nucleus contained in the plane of symmetry σ_v, 2 is obtained as the element of the product

$$(2I_a+1)=2 \text{ and } (-1)^{(2I_a)(n_a-1)}=(-1)^{(0)}=1$$

and for the group of two H nuclei that are exchanged, -2 is obtained as the element of the product

$$(2I_a+1)=2 \text{ and } (-1)^{(2I_a)(n_a-1)}=(-1)^{(1)}=-1$$

so that the product (1 x 3 x 2 x -2) yields $\chi_{sp}(\sigma_v)=-12$.

This representation is reduced by applying the reduction formula [1.23], given in Chapter 1.

The irreducible representations are obtained as: $12A_2 \oplus 6\,E$.

CDH$_3$		
K	*J*	Γr
0	Even	A_1
0	Odd	A_2
3n+1	J	$A_1 \oplus A_1$
3n±1	J	E

a)

CDH$_3$ (C_{3v})			
Γ_{rve}	Γ_{ns}	Γ_{int}	g_{ns}
A_1	$12A_1$	A_1	12
A_2	$12A_1$	A_2	12
E	6E	A_1 A_2	12

b)

Table 3.28. *Symmetries of rotational levels and statistical weights of rovibronic states for CDH$_3$ without inversion C_{3v}*

In the case of the CDH$_3$ molecule, the irreducible representations are given by: $20A_1 \oplus 2A_2 \oplus 16E$.

Similarly to NH$_3$, the symmetry of the complete wave function being of type *A* symmetric or antisymmetric, the statistical weights of the rotational levels of type *A* or *E* for a fully symmetric vibronic state can be determined as summarized in Table 3.28(a) and (b) for CDH$_3$.

In the case of the methane CH$_4$ molecule, which belongs to the symmetry group T_d, the spin of the C nucleus is equal to 0 and that of the H nucleus is equal to 1/2. The formula of equation [3.68] can be used to determine the reducible character for each symmetry operation of the group T_d. The group is composed of the identity operation *E*, the operation $8C_3$, the operation $3C_2$, the operation $6\sigma_d$ and the operation $6S_4$.

In an identity operation *E*, the five nuclei preserve their position. Each nucleus constitutes one group; therefore, the values of *a* range from 1 to 5. For the C nucleus, when applying the identity transformation *E*, 1 is obtained as the element of the product

$$(2I_a+1)=1 \text{ and } (-1)^{(2I_a)(n_a-1)}=(-1)^{(0)}=1$$

and for each H nucleus, 2 is obtained as the element of the product

$$(2I_a+1)=2 \text{ and } (-1)^{(2I_a)(n_a-1)}=(-1)^{(0)}=1$$

so that the product (1 x 2 x 2 x 2 x 2) yields $\chi_{sp}(E)=16$.

Similarly, when the transformation C_3 is applied, three groups of atoms are identified: C, which is invariant, one H nucleus, which is invariant and three other H nuclei, which are exchanged. For the C nucleus, when the transformation C_3 is applied, 1 is obtained as the element of the product

$$(2I_a+1)=1 \text{ and } (-1)^{(2I_a)(n_a-1)}=(-1)^{(0)}=1$$

for the group of one H nucleus, 2 is obtained as the element of the product

$$(2I_a+1)=2 \text{ and } (-1)^{(2I_a)(n_a-1)}=(-1)^{(0)}=1$$

and for the group of three H nuclei, 2 is obtained as the element of the product

$$(2I_a+1)=2 \text{ and } (-1)^{(2I_a)(n_a-1)}=(-1)^{(2)}=1$$

so that the product (1 x 2 x 2) yields $\chi_{sp}(C_3)=4$.

Similarly, when transformation C_2 is applied, three groups of atoms are identified: C, which is invariant, two H nuclei, which are exchanged and two other H nuclei, which are also exchanged.

For the C nucleus, when the transformation C_2 is applied, 1 is obtained as the element of the product

$$(2I_a+1)=1 \text{ and } (-1)^{(2I_a)(n_a-1)}=(-1)^{(0)}=1$$

for the group of two H nuclei, –2 is obtained as the element of the product

$$(2I_a+1)=2 \text{ and } (-1)^{(2I_a)(n_a-1)}=(-1)^{(1)}=-1$$

and for the group of the other two H nuclei, –2 is obtained as the element of the product

$$(2I_a+1)=2 \text{ and } (-1)^{(2I_a)(n_a-1)}=(-1)^{(1)}=-1$$

so that the product (1 x -2 x -2) yields $\chi_{sp}(C_2)=4$.

Similarly, when the transformation σ_d is applied, four groups of atoms are identified: C, which is invariant, one H nucleus, also in the plane of symmetry, and a second H nucleus, also in the plane of symmetry, and which does not change its place under the operation and the two remaining H nuclei, which are exchanged. For the C nucleus, when the identity transformation σ_d is applied, 1 is obtained as the element of the product

$$(2I_a+1)=1 \text{ and } (-1)^{(2I_a)(n_a-1)}=(-1)^{(0)}=1$$

for the two H nuclei contained in the plane of symmetry σ_d, 2 is obtained for each as the element of the product

$$(2I_a+1)=2 \text{ and } (-1)^{(2I_a)(n_a-1)}=(-1)^{(0)}=1$$

and for the group of two H nuclei that are exchanged, -2 is obtained as the element of the product

$$(2I_a+1)=2 \text{ and } (-1)^{(2I_a)(n_a-1)}=(-1)^{(1)}=-1$$

so that the product (1 x 2 x 2 x –2) yields $\chi_{sp}(\sigma_d)=-\,8$.

Similarly, when the transformation S_4 is applied, two groups of atoms are identified: C, which is invariant and the four H nuclei, which are exchanged. For the C nucleus, when applying the S_4 transformation, 1 is obtained as the element of the product

$$(2I_a+1)=1 \text{ and } (-1)^{(2I_a)(n_a-1)}=(-1)^{(0)}=1$$

and for the group of four H nuclei, -2 is obtained as the element of the product

$$(2I_a+1)=2 \text{ and } (-1)^{(2I_a)(n_a-1)}=(-1)^3=-1$$

so that the product (1 x –2) yields $\chi_{sp}(S_4)=-2$.

This representation is reduced by applying the reduction formula [1.23] of Chapter 1.

The irreducible representations are given by: $5A_2 \oplus E \oplus 3T_1$ that correspond to the total spin 2 (meta), 0 (para) and 1(ortho). Table 3.29 summarizes the statistical weights of the rotation levels for $J = 0$ to $J = 4$ taking into account the spin function with the notations of group T_d.

CH_4 (T_d) (J = 0, 1, 2, 3, 4)					
J	**Γ_{rve}**	**Γ_{ns}**	**Γ_{int}**	**g_{ns}**	**g_J**
0	**A_1**	$5A_2$	A_2	5	1
1	**T_1**	$3T_1$	A_1	3	3
2	**E**	E	$A_1 \oplus A_2$	2	5
2	**T_2**	$3T_1$	A_2	3	5
3	**A_1**	$5A_2$	A_2	5	7
3	**T_1**	$3T_1$	A_1	3	7
3	**T_2**	$3T_1$	A_2	3	7
4	**A_1**	$5A_2$	A_2	5	9
4	**E**	E	$A_1 \oplus A_2$	2	9
4	**T_1**	$3T_1$	A_1	3	9
4	**T_2**	$3T_1$	A_2	3	9

Table 3.29. *Statistical weights of the rovibronic states of CH_4 for J = 0 to J = 4*

The rotation subgroup of group T_d being the group T, the notation of group T is sometimes used to label the eigenstates of CH_4 [HER 45]. In this case, the irreducible representations are given by: $5A \oplus E \oplus 3T$. As a result of considering the spin states, the complete wave function may be of type A for all the rotation sub-levels. If the rotational level is of type A, the product should use the spin function of type A, which leads to a statistical weight of 5 (A appears five times). If the rotational level is of type E, the product should use the spin function of type E, which leads to a statistical weight of 2, because $E \otimes E = 2A \oplus E$, the type A appearing twice and E appearing just once. If the rotational level is of type T, the product should use the spin function of type T, which yields a statistical weight of 3, as $T \otimes T = A \oplus E \oplus 2T$, the type A appearing just once and T appearing three times.

In the case of the CD_4 molecule, the irreducible representations are given by: $15A_1 \oplus 6E \oplus 3T_1 \oplus 15T_3$. The statistical weights of the rotation levels x are determined following the same procedure as the one used for CH_4.

3.8.5. *IR and Raman selection rules for the rotational levels*

For the spherical top, for which the rotational energy of the rigid rotator is a function of J only with a degeneracy of $2J + 1$ in K (see equation [1.11]), a decomposition along the rules of the vector addition model in group K (or SO(3)) can be followed, using the fact that the electric dipole moment operator is a vector operator, that is, a first-order spherical tensor. It is possible to obtain the selection rules for the electric dipole moment operator ($D^{(1)}$) active in IR and which induces a transition between the level characterized by J' and that characterized by J'' from the following:

$$D^{(J'')} \otimes D^{(1)} \otimes D^{(J')} \supset D^{(0)} \qquad [3.69]$$

provided that $J'' = J'$ except for $J'' = J' = 0$ or $J'' = J' \pm 1$. In the case of a symmetric top, the selection rule of the quantum number K, projection of $\boldsymbol{J}$ on the Oz axis of the mobile reference frame, should be taken into account.

The electric polarizability operator is built as a dyadic product of polar vectors leading to a reducible Cartesian tensor of rank 2. It can be decomposed into a scalar part (a scalar product invariant under rotation), a vector part (a vector product or antisymmetric tensor) and a part that corresponds to a zero trace tensor, giving three components that behave differently under a rotation operation. The second-order tensor can thus be reduced in the form of irreducible spherical tensors of rank 0 (symmetric), rank 1 (antisymmetric) and rank 2 (symmetric). The selection rules for the electrical polarizability operator $\left(D^{(0)} \otimes D^{(1)} \otimes D^{(2)}\right)$ that is active in Raman spectroscopy and that induces a transition between the level characterized by J' and that characterized by J'' are established from the following relation:

$$D^{(J'')} \otimes \left(D^{(0)} \oplus D^{(1)} \oplus D^{(2)}\right) \otimes D^{(J')} \supset D^{(0)} \qquad [3.70]$$

which leads to $J'' = J'$ except for $J'' = J' = 0$ or $J'' = J' \pm 1$ or $J'' = J' \pm 3$.

3.9. Conclusion

The group theory methods were applied in order to determine the functions generating irreducible representations of the symmetry group of the symmetric top and spherical top molecules by reference to NH_3 and CH_4 molecules and to their isotopes for various degrees of freedom: electronic, vibrational, rotational and electronic and nuclear spins. This was conducted within the framework of an approximate Hamiltonian, which allows us to express the complete wave function as a product of wave functions, assuming that the vibration–rotation interaction is negligible.

The statistical weights of the levels were calculated with the generic formula given by Landau [LAN 75] and the inversion phenomena described within the framework of the molecular symmetry group according to the Bunker and Jensen [BUN 98] approach.

3.10. Appendix: Absorption and emission of a molecule in the gas phase

This appendix reviews the theoretical model used for calculating the absorption coefficient of a molecule in the gas phase. Gas molecules are exposed to a radiation of energy density $\varrho(\sigma)$ in an infinitesimal bandwidth of wave number $d\sigma$ around the wave number σ. The molecules are thus undergoing induced absorption and induced (or stimulated) emission.

The probability for a molecule to undergo a transition from the lower level $|i\rangle$ to a higher level $|f\rangle$ by absorbing a photon in a unit of time is given by the equation:

$$\left(\frac{dP_{if}}{dt}\right)_{abs} = B_{if}\rho(\sigma) \qquad [3.71]$$

Similarly, the probability for a molecule to undergo a transition from the higher level $|f\rangle$ to a lower level $|i\rangle$ by emitting a photon in a unit of time is given by the equation:

$$\left(\frac{dP_{if}}{dt}\right)_{emis} = B_{fi}\rho(\sigma) \qquad [3.72]$$

Spontaneous emission of an excited level as a result of vacuum fluctuations in the absence of radiation should also be considered. Its probability per unit of time is given by:

$$\left(\frac{dP_{if}}{dt}\right)_{spont} = A_{fi} \qquad [3.73]$$

At thermodynamic equilibrium of the radiation–molecule system, the following equality is verified:

$$N_f\left(B_{fi}\rho(\sigma)+A_{fi}\right)=N_iB_{if}\rho(\sigma) \qquad [3.74]$$

since the emission rate must be equal to the absorption rate, with the system being in a steady state for which N_α ($\alpha = i$ or f) is the number of molecules on the level of energy E_α.

Moreover, for a system subject to the Maxwell–Boltzmann statistics at temperature T, N_α is given by:

$$\frac{N_\alpha}{N} = g_\alpha e^{-\frac{E_\alpha}{k_BT}} \qquad [3.75]$$

where g_α is the degeneracy of the level of energy E_α and N is the total number of molecules.

Including [3.75] in [3.74] leads to:

$$\rho(\sigma)=\frac{A_{fi}/B_{fi}}{g_iB_{if}/g_fB_{fi}}\frac{1}{e^{hc\sigma/k_BT}-1} \qquad [3.76]$$

and the comparison with Planck's law:

$$\rho(\sigma)=8\pi h\sigma^3\frac{1}{e^{hc\sigma/k_BT}-1} \qquad [3.77]$$

leads to the following relations:

$$A_{fi}/B_{fi}=8\pi h\sigma^3 \text{ and } g_iB_{if}/g_fB_{fi}=1 \qquad [3.78]$$

As indicated in section 1.1 of Volume 1 [DAH 17], in the case of a transition between the states $|i\rangle$ and $|f\rangle$, the probability of radiative transition is given by Einstein's coefficient for spontaneous emission:

$$A_{if} = \tau^{-1} = \frac{64\pi^4}{h}\frac{g_i}{g_f}\nu_{if}^3\left|R_{if}\right|^2 \qquad [3.79]$$

where τ^{-1} is the radiative lifetime, g_i and g_f are the degeneracies of levels $|i\rangle$ and $|f\rangle$, ν_{if} is the frequency of the considered transition and R_{if} is the matrix element of the operator inducing the transition. This leads, from the expression of Einstein's coefficient for spontaneous emission, to those of the induced absorption coefficient B_{if} or induced emission coefficient B_{fi} (see equation [3.78]) such that:

$$A_{if} = \frac{8\pi h\nu_{if}^3}{c^3}B_{if} = \frac{8\pi h}{\lambda_{if}^3}B_{if} = 8\pi h\sigma_{if}^3 B_{if} \qquad [3.80]$$

where $\frac{B_{if}}{g_f} = \frac{B_{fi}}{g_i}$ (see equation [3.78]) and λ_{if} is the wavelength and σ_{if} (in cm^{-1}) is the wave number associated with the transition between the two levels.

The number of transitions N_{tr} per unit of volume, per unit of time of the initial lower level of energy E_i towards a final higher level of energy E_f is given by:

$$N_{tr} = \rho(\sigma)\left(N_i B_{if} - N_f B_{fi}\right) = \rho(\sigma) N_i B_{if}\left(1 - e^{hc\sigma_{if}/k_B T}\right) \qquad [3.81]$$

where σ_{if} is the wave number of the transition.

The intensity I of an isotropic radiation in a given direction is the energy flux per unit of time through a unit of surface perpendicular to the given direction. It is given by $I(\sigma) = c\rho(\sigma)$. As described in Chapter 3 of Volume 1, the Bouguer–Lambert law is used to determine the variation of this intensity when the radiation travels through a length l of a gas at pressure P using the expression:

$$dI(\sigma) = -k(\sigma,P)I(\sigma)dl \qquad [3.82]$$

where $k(\sigma,P)$ is the coefficient of absorption at wave number σ and at pressure P. From the intensity transmitted by a length l of gas: $I_t(\sigma) = I_0(\sigma)\exp\left(-k(\sigma,P)l\right)$ obtained by integration of [3.82], $k(\sigma,P)$ is measured using the formula:

$$k(\sigma,P)=\frac{1}{l}Log\left(\frac{I_0(\sigma)}{I_t(\sigma)}\right) \quad [3.83]$$

The quantity defined as:

$$S=\int_0^{\infty}k(\sigma,P)d\sigma \quad [3.84]$$

is termed as the integrated absorption coefficient of the absorption line that is written in the form $S = S^0P$, for moderate pressures (below several terrestrial atmospheres), where S^0 defines the intensity of the absorption line at temperature T.

From the balance equation between the transition levels and taking into account both the populations of each level:

$$N_i = N_0\exp\left(-\frac{E_i}{k_BT}\right) = N_0\exp\left(-\frac{hc\sigma_i}{k_BT}\right), \quad [3.85]$$

and the profile of the absorption line (Gaussian by Doppler effect, Lorenztian by pressure broadening), the intensity of a vibration–rotation absorption line in the gas phase in the IR range is given by:

$$S_{if}^g=\int_{\sigma_{\min}}^{\sigma_{\max}}\frac{8\pi^3}{3hc}\sigma_{if}f(\sigma-\sigma_{if})N_0\left[1-e^{\left(-\frac{hc\sigma_{if}}{k_BT}\right)}\right]g_i\frac{e^{\left(-\frac{hc\sigma_i}{k_BT}\right)}}{Q(T)}R_{if}^2d\sigma \quad [3.86]$$

where σ_{if} is the wave number (in cm^{-1}) of the given rovibrational transition, N_0 is the number of molecules per unit volume, $hc\sigma_i$ is the energy of the initial level, $Q(T)=\sum_i g_i\exp\left(-E_i/k_BT\right)$ designates the total partition function of the molecule at temperature T, R_{if} is the moment of rovibrational transition and finally g_i is the degeneracy of the initial level, due to the nuclear spin statistics. In this formula, the profile of the line is a Voigt function, the convolution product of a Gaussian by a Lorenztian, since the Doppler effect and the pressure broadening are simultaneous, and its integral over the absorption domain is equal to:

$$1\left(\int_{\sigma_{\min}}^{\sigma_{\max}}f(\sigma-\sigma_{if})d\sigma=1\right)$$

In the solid phase, a corrective term due to the effect of the refractive index n (from the polarization of the medium) of the solid rare gas is required. If the vibration–rotation interaction is considered negligible and if the fundamental state is not degenerate ($g_0 = 1$), the intensity of a vibrational line at a low temperature for a molecule can be written as:

$$S_{if}^{m} = \int_{\sigma_{\min}}^{\sigma_{\max}} \frac{1}{n}\left(\frac{n^2+1}{3}\right)^2 \frac{8\pi^3}{3hc}\sigma_{if} f(\sigma-\sigma_{if})\left|R_{if}\right|^2 \mathrm{d}\sigma \qquad [3.87]$$

where R_{if} designates the vibrational transition moment. Written in this form, relation [3.87] brings essential information on the dipole transition moments of the vibration modes in a rare gas matrix. These quantities can be experimentally obtained from the integrated intensity of the absorption spectrum:

$$S_{if}^{m} = \frac{1}{2lN}\int Ln\frac{I_0(\sigma-\sigma_{if})}{I_t(\sigma-\sigma_{if})}d\sigma = \frac{1}{2lN}I_{if} \qquad [3.88]$$

where N is the number of molecules per cm^3, $2l$ is the length traveled in the sample and the term under integral is the integrated absorption measured over the IR absorption spectrum domain.

4

Energy Levels of Symmetric Tops in the Gas Phase

One of the major problems in the interpretation of infrared (IR) and Raman spectra of a polyatomic molecule is the determination of its vibrational potential function. This must contain all the information on the forces between the atoms of the molecule, and represent a starting point of the perturbation calculations which allows a fine analysis of the interactions between these atoms in order to interpret or predict the detailed spectra of the molecule.

In the gas phase, the molecular spectra result from the interaction between an electromagnetic radiation and the nuclei and electrons composing the molecule, through its dipole moment or its induced dipole moment. As a result of this interaction, the transitions between energy levels lead to absorption and/or emission spectra of the molecule or to its Raman spectrum. Therefore, knowledge of the gas phase molecular spectra offers the possibility to use these molecules as tracers (probes) to characterize the physical and chemical properties of the medium, for example, a condensed medium, in which they could evolve.

More specifically, infrared absorption spectroscopy can be used to observe the cold bands, i.e. the lines corresponding to vibration–rotation transitions (near IR) or to rotation transitions (far IR) starting from the fundamental level, assuming that the molecule is in its fundamental electronic state.

Unlike diatomic molecules, which have only one internal vibration mode, characterized by only one frequency, polyatomic molecules have several vibrational modes, some of which may be degenerate, characterized by very different frequencies.

This chapter focuses on the study of IR spectra of symmetric top molecules, more specifically the pyramidal molecules of type XY_3 such as NH_3, PH_3, AsH_3, PF_3, PCl_3 and AsF_3. In the following, the studied molecules are assumed to remain in their fundamental electronic state.

4.1. Introduction

The interpretation of a molecular spectrum relies on the modeling of the interaction between the molecule and the electromagnetic radiation which, through the Hamiltonian operator associated with the different internal and external motions of the molecule, leads to transitions between the states of the molecule. In general, the Schrödinger eigenvalue equation associated with the Hamiltonian operator does not have simple analytical solutions. The Born–Oppenheimer (BO) approximation is applied to decouple the rapid motions of electrons around the nuclei from those of the heavier nuclei (proton to electronic mass ratio of 1836.152), which are far slower. Indeed, for each electronic state, the nuclei evolve in an average potential that depends only on their coordinates and their spin in a reference frame attached to the equilibrium configuration of the molecule.

A pyramidal symmetric top molecule of type XY_3 is a stable assembly of four atoms connected by electromagnetic forces and forming, in its equilibrium configuration, a regular pyramidal structure of height h^e, with nucleus X at its top, and the sides which are the covalent bonds between the nucleus X and the three nuclei Y all of the same length r^e making an angle β^e with the threefold symmetry axis of the molecule denoted by C_3 (see Figure 4.1). This type of molecule belongs to the symmetry group C_{3v}, and can be oblate or prolate depending on the values of their moments of inertia.

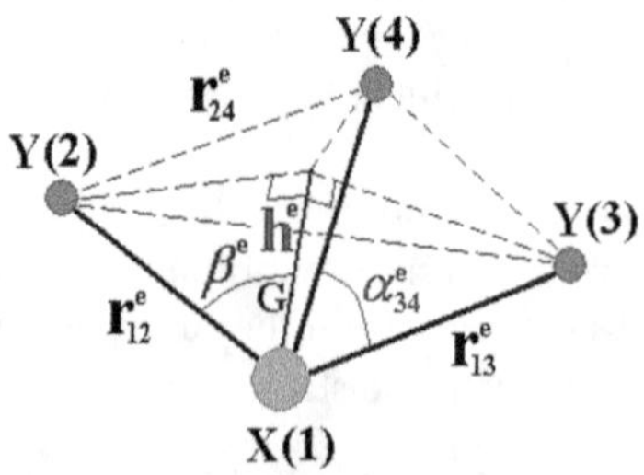

Figure 4.1. *Geometrical characteristics of a symmetric top molecule of type XY_3 in its equilibrium configuration. For a color version of this figure, see www.iste.co.uk/dahoo/infrared.zip*

The relations between the geometrical characteristics of the XY_3 molecule in its equilibrium configuration are given as follows:

$$
\begin{aligned}
&r_{12}^{e} = r_{13}^{e} = r_{14}^{e} = r^{e} \\
&\beta_2^{e} = \beta_3^{e} = \beta_4^{e} = \beta^{e} \\
&r_{23}^{e} = r_{24}^{e} = r_{34}^{e} = \sqrt{3} \times r^{e} \times \sin\beta^{e} \\
&\alpha_{23}^{e} = \alpha_{23}^{e} = \alpha_{23}^{e} = \alpha^{e} \\
&\sin\frac{\alpha^{e}}{2} = \frac{\sqrt{3}}{2}\sin\beta^{e} \\
&h^{e} = r^{e}\cos\beta^{e}
\end{aligned}
$$

4.2. Vibrational–rotational motions of an isolated symmetric top

The study of the vibrational–rotational motions of a molecule requires a decoupling of the latter from the translational motion of the molecule. This decoupling can be rigorously obtained by choosing a reference frame $(G, \vec{X}, \vec{Y}, \vec{Z})$ in uniform translational motion with the molecule. The origin of the reference frame coincides with the center of mass G of the molecule and its axes are permanently parallel to those of the laboratory fixed direct orthonormal reference frame $(O, \vec{X}, \vec{Y}, \vec{Z})$ (first Eckart condition [ECK 35], see Chapter 1, Volume 2 [DAH 19]) as shown in Figure 4.2.

However, the vibrational–rotational motions cannot be separated in an exact manner. But, given the wide energy difference between the rapid and low-amplitude vibrational motions and the slow rotational motion, it is, assuming that the Born–Oppenheimer approximation is valid, possible to study them separately and then make the corrections due to their interaction.

In order to study the vibrational motions, a reference frame that is generally direct orthonormal $(G, \vec{x}, \vec{y}, \vec{z})$ is associated with the molecule, whose origin lies in its center of mass and whose axes are judiciously chosen along the principal directions of inertia of its equilibrium configuration (see Figure 4.2). Indeed, the use of such a reference frame $(G, \vec{x}, \vec{y}, \vec{z})$ makes it possible to minimize the vibration–rotation coupling, and the vibrational motions can be approached as a set of harmonic oscillators and the rotational motion as that of a rigid rotor. The said coupling as well as the anharmonic terms of the vibrational potential can then be accounted for using perturbation methods.

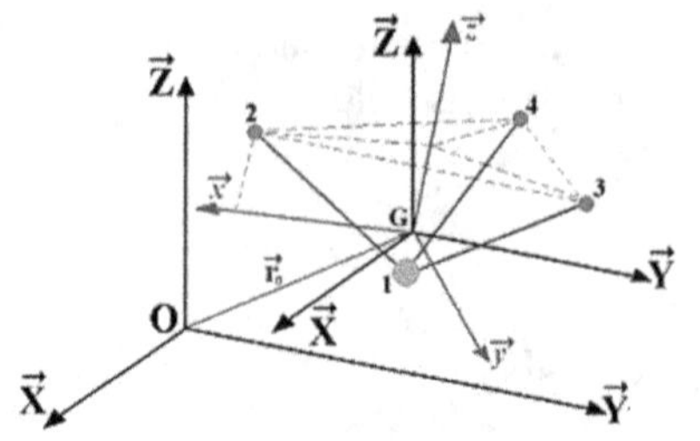

Figure 4.2. *Definition of the fixed direct orthonormal systems of axes:* $(O,\vec{X},\vec{Y},\vec{Z})$ *associated with the laboratory reference frame and assumed Galilean* $(G,\vec{x},\vec{y},\vec{z})$ *attached to the equilibrium configuration of the symmetric top molecule and* $(G,\vec{X},\vec{Y},\vec{Z})$ *attached to the molecule in its translational motion. For a color version of this figure, see www.iste.co.uk/dahoo/infrared.zip*

Generally, in a molecule composed of N atoms, the nuclei undergo motions in the three directions of the space, which are described by *3N* degrees of freedom. Among these *3N* degrees of freedom, 3 characterize the overall translational motion of the molecule represented by the position vector $\vec{r}_0(X_0,Y_0,Z_0)$ of its center of mass, which will be ignored in what follows in this chapter according to Eckart conditions [ECK 35], and 3 correspond to the rotational motion of the molecule represented by Euler angles, φ for precession, θ for nutation and χ for intrinsic rotation, associated at each instant with the orientation of its equilibrium configuration reference frame $(G,\vec{x},\vec{y},\vec{z})$ with respect to the reference frame $(G,\vec{X},\vec{Y},\vec{Z})$. The rotation matrix allowing the $(G,\vec{X},\vec{Y},\vec{Z})$ reference frame to change into the $(G,\vec{x},\vec{y},\vec{z})$ reference frame was established in section 2.7.1 of Volume 2 [DAH 19].

Finally, the remaining *3N-6* degrees of freedom correspond to the vibrational modes of the molecule through which the nuclei vibrate about their equilibrium position at the same frequencies. In the case of the pyramidal symmetric top molecules of type XY_3, the number of vibrational modes is: $3\times 4-6=6$.

The application of group theory makes it possible to determine the vibrational modes of a molecule and their species. When a pyramidal molecule of type XY_3 is subjected to the symmetry operations of the group C_{3v}, four vibrational modes corresponding to the irreducible representations of type A_1 and E are depicted:

– two fully symmetric non-degenerate modes ν_1 and ν_2 of species A_1;

– two doubly degenerate antisymmetric bending-deformation modes ν_3 and ν_4 of species E (also denoted by ν_{3a}, ν_{3b}, ν_{4a} and ν_{4b}).

These two species are mutually orthogonal.

Figure 4.3 presents the vibrational modes of a pyramidal symmetric top molecule of type XY_3. In this figure, the arrows only represent the direction of the nuclei displacement during the vibration motions and not their amplitude.

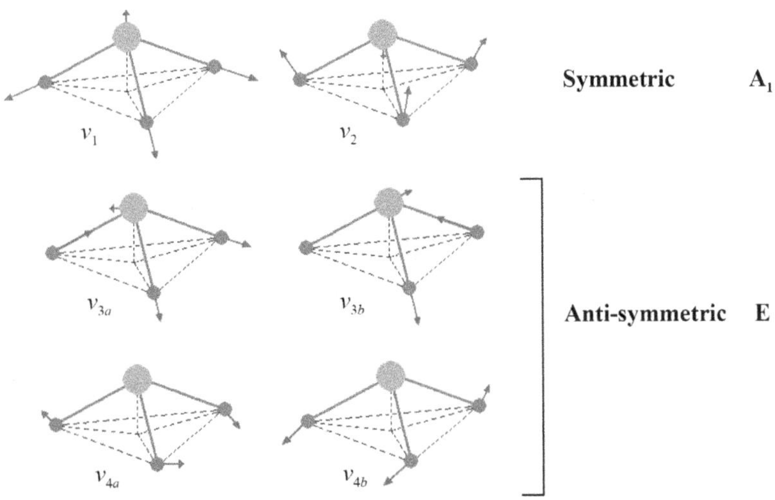

Figure 4.3. *Vibration modes of a symmetric top molecule of type XY_3. The arrows indicate only the direction of the nuclei displacement. For a color version of this figure, see www.iste.co.uk/dahoo/infrared.zip*

The energies and wave functions of the vibration–rotation states of a molecule, assumed to be and to remain in its fundamental electronic state, are the eigenvalues and eigenvectors of the nuclear Hamiltonian operator, solutions of the Schrödinger's time independent equation:

$$H_N \Psi_N = E_N \Psi_N \qquad [4.1]$$

where H_N is the sum of the nuclear (vibration–rotation) kinetic energy operator T_N and the vibrational potential energy operator V_{vib} of the molecule:

$$H_N = T_N + V_{\text{vib}} \qquad [4.2]$$

Figure 4.4 presents a symmetric top molecule of type XY_3 with its equilibrium configuration reference frame $(\text{G}, \vec{x}, \vec{y}, \vec{z})$. The Cartesian coordinates of the nuclei of the rigid molecule are presented in Table 4.1.

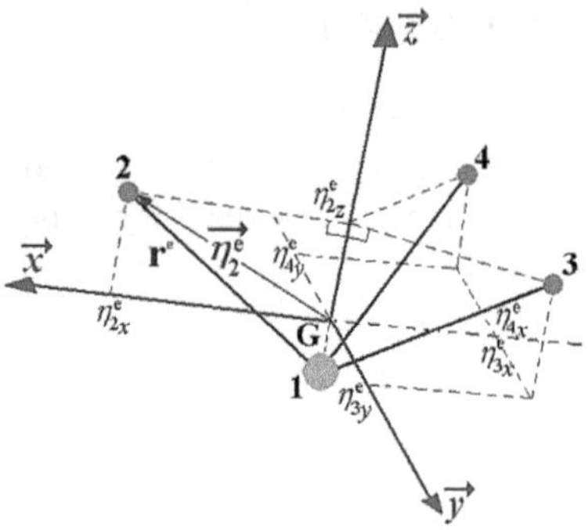

Figure 4.4. *System of axes* $(\mathrm{G}, \vec{x}, \vec{y}, \vec{z})$ *attached to the equilibrium configuration of a symmetric top molecule XY*$_3$*; and position vector of a nucleus* (nucleus *Y(2)*, *for example*) *in this system. For a color version of this figure, see www.iste.co.uk/dahoo/infrared.zip*

Nucleus coordinates	$\eta^{e}_{i_x}$	$\eta^{e}_{i_y}$	$\eta^{e}_{i_z}$
X (i = 1)	0	0	$\frac{-3m_Y}{m_X+3m_Y}\mathrm{r}^e\cos\beta^e$
Y (i = 2)	$\mathrm{r}^e\sin\beta^e$	0	$\frac{m_X}{m_X+3m_Y}\mathrm{r}^e\cos\beta^e$
Y (i = 3)	$-\frac{1}{2}\mathrm{r}^e\sin\beta^e$	$\frac{\sqrt{3}}{2}\mathrm{r}^e\sin\beta^e$	$\frac{m_X}{m_X+3m_Y}\mathrm{r}^e\cos\beta^e$
Y (i = 4)	$-\frac{1}{2}\mathrm{r}^e\sin\beta^e$	$-\frac{\sqrt{3}}{2}\mathrm{r}^e\sin\beta^e$	$\frac{m_X}{m_X+3m_Y}\mathrm{r}^e\cos\beta^e$

Table 4.1. *Cartesian coordinates of the nuclei of a symmetric top molecule of type XY*$_3$ *in the system of axes* $(\mathrm{G}, \vec{x}, \vec{y}, \vec{z})$ *attached to its equilibrium configuration.* m_X *and* m_Y *are, respectively, the mass of nucleus X and that of nuclei Y*

Consider a nucleus i of mass m_i, of position vector $\vec{\eta}_i$ defined, in the reference frame $(\mathrm{G}, \vec{\mathrm{X}}, \vec{\mathrm{Y}}, \vec{\mathrm{Z}})$ attached to the molecule, by its Cartesian components $(\eta_{i_X}, \eta_{i_Y}, \eta_{i_Z})$. Over time, this nucleus undergoes, relative to this frame, low-amplitude motions around its equilibrium position defined by $\vec{\eta}_i^{\,e}$, with a velocity vector $\dot{\vec{\eta}}_i$ of components $(\dot{\eta}_{i_X}, \dot{\eta}_{i_Y}, \dot{\eta}_{i_Z})$.

In classical mechanics, the nuclear kinetic energy T_N of a molecule is the sum of the kinetic energies of its nuclei. It is written as:

$$T_N = \frac{1}{2}\sum_{i=1}^{4} m_i \dot{\vec{\eta}}_i \cdot \dot{\vec{\eta}}_i \qquad [4.3]$$

Nevertheless, the motion of the set of nuclei of the molecule is composed of:

– a block translational motion, which is a uniform rectilinear motion of the center of mass of the molecule with respect to a fixed reference frame $(\mathrm{O}, \vec{\mathrm{X}}, \vec{\mathrm{Y}}, \vec{\mathrm{Z}})$. This motion is not taken into consideration, as the study is conducted in the reference frame $(\mathrm{G}, \vec{\mathrm{X}}, \vec{\mathrm{Y}}, \vec{\mathrm{Z}})$ attached to the molecule;

– a motion around the center of mass of the molecule: the rotational motion;

– motions of the nuclei, each one with respect to the others, leading to molecule deformation: the vibrational motions.

In the reference frame $(\mathrm{G}, \vec{\mathrm{X}}, \vec{\mathrm{Y}}, \vec{\mathrm{Z}})$ attached to the molecule, using the derivative basis change rule, the velocity vector $\dot{\vec{\eta}}_i$ is written as:

$$\dot{\vec{\eta}}_i = \dot{\vec{\mathrm{r}}}_i + \vec{\omega} \times \vec{\eta}_i \qquad [4.4]$$

where $\vec{\omega}$ is the rotation velocity vector of the reference frame $(\mathrm{G}, \vec{x}, \vec{y}, \vec{z})$ with respect to the reference frame $(\mathrm{G}, \vec{\mathrm{X}}, \vec{\mathrm{Y}}, \vec{\mathrm{Z}})$.

After multiple transformations of the nuclear kinetic energy of the molecule and following the quantization of the vibrational–rotational (or rovibrational) motions, Wilson and Howard [WIL 36] established an expression of the nuclear Hamiltonian H_N, which was reformulated by Darling and Denisson [DAR 40] and simplified by Watson [WAT 68a]. The reader is referred to section 2.2 of Volume 2 [DAH 19].

Finally, the nuclear Hamiltonian can be written as the sum of: (1) the Hamiltonian of vibrations $T_{\mathrm{vib}} + V_{\mathrm{vib}}$ relative to the reference frame of the equilibrium configuration $(\mathrm{G}, \vec{x}, \vec{y}, \vec{z})$; (2) the rotational Hamiltonian of the molecule assumed rigid T_{rot} (rigid rotor); and (3) a Hamiltonian, which is far smaller in magnitude than the previous ones, containing vibration–rotation interaction (coupling).

In order to determine the eigenstates of the nuclear Hamiltonian H_N, the first step is to consider the term of vibration–rotation interaction as being negligible in order to separate the vibrational motions from the rotational motion. The energy E_N and the wave function Ψ_N are written as:

$$
\begin{aligned}
E_N &= E_{\text{vib}} + E_{\text{rot}} \\
\Psi_N &= \Psi_{\text{vib}} \Psi_{\text{rot}}
\end{aligned}
\qquad [4.5]
$$

The interaction term can then be introduced by perturbation calculation on the basis of eigenenergies and eigenfunctions of expressions [4.5].

4.3. Vibrational motions of an isolated pyramidal symmetric top

4.3.1. *Kinetic and potential energy functions*

Molecular vibrations are a standard problem of small movements in mechanics. Their study requires good knowledge of kinetic and potential energy functions, and fundamental functions of mechanics.

To solve this problem, the nuclei in the molecule are assumed to execute only small-amplitude motions in the vicinity of their equilibrium position.

4.3.1.1. *Cartesian coordinates*

In the reference frame $(\mathrm{G}, \vec{x}, \vec{y}, \vec{z})$ of the equilibrium configuration of the molecule, the position vector $\vec{\eta}_i$ of the nucleus i can be written as $\vec{\eta}_i = \vec{\eta}_i^{\mathrm{e}} + \vec{\mathrm{r}}_i$, where $\vec{\mathrm{r}}_i$ is the instantaneous displacement vector (small displacements relative to internuclear distances) of the nucleus in the vicinity of its equilibrium position, of Cartesian components (x_i, y_i, z_i) termed as "vibrational coordinates". Its velocity vector is then $\dot{\vec{\eta}}_i = \dot{\vec{\mathrm{r}}}_i$, and the vibrational kinetic energy T_{vib} of the molecule is written as:

$$
T_{\text{vib}} = \frac{1}{2} \sum_{i=1}^{4} m_i \left(\dot{x}_i^2 + \dot{y}_i^2 + \dot{z}_i^2 \right) \qquad [4.6]
$$

Moreover, the vibrational potential energy V_{vib} is a function of Cartesian vibrational coordinates α_i of the nuclei and has a shape of multidimensional well, whose minimum corresponds to the potential energy of the equilibrium configuration of the molecule. Moreover, the amplitudes of the motions of the nuclei being small, the vibrational potential energy function can be represented by a series expansion about the equilibrium position:

$$V_{\text{vib}} = V_0 + \sum_{i=1}^{4} \sum_{\alpha=x,y,z} \left(\frac{\partial V_{\text{vib}}}{\partial \alpha_i} \right)_0 \alpha_i + \frac{1}{2!} \sum_{i,j=1}^{4} \sum_{\alpha,\beta} \left(\frac{\partial^2 V_{\text{vib}}}{\partial \alpha_i \partial \beta_j} \right)_0 \alpha_i \beta_j$$
$$+ \frac{1}{3!} \sum_{i,j,k=1}^{4} \sum_{\alpha,\beta,\gamma} \left(\frac{\partial^3 V_{\text{vib}}}{\partial \alpha_i \partial \beta_j \partial \gamma_k} \right)_0 \alpha_i \beta_j \gamma_k + \ldots \quad [4.7]$$

where V_0 is the potential energy of the equilibrium configuration (minimum energy), a constant that can be arbitrarily chosen as origin.

– The second set of terms (first order) contains the corrective forces acting on the nuclei at their equilibrium position. These forces are therefore equal to zero.

– The third set of terms (second order) constitutes the harmonic part of the vibrational potential energy of the nuclei:

$$V_{\text{vib}}^{\text{har}} = \frac{1}{2} \sum_{i,j=1}^{4} \sum_{\alpha,\beta} k_{\alpha\beta}^{ij} \alpha_i \beta_j \quad [4.8]$$

where $k_{\alpha\beta}^{ij} = \left(\frac{\partial^2 V_{\text{vib}}}{\partial \alpha_i \partial \beta_j} \right)_0$ are the second derivatives of the potential energy function, termed as "force constants". This is the most important part of V_{vib} and contains the interaction terms (cross products) $\alpha_i \beta_j$ of the Cartesian vibration coordinates of the nuclei, whose force constants verify the relation $k_{\alpha\beta}^{ij} = k_{\beta\alpha}^{ji}$.

– Finally, the following terms (of orders 3, 4, etc.), $\Delta V_{\text{vib}}^{\text{anhar}}$, characterize the anharmonicity of the vibrational motions, and they are considered as perturbation terms.

Due to their small amplitudes, the vibrational motions can be treated in the harmonic oscillator approximation, the anharmonic terms being then introduced by perturbation methods.

4.3.2. *Harmonic oscillators – classical approach*

In order to determine the set of vibration coordinates functions $\{\alpha_i(t)\}$ of the molecule and their associated frequencies (or wave numbers) $\{\omega_0\}$, the Lagrange equations of the system must be solved:

$$\frac{d}{dt}\left(\frac{\partial L_{\text{vib}}}{\partial \dot{\alpha}_i}\right) - \frac{\partial L_{\text{vib}}}{\partial \alpha_i} = 0 \quad [4.9a]$$

where L_{vib} is the Lagrange function defined as: $L_{\text{vib}} = T_{\text{vib}} - V_{\text{vib}}$.

Knowing that a free molecule in space is a conservative system, T_{vib} depends only on the components of the velocity vectors of the nuclei motion, and V_{vib} depends only on the Cartesian vibration coordinates of these nuclei (time independent), Lagrange equations then become:

$$\frac{d}{dt}\left(\frac{\partial T_{\text{vib}}}{\partial \dot{\alpha}_i}\right) + \frac{\partial V_{\text{vib}}}{\partial \alpha_i} = 0 \quad [4.9b]$$

In the harmonic approximation, the resolution of these equations leads to the resolution of a symmetric secular equation (determinant equation), of dimensions $3\text{N}\times3\text{N}$ of the following form:

$$\begin{vmatrix} k_{xx}^{11}-m_1\lambda & k_{xy}^{11} & k_{xz}^{11} & k_{xx}^{12} & \cdots\cdots & k_{xz}^{1\text{N}} \\ k_{yx}^{11} & k_{yy}^{11}-m_1\lambda & k_{yz}^{11} & k_{yx}^{12} & \cdots\cdots & k_{yz}^{1\text{N}} \\ k_{zx}^{11} & k_{zy}^{11} & k_{zz}^{11}-m_1\lambda & k_{zx}^{12} & \cdots\cdots & k_{zz}^{1\text{N}} \\ k_{xx}^{21} & k_{xy}^{21} & k_{xz}^{21} & k_{xx}^{22}-m_2\lambda & \cdots\cdots & k_{xz}^{2\text{N}} \\ \cdots\cdots & \cdots\cdots & \cdots\cdots & \cdots\cdots & \cdots\cdots & \cdots\cdots \\ k_{zx}^{\text{N}1} & k_{zy}^{\text{N}1} & k_{zz}^{\text{N}1} & k_{zx}^{\text{N}2} & \cdots\cdots & k_{zz}^{\text{NN}}-m_{\text{N}}\lambda \end{vmatrix} = 0$$

[4.10]

where, as previously seen, $k_{\alpha\beta}^{ij} = k_{\beta\alpha}^{ji}$ and $\lambda = \omega^2 = 4\pi^2\nu^2$.

3N roots are thus obtained, out of which 6 are equal to zero and correspond to the translational and rotational motions of the molecule. The remaining 3N-6 roots – real and non-zero – are the frequencies (wave numbers) associated with the normal vibrational modes of the molecule [WHI 44]. During a normal mode, each nucleus moves about its equilibrium position, in a synchronized manner, the center of mass of the molecule remaining fixed.

The frequencies are not necessarily distinct. If one frequency is unique, the corresponding vibrational mode is termed as "non-degenerate". If two (respectively three) frequencies are identical, the corresponding vibrational modes are termed as "doubly (respectively triply) degenerate".

It is important to note that the determinant [4.10] generally contains many non-zero elements. Its analytical resolution is difficult.

Moreover, it is difficult to determine various force constants of the expression [4.7]. Indeed, a force constant $k^{ij}_{\alpha\beta}$ characterizes the strength of the force applied on nucleus *i* moving in direction α by nucleus *j* moving in direction β. Therefore, a large number of constants must be determined.

In fact, in practice, the force constants are determined by *ab initio* methods based on information deduced from the observed IR spectra, associated with vibrational–rotational motions of the molecule.

A new set of coordinates that are better adapted need to be introduced: the "internal coordinates", in order to simplify the kinetic and potential functions and reduce the number of constants to be determined.

4.3.2.1. *Internal coordinates*

The geometry of a molecule is essentially defined by the relative position of the nuclei with respect to one another: the internuclear distances r_{ij} and the valence angles between internuclear bonds α_{ij} (see Figure 4.5). The variations $q_{ij} = \mathrm{r}_{ij} - \mathrm{r}^{\mathrm{e}}_{ij}$ and $\Delta\alpha_{ij} = \alpha_{ij} - \alpha^{\mathrm{e}}_{ij}$ of these distances and angles constitute the set of internal coordinates describing the strain of the molecule. In this system, the overall translational and rotational degrees of freedom of the molecule are eliminated from the secular equation whose dimension becomes $(3N\text{-}6)\times(3N\text{-}6)$ associated only with the vibrational motions; for example, 6×6 in the case of a nonlinear molecule of four atoms.

Moreover, the use of the molecule symmetry reduces the number of force constants to be determined. Indeed, the vibrational potential energy function is invariant when the molecule is subjected to symmetry operations associated with the elements of its symmetry group.

Two sets of internal coordinates are possible [WIL 38].

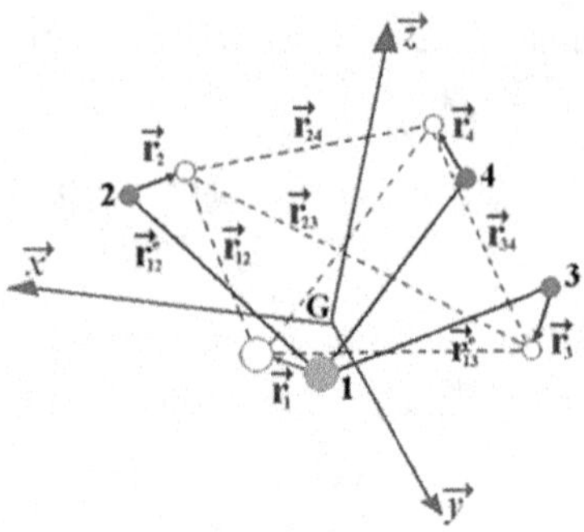

Figure 4.5. *Displacement vectors of the nuclei, at an instant t, during a vibrational motion of the molecule XY₃. For a color version of this figure, see www.iste.co.uk/dahoo/infrared.zip*

4.3.2.2. *Hypothesis of central forces*

According to this hypothesis, only central forces act between the nuclei of the molecule. For example, the force acting on the nucleus i is the resultant of the attraction and repulsion forces exerted on it by all the other nuclei j. These forces are directed along the internuclear line i-j between the two given nuclei and depends only on the distance between the latter; the chosen internal coordinates are then the variations of the internuclear distances q_{ij} [SCH 39a], where q_{1j} ($j = 2, 3, 4$) is the variation of the length of the bond between nucleus X and nucleus Y and q_{ij} $(i, j = 2, 3, 4)$ is the variation of the distance between Y nuclei. Moreover, the vibrational motions being small-amplitude motions, the internal coordinates may be assumed to vary linearly with the Cartesian coordinates. In the equilibrium configuration reference frame $(\mathrm{G}, \vec{x}, \vec{y}, \vec{z})$ of the molecule, the following expressions are thus obtained:

$$
\begin{aligned}
q_{12} &= \sin\beta^{e}(x_2 - x_1) + \cos\beta^{e}(z_2 - z_1) \\
q_{13} &= -\frac{1}{2}\sin\beta^{e}(x_3 - x_1) + \frac{\sqrt{3}}{2}\sin\beta^{e}(y_3 - y_1) + \cos\beta^{e}(z_3 - z_1) \\
q_{14} &= -\frac{1}{2}\sin\beta^{e}(x_4 - x_1) - \frac{\sqrt{3}}{2}\sin\beta^{e}(y_4 - y_1) + \cos\beta^{e}(z_4 - z_1) \\
q_{23} &= -\frac{\sqrt{3}}{2}(x_3 - x_2) + \frac{1}{2}(y_3 - y_2) \\
q_{24} &= -\frac{\sqrt{3}}{2}(x_4 - x_2) - \frac{1}{2}(y_4 - y_2) \\
q_{34} &= -(y_4 - y_3)
\end{aligned}
\qquad [4.11]
$$

In order to establish the expressions of the kinetic and potential energies in terms of internal coordinates, the Cartesian coordinates α_i (x_i, y_i, z_i) must be expressed as a function of the internal coordinates q_{ij} reversing the system comprising equations [4.11] and the six Eckart relations.

The expression [4.6] of the kinetic energy can then be written as:

$$T_{\text{vib}} = \frac{1}{2}a_1(\dot{q}_{12}^2 + \dot{q}_{13}^2 + \dot{q}_{14}^2) + \frac{1}{2}a_2(\dot{q}_{23}^2 + \dot{q}_{24}^2 + \dot{q}_{34}^2) + \frac{1}{2}\sum a_{ij}^{i'j'}\dot{q}_{ij}\dot{q}_{i'j'} \qquad [4.12]$$

where the constants a have the dimension of a mass and $\dot{q}$ are the velocities associated with the internal vibration coordinates q.

Moreover, expression [4.6] of the potential energy function, considering the symmetry properties of group C_{3v} to which the pyramidal symmetric top molecules of type XY_3 belong, is written as

$$V_{\text{vib}} = V_{\text{vib}}^{\text{har}} + \Delta V_{\text{vib}}^{\text{anhar}}$$

with:

$$V_{\text{vib}}^{\text{har}} = \frac{1}{2}f_1(q_{12}^2 + q_{13}^2 + q_{14}^2) + \frac{1}{2}f_2(q_{23}^2 + q_{24}^2 + q_{34}^2) + \left\{ \begin{aligned} &+f_3(q_{12}q_{13} + q_{12}q_{14} + q_{13}q_{14}) \\ &+f_4(q_{23}q_{24} + q_{23}q_{34} + q_{24}q_{34}) \\ &+f_5(q_{12}q_{34} + q_{13}q_{24} + q_{14}q_{23}) \\ &+f_6(q_{12}q_{23} + q_{12}q_{24} + q_{13}q_{23} \\ &\qquad + q_{13}q_{34} + q_{14}q_{24} + q_{14}q_{34}) \end{aligned} \right\} \qquad [4.13]$$

In expression [4.13], the 3rd, 4th, 5th and 6th terms (between curly brackets) of $V_{\text{vib}}^{\text{har}}$ represent the interactions between internal coordinates (simultaneous variation of two internuclear distances) of the respective force constants f_3, ..., f_6, and $\Delta V_{\text{vib}}^{\text{anhar}}$ contain the much smaller terms of orders 3, 4, ... , characterizing the anharmonicity of the vibrational motions.

Nevertheless, in expression [4.12] of the kinetic energy, the present velocity interaction terms (cross products) render this expression more complicated, which further adds to the difficulty in solving the secular equation associated with the vibrational motions of the molecule.

4.3.2.3. *Hypothesis of valence forces*

According to this hypothesis, the forces acting on the nuclei are binding forces between the nucleus X and the nuclei Y, on the one hand, and the forces due to the variations of angles between these bonds, on the other hand. In this case, the internal coordinates are the variations of lengths of bonds q_{1j} and the variations of angles $\Delta\alpha_{ij}$ between these bonds. Hence, assuming a linear variation of coordinates q_{1j} and $\Delta\alpha_{ij}$ in terms of the Cartesian coordinates, the expressions are obtained as:

$$
\begin{aligned}
q_{12} &= \sin\beta^e(x_2 - x_1) + \cos\beta^e(z_2 - z_1) \\
q_{13} &= -\frac{1}{2}\sin\beta^e(x_3 - x_1) + \frac{\sqrt{3}}{2}\sin\beta^e(y_3 - y_1) + \cos\beta^e(z_3 - z_1) \\
q_{14} &= -\frac{1}{2}\sin\beta^e(x_4 - x_1) - \frac{\sqrt{3}}{2}\sin\beta^e(y_4 - y_1) + \cos\beta^e(z_4 - z_1) \\
\Delta\alpha_{23} &= \frac{1}{r^e \sin\alpha^e}\left\{\begin{aligned} &\frac{3}{4}\sin^3\beta^e(x_3 + x_1 - 2x_2) - \frac{3}{2}\sin\beta^e(x_3 - x_2) \\ &-\frac{3\sqrt{3}}{4}\sin^3\beta^e(y_3 - y_1) + \frac{\sqrt{3}}{2}\sin\beta^e(y_3 - y_2) \\ &-\frac{3}{2}\sin^2\beta^e\cos\beta^e(z_3 + z_2 - 2z_1) \end{aligned}\right\} \\
\Delta\alpha_{24} &= \frac{1}{r^e \sin\alpha^e}\left\{\begin{aligned} &\frac{3}{4}\sin^3\beta^e(x_4 + x_1 - 2x_2) - \frac{3}{2}\sin\beta^e(x_4 - x_2) \\ &+\frac{3\sqrt{3}}{4}\sin^3\beta^e(y_4 - y_1) - \frac{\sqrt{3}}{2}\sin\beta^e(y_4 - y_2) \\ &-\frac{3}{2}\sin^2\beta^e\cos\beta^e(z_4 + z_2 - 2z_1) \end{aligned}\right\} \\
\Delta\alpha_{34} &= \frac{1}{r^e \sin\alpha^e}\left\{\begin{aligned} &\frac{3}{4}\sin^3\beta^e(x_4 + x_3 - 2x_1) - \sqrt{3}\sin\beta^e(x_4 - x_3) \\ &+\frac{3\sqrt{3}}{4}\sin^3\beta^e(y_4 - y_3) \\ &-\frac{3}{2}\sin^2\beta^e\cos\beta^e(z_4 + z_3 - 2z_1) \end{aligned}\right\}
\end{aligned}
\qquad [4.14]
$$

As previously shown, by reversing the system: "equations [4.13] + the six Eckart relations", the expressions of Cartesian coordinates α_i can be deduced as a function of the internal coordinates q_{1j} and $\Delta\alpha_{i'j'}$, which are replaced in expressions [4.5] and [4.7].

Taking into account the symmetry properties of group C_{3v} to which the molecules of type XY_3 belong, the potential energy function is written as:

$$V_{\text{vib}} = V_{\text{vib}}^{\text{har}} + \Delta V_{\text{vib}}^{\text{anhar}}$$

with:

$$V_{\text{vib}}^{\text{har}} = \frac{1}{2} f_1^{'}(q_{12}^2 + q_{13}^2 + q_{14}^2) + \frac{1}{2} f_2^{'}(\Delta\alpha_{23}^2 + \Delta\alpha_{24}^2 + \Delta\alpha_{34}^2) + \left\{ \begin{array}{l} + f_3^{'}(q_{12}q_{13} + q_{12}q_{14} + q_{13}q_{14}) \\ + f_4^{'}(\Delta\alpha_{23}\Delta\alpha_{24} + \Delta\alpha_{23}\Delta\alpha_{34} + \Delta\alpha_{24}\Delta\alpha_{34}) \\ + f_5^{'}(q_{12}\Delta\alpha_{34} + q_{13}\Delta\alpha_{24} + q_{14}\Delta\alpha_{23}) \\ + f_6^{'}(q_{12}\Delta\alpha_{23} + q_{12}\Delta\alpha_{24} + q_{13}\Delta\alpha_{23} \\ \qquad + q_{13}\Delta\alpha_{34} + q_{14}\Delta\alpha_{24} + q_{14}\Delta\alpha_{34}) \end{array} \right\} \quad [4.15]$$

where the 3rd, 4th, 5th and 6th terms (between curly brackets) of $V_{\text{vib}}^{\text{har}}$ represent the interactions between internal coordinates (simultaneous variation of two coordinates) (1) of bonds $q.q'$; (2) angular $\Delta\alpha.\Delta\alpha'$; and (3) mixed $q.\Delta\alpha$.

Finally, the kinetic energy is written as:

$$T_{\text{vib}} = \frac{1}{2} a_1^{'}(\dot{q}_{12}^2 + \dot{q}_{13}^2 + \dot{q}_{14}^2) + \frac{1}{2} a_2^{'}(\Delta\dot{\alpha}_{23}^2 + \Delta\dot{\alpha}_{24}^2 + \Delta\dot{\alpha}_{34}^2) + \frac{1}{2}\sum a'^{i'j'}_{ij}\,\dot{q}_{ij}\dot{q}_{i'j'} + \frac{1}{2}\sum a'^{i'j'}_{ij}\,\Delta\dot{\alpha}_{ij}\Delta\dot{\alpha}_{i'j'} + \frac{1}{2}\sum a'^{i'j'}_{ij}\,\dot{q}_{ij}\Delta\dot{\alpha}_{i'j'} \quad [4.16]$$

where a' are the constants and where interaction terms (cross products) of the velocities of the variations of the internal coordinates are still present. Again, this approach does not reduce the difficulty in solving the secular equation associated with the molecular vibrational motions.

Finally, in order to reconcile: (1) the search for simple expressions for T_{vib} and $V_{\text{vib}}^{\text{har}}$; (2) the results of the application of the elements of the symmetry group to which the molecule belongs; and (3) the results of *ab initio* calculations of internal force constants, *new vibrational coordinates must be introduced: the "symmetry coordinates"*.

4.3.2.4. *Symmetry coordinates*

In the case of symmetric molecules, the best way to determine the normal modes of vibrations is to use the symmetry coordinates S_v, components of a vector $\boldsymbol{S}$. These coordinates are generally chosen as much as possible to be in the direction of the internuclear bonds or in a perpendicular direction or along an axis of the equilibrium configuration reference frame.

In the case of pyramidal symmetric top molecules XY_3, the symmetry coordinates introduced by Howard and Wilson [HOW 34] are presented in Figure 4.6 [HER 45]. They characterize the motions of the Y nuclei of the molecule.

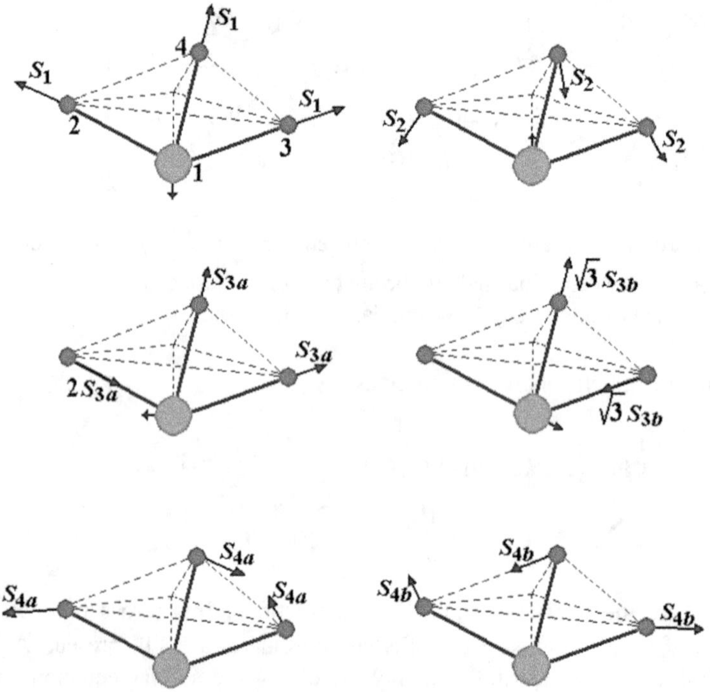

Figure 4.6. *Definition of the system of symmetry coordinates of a symmetric top molecule XY_3 [HER 45]. The factor $\sqrt{3}$ of mode S_{3b} is introduced to normalize the latter with the mode S_{3a}. For a color version of this figure, see www.iste.co.uk/dahoo/infrared.zip*

The expressions [4.17] below give the Cartesian vibrational coordinates of the nuclei of the molecule as a function of the symmetry coordinates defined above in the equilibrium configuration reference frame of the molecule, unlike the approach in reference [HER 45] where each nucleus features a reference frame that differs from those of the other nuclei.

$$
\begin{aligned}
x_1 &= 3\frac{m_{\mathrm{Y}}}{m_{\mathrm{X}}}\sin\beta^{\mathrm{e}}\cdot S_{3a}\\
y_1 &= 3\frac{m_{\mathrm{Y}}}{m_{\mathrm{X}}}\sin\beta^{\mathrm{e}}\cdot S_{3b}\\
z_1 &= -3\frac{m_{\mathrm{Y}}}{m_{\mathrm{X}}}\cos\beta^{\mathrm{e}}\cdot S_1+3\frac{m_{\mathrm{Y}}}{m_{\mathrm{X}}}\sin\beta^{\mathrm{e}}\cdot S_2\\
x_2 &= \sin\beta^{\mathrm{e}}\cdot S_1+\cos\beta^{\mathrm{e}}\cdot S_2-2\sin\beta^{\mathrm{e}}\cdot S_{3a}+S_{4a}\\
y_2 &= -S_{4b}\\
z_2 &= \cos\beta^{\mathrm{e}}\cdot S_1-\sin\beta^{\mathrm{e}}\cdot S_2-2\cos\beta^{\mathrm{e}}\cdot S_{3a}\\
x_3 &= -\frac{1}{2}\sin\beta^{\mathrm{e}}\cdot S_1-\frac{1}{2}\cos\beta^{\mathrm{e}}\cdot S_2-\frac{1}{2}\sin\beta^{\mathrm{e}}\cdot S_{3a}\\
&\quad+\frac{\sqrt{3}}{2}\sin\beta^{\mathrm{e}}\cdot S_{3b}-\frac{1}{2}\cdot S_{4a}-\frac{\sqrt{3}}{2}\cdot S_{4b}\\
y_3 &= \frac{\sqrt{3}}{2}\sin\beta^{\mathrm{e}}\cdot S_1+\frac{\sqrt{3}}{2}\cos\beta^{\mathrm{e}}\cdot S_2+\frac{\sqrt{3}}{2}\sin\beta^{\mathrm{e}}\cdot S_{3a}\\
&\quad-\frac{3}{2}\sin\beta^{\mathrm{e}}\cdot S_{3b}-\frac{\sqrt{3}}{2}\cdot S_{4a}+\frac{1}{2}\cdot S_{4b}\\
z_3 &= \cos\beta^{\mathrm{e}}\cdot S_1-\sin\beta^{\mathrm{e}}\cdot S_2+\cos\beta^{\mathrm{e}}\cdot S_{3a}-\sqrt{3}\cos\beta^{\mathrm{e}}\cdot S_{3b}\\
x_4 &= -\frac{1}{2}\sin\beta^{\mathrm{e}}\cdot S_1-\frac{1}{2}\cos\beta^{\mathrm{e}}\cdot S_2-\frac{1}{2}\sin\beta^{\mathrm{e}}\cdot S_{3a}\\
&\quad-\frac{\sqrt{3}}{2}\sin\beta^{\mathrm{e}}\cdot S_{3b}-\frac{1}{2}\cdot S_{4a}+\frac{\sqrt{3}}{2}\cdot S_{4b}\\
y_4 &= -\frac{\sqrt{3}}{2}\sin\beta^{\mathrm{e}}\cdot S_1-\frac{\sqrt{3}}{2}\cos\beta^{\mathrm{e}}\cdot S_2-\frac{\sqrt{3}}{2}\sin\beta^{\mathrm{e}}\cdot S_{3a}\\
&\quad-\frac{3}{2}\sin\beta^{\mathrm{e}}\cdot S_{3b}+\frac{\sqrt{3}}{2}\cdot S_{4a}+\frac{1}{2}\cdot S_{4b}\\
z_4 &= \cos\beta^{\mathrm{e}}\cdot S_1-\sin\beta^{\mathrm{e}}\cdot S_2+\cos\beta^{\mathrm{e}}\cdot S_{3a}+\sqrt{3}\cos\beta^{\mathrm{e}}\cdot S_{3b}
\end{aligned}
\qquad [4.17]
$$

Therefore, the substitution of the Cartesian coordinates by their expressions as a function of the symmetry coordinates introduced above [4.17] in the expressions [4.11] and [4.14] leads to the expressions of the internal vibration coordinates as follows:

$$
\begin{aligned}
q_{12} &= p.S_1 - \frac{3m_{\mathrm{Y}}}{m_{\mathrm{X}}}\sin\beta^{\mathrm{e}}\cos\beta^{\mathrm{e}}.S_2 \\
&\quad -(1+u).S_{3a} + \sin\beta^{\mathrm{e}}.S_{4a} \\
q_{13} &= p.S_1 - \frac{3m_{\mathrm{Y}}}{m_{\mathrm{X}}}\sin\beta^{\mathrm{e}}\cos\beta^{\mathrm{e}}.S_2 \\
&\quad +\frac{1}{2}(1+u).S_{3a} - \frac{\sqrt{3}}{2}(1+u).S_{3b} \\
&\quad -\frac{1}{2}\sin\beta^{\mathrm{e}}.S_{4a} + \frac{\sqrt{3}}{2}\sin\beta^{\mathrm{e}}.S_{4b} \\
q_{14} &= p.S_1 - \frac{3m_{\mathrm{Y}}}{m_{\mathrm{X}}}\sin\beta^{\mathrm{e}}\cos\beta^{\mathrm{e}}.S_2 \\
&\quad +\frac{1}{2}(1+u).S_{3a} + \frac{\sqrt{3}}{2}(1+u).S_{3b} \\
&\quad -\frac{1}{2}\sin\beta^{\mathrm{e}}.S_{4a} - \frac{\sqrt{3}}{2}\sin\beta^{\mathrm{e}}.S_{4b} \\
q_{23} &= \sqrt{3}\sin\beta^{\mathrm{e}}.S_1 + \sqrt{3}\cos\beta^{\mathrm{e}}.S_2 - \frac{\sqrt{3}}{2}\sin\beta^{\mathrm{e}}.S_{3a} - \frac{3}{2}\sin\beta^{\mathrm{e}}.S_{3b} \\
&\quad +\frac{\sqrt{3}}{2}.S_{4a} + \frac{3}{2}.S_{4b} \\
q_{24} &= \sqrt{3}\sin\beta^{\mathrm{e}}.S_1 + \sqrt{3}\cos\beta^{\mathrm{e}}.S_2 - \frac{\sqrt{3}}{2}\sin\beta^{\mathrm{e}}.S_{3a} + \frac{3}{2}\sin\beta^{\mathrm{e}}.S_{3b} \\
&\quad +\frac{\sqrt{3}}{2}.S_{4a} - \frac{3}{2}.S_{4b} \\
q_{34} &= \sqrt{3}\sin\beta^{\mathrm{e}}.S_1 + \sqrt{3}\cos\beta^{\mathrm{e}}.S_2 + \sqrt{3}\sin\beta^{\mathrm{e}}.S_{3a} - \sqrt{3}.S_{4a}
\end{aligned}
\qquad [4.18]
$$

and

$q_{12} =$ ... see expression [4.18]

$q_{13} =$ ... see expression [4.18]

$q_{14} =$ ... see expression [4.18]

$$\Delta\alpha_{23} = \frac{1}{r^e \sin\alpha^e}\left\{\begin{array}{l} -\frac{9m_Y}{m_X}\sin^2\beta^e\cos^2\beta^e.S_1 \\ +3u.\sin\beta^e\cos\beta^e.S_2 \\ +\frac{9m_Y}{4m_X}\sin^4\beta^e.S_{3a} + \frac{9\sqrt{3}m_Y}{4m_X}\sin^4\beta^e.S_{3b} \\ +\frac{3}{2}\sin\beta^e\left(1-\frac{1}{2}\sin^2\beta^e\right).S_{4a} \\ +\frac{3\sqrt{3}}{2}\sin\beta^e\left(1-\frac{1}{2}\sin^2\beta^e\right).S_{4b} \end{array}\right\}$$

$$\Delta\alpha_{24} = \frac{1}{r^e \sin\alpha^e}\left\{\begin{array}{l} -\frac{9m_Y}{m_X}\sin^2\beta^e\cos^2\beta^e.S_1 \\ +3u.\sin\beta^e\cos\beta^e.S_2 \\ +\frac{9m_Y}{4m_X}\sin^4\beta^e.S_{3a} - \frac{9\sqrt{3}m_Y}{4m_X}\sin^4\beta^e.S_{3b} \\ +\frac{3}{2}\sin\beta^e\left(1-\frac{1}{2}\sin^2\beta^e\right).S_{4a} \\ -\frac{3\sqrt{3}}{2}\sin\beta^e\left(1-\frac{1}{2}\sin^2\beta^e\right).S_{4b} \end{array}\right\} \qquad [4.19]$$

$$\Delta\alpha_{34} = \frac{1}{r^e \sin\alpha^e}\left\{\begin{array}{l} -3p.\sin^2\beta^e.S_1 \\ +\frac{9m_Y}{m_X}\sin^3\beta^e\cos\beta^e.S_2 \\ -\frac{3}{2}\sin^2\beta^e(1+u).S_{3a} \\ +3\sin^2\beta^e.S_{3b} + \frac{3}{2}\sin^3\beta^e.S_{4a} \\ -3\sin\beta^e.S_{4b} \end{array}\right\}$$

where $p = 1 + \dfrac{3\mathrm{m_Y}}{\mathrm{m_X}} \cos^2 \beta^e$ and $u = 1 + \dfrac{3\mathrm{m_Y}}{\mathrm{m_X}} \sin^2 \beta^e$.

A formal calculation software (open source Xcas, mathematica, etc.) can be used to write the vibrational kinetic energy T_{vib} (expression [4.6]), in terms of the set of symmetry coordinates as:

$$T_{\text{vib}} = \frac{1}{2} d_{11}.\dot{S}_1^2 + d_{12}.\dot{S}_1 \dot{S}_2 + \frac{1}{2} d_{22}.\dot{S}_2^2 + \frac{1}{2} d_{33} \left(\dot{S}_{3a}^2 + \dot{S}_{3b}^2 \right)$$
$$+ d_{34} \left(\dot{S}_{3a} \dot{S}_{4a} + \dot{S}_{3b} \dot{S}_{4b} \right) + \frac{1}{2} d_{44} \left(\dot{S}_{4a}^2 + \dot{S}_{4b}^2 \right) \quad [4.20]$$

with the constants d expressed as:

$$d_{11} = 3\mathrm{m_Y} p,\ d_{12} = -\frac{9\mathrm{m_Y^2}}{\mathrm{m_X}} \sin \beta^e \cos \beta^e,\ d_{22} = 3\mathrm{m_Y} u$$

$$d_{33} = 3\mathrm{m_Y} (1+u),\ d_{34} = -3\mathrm{m_Y} \sin \beta^e,\ d_{44} = 3\mathrm{m_Y}$$

It can be noted that coefficients d_{12} and d_{34} are not equal to zero. This is due to the non-orthogonality of symmetry coordinates S_1, S_2, on the one hand, and S_3 and S_4, on the other hand.

Moreover, the vibrational potential energy $V_{\text{vib}}^{\text{har}}$ has the following forms:

Hypothesis of central forces:

$$V_{\text{vib}}^{\text{har}} = \frac{1}{2} k_{11}.S_1^2 + k_{12}.S_1 S_2 + \frac{1}{2} k_{22}.S_2^2$$

$$+ \frac{1}{2} k_{33}.\left(S_{3a}^2 + S_{3b}^2 \right) + k_{34}.\left(S_{3a} S_{4a} + S_{3b} S_{4b} \right) + \frac{1}{2} k_{44}.\left(S_{4a}^2 + S_{4b}^2 \right) \quad [4.21]$$

Hypothesis of valence forces:

$$V_{\text{vib}}^{\text{har}} = \frac{1}{2} \sum_{v} k'_{vv}.S_v^2 + \sum_{v,v'}^{v<v'} k'_{vv'}.S_v S_{v'} \quad [4.22]$$

where v and v' take the values 1, 2, 3a, 3b, 4a and 4b, and $k'_{vv'} = k'_{v'v}$.

The dimension of the secular equation to be solved is 6×6 and can be presented in matrix form as:

$$|\boldsymbol{k}-\boldsymbol{d}\lambda_0|=\mathbf{0} \qquad [4.23]$$

where $\boldsymbol{k}$ ($\boldsymbol{k'}$) and $\boldsymbol{d}$ are the force constants and the mass matrices of elements $k_{vv'}$ ($k'_{vv'}$) and $d_{vv'}$, respectively.

The expressions of the new force constants $k_{vv'}$ [4.21] and $k'_{vv'}$ [4.22] as a function of internal force constants f of [4.13] and f' of [4.15] are presented in section 4.7.2.

In the expression of the vibrational kinetic energy [4.20], it can be noted that the vibrational modes ν_1 and ν_2 of species A_1 are fully separated from modes ν_3 and ν_4 of species E, in accordance with group theory. This property is also verified in expression [4.21] of the vibrational potential energy according to the hypothesis of central forces. Moreover, these doubly degenerate vibrational modes ν_3 and ν_4 of species E are also separated into two sets ν_{3a} and ν_{4a}, on the one hand, and ν_{3b} and ν_{4b}, on the other hand. The 6×6 secular equation to be solved is written as:

$$\begin{vmatrix} k_{11}-d_{11}\lambda_0 & k_{12}-d_{12}\lambda_0 & 0 & 0 & 0 & 0 \\ k_{12}-d_{12}\lambda_0 & k_{22}-d_{22}\lambda_0 & 0 & 0 & 0 & 0 \\ 0 & 0 & k_{3a3a}-d_{33}\lambda_0 & k_{3a4a}-d_{34}\lambda_0 & 0 & 0 \\ 0 & 0 & k_{3a4a}-d_{34}\lambda_0 & k_{4a4a}-d_{44}\lambda_0 & 0 & 0 \\ 0 & 0 & 0 & 0 & k_{3b3b}-d_{33}\lambda_0 & k_{3b4b}-d_{34}\lambda_0 \\ 0 & 0 & 0 & 0 & k_{3b4b}-d_{34}\lambda_0 & k_{4b4b}-d_{44}\lambda_0 \end{vmatrix}=0 \qquad [4.24]$$

which leads to solving three secular equations of $\dim 2\times 2$, two of which are identical:

$$|\boldsymbol{k}-\boldsymbol{d}\lambda_0|_{A_1}=\mathbf{0} \qquad \text{and} \qquad |\boldsymbol{k}-\boldsymbol{d}\lambda_0|_{E}=\mathbf{0} \qquad [4.25]$$

associated respectively with species A_1 and species E, and thus determine the frequencies ω_0 of the vibrational modes of the molecule.

In contrast, the orthogonality of the species A_1 and E is not verified in the vibrational potential energy function when the angular internal coordinates are considered under the hypothesis of valence forces. Indeed, the terms characterizing the interactions between symmetry coordinates are present in expression [4.22]. The secular equation of $\dim 6\times 6$ is written as:

$$\begin{vmatrix} k'_{11}-d_{11}\lambda_0 & k'_{12}-d_{12}\lambda_0 & k'_{13a} & k'_{14a} & k'_{13b} & k'_{14b} \\ k'_{12}-d_{12}\lambda_0 & k'_{22}-d_{22}\lambda_0 & k'_{23a} & k'_{24a} & k'_{23b} & k'_{24b} \\ k'_{13a} & k'_{23a} & k'_{3a3a}-d_{33}\lambda_0 & k'_{3a4a}-d_{34}\lambda_0 & k'_{3a3b} & k'_{3a4b} \\ k'_{14a} & k'_{24a} & k'_{3a4a}-d_{34}\lambda_0 & k'_{4a4a}-d_{44}\lambda_0 & k'_{3b4a} & k'_{4a4b} \\ k'_{13b} & k'_{23b} & k'_{3a3b} & k'_{3b4a} & k'_{3b3b}-d_{33}\lambda_0 & k'_{3a4b}-d_{34}\lambda_0 \\ k'_{14b} & k'_{24b} & k'_{3a4b} & k'_{4a4b} & k'_{3a4b}-d_{34}\lambda_0 & k'_{4b4b}-d_{44}\lambda_0 \end{vmatrix} = 0 \quad [4.26]$$

However, depending on the nature of the nuclei in the studied XY_3 molecule, the values of the internal force constants f' in expression [4.15] can be used with a good approximation to find the orthogonality property and obtain a secular equation of the form [4.24].

4.3.3. *Separation of vibrational modes*

In the harmonic approximation, to calculate separately the vibrational modes of the molecule, we need to introduce six new coordinates, termed "vibrational normal coordinates Q_v", each of which represents a vibrational mode v. These coordinates are the components of a vector $\boldsymbol{Q}$.

These norm coordinates are such that the total vibrational energy is written in the simple form:

$$T_{\text{vib}} + V_{\text{vib}}^{\text{har}} = \frac{1}{2}\sum_v \dot{Q}_v^2 + \frac{1}{2}\sum_v \lambda_{0_v} Q_v^2 \quad [4.27]$$

which corresponds to a set of independent harmonic oscillators. These coordinates are then linked to the previously introduced symmetry coordinates by the relation:

$$\boldsymbol{S} = \boldsymbol{LQ} \quad [4.28]$$

where $\boldsymbol{L}$ is a 6×6 matrix whose elements have a dimension homogeneous to that of the inverse of the square root of a mass. Moreover, this matrix defines an orthogonal transformation, since the axes of the reference frame $(G,\vec{x},\vec{y},\vec{z})$ related to the equilibrium configuration of the molecule are the principal axes of inertia.

If the left term of the secular equation [4.23] is multiplied by the matrix $\boldsymbol{d}^{-1}$, inverse of matrix $\boldsymbol{d}$, the result is:

$$\left|\boldsymbol{d}^{-1}\boldsymbol{k} - \boldsymbol{I}\boldsymbol{\lambda}_0\right| = 0 \qquad [4.29]$$

where $\boldsymbol{I}$ is the unitary matrix and $\boldsymbol{\lambda}_0$ is the diagonal matrix containing the eigenvalues λ_{0_v} (expressed in s^{-2}) related to frequencies of vibrations ω_{0_v} (expressed in cm^{-1}) by $\lambda_{0_v} = \omega_{0_v}^2$.

Finally, introducing transformation [4.28] in equation [4.29], the latter becomes:

$$\boldsymbol{d}^{-1}\boldsymbol{k}\boldsymbol{L} = \boldsymbol{L}\boldsymbol{\lambda}_0 \qquad [4.30]$$

an eigenvalue equation whose solutions are obtained by the diagonalization of the 6 x 6 matrix $\left[\boldsymbol{d}^{-1}\boldsymbol{k}\right]$. If the potential energy $V_{\text{vib}}^{\text{har}}$ is expressed according to the central force hypothesis, this amounts to the diagonalization of two 2×2 matrices: $\left[\boldsymbol{d}^{-1}\boldsymbol{k}\right]_{A_1}$ for the non-degenerate vibrational modes ν_1 and ν_2, and $\left[\boldsymbol{d}^{-1}\boldsymbol{k}\right]_E$ for the doubly degenerate vibrational modes ν_3 and ν_4.

4.3.4. *Harmonic oscillators – quantum approach*

When a molecule is subjected to an electromagnetic field in the IR range, it absorbs or emits one or several photons of well-defined energies or frequencies. The resulting absorption or emission spectrum has well-identified lines. The vibrational motions of the molecule are called "quantized".

In quantum mechanics, the vibrational state of the molecule, at a given instant, is fully determined by the vibrational wave function Ψ_{vib}, which determines its state at all subsequent instants. However, assuming that the molecule is isolated, its vibrational wave function is time-independent and depends only on the vibrational coordinates. The vibrational states of the molecule are termed as: "stationary states". Under the harmonic approximation, their wave functions Ψ_v^0 and energies E_v^0, also known as eigenvectors and eigenvalues, are determined by solving Schrödinger's equation:

$$H^0_{\text{vib}} \Psi^0_{\text{v}} = E^0_{\text{v}} \Psi^0_{\text{v}} \qquad [4.31]$$

where H^0_{vib} is the zeroth-order vibrational Hamiltonian operator, which is defined by:

$$H^0_{\text{vib}} = T_{\text{vib}} + V^{\text{har}}_{\text{vib}} = -\frac{1}{2}\hbar^2 \sum_{v} \frac{\partial^2}{\partial Q_v^2} + \frac{1}{2} \sum_{v} \lambda_{0_v} Q_v^2 \qquad [4.32]$$

T_{vib} and $V^{\text{har}}_{\text{vib}}$ are the kinetic energy and potential energy operators, respectively, and $\hbar = h/2\pi$, with h being the Planck constant. This Hamiltonian operator H^0_{vib} characterizes a set of independent harmonic oscillators, each of which is associated with a vibrational mode of the molecule.

Moreover, according to the harmonic approximation and using the methods of group theory, in order to find a simple expression of the wave functions and of the vibrational energies, it is possible to introduce a basis of dimensionless normal coordinates q_v each characterizing a normal mode of vibration (harmonic oscillator) of the molecule and defined by:

$$q_v = \left(\frac{d_{vv} \sqrt{\lambda_{0_v}}}{\hbar} \right)^{1/2} Q_v \qquad [4.33]$$

and also by their conjugate moment operator $p_v = -i\frac{\partial}{\partial q_v}$. These operators are Hermitian and verify the canonical commutation relation:

$$[q_v, p_v] = q_v p_v - p_v q_v = i \qquad [4.34]$$

These operators act on a space of quantum states of square integrable functions, the "Hilbert space".

Finally, by introducing the dimensionless normal coordinates q_v in expression [4.32], the Schrödinger's equation [4.31] associated with the set of harmonic oscillators is written as:

$$H_{\text{vib}}^{0}\Psi_{\text{v}}^{0}=-\frac{1}{2}\sum_{\nu}\hbar\omega_{0_\nu}\left(\frac{\partial^2}{\partial q_\nu^2}-q_\nu^2\right)\Psi_{\text{v}}^{0}=E_{\text{v}}^{0}\Psi_{\text{v}}^{0} \quad [4.35]$$

Solving this equation leads to determining the eigenenergies:

$$E_{\text{v}}^{0}=\sum_{\nu}\hbar\omega_{0_\nu}\left(\text{v}_\nu+\frac{1}{2}\right) \quad [4.36]$$

and the eigenvectors:

$$\Psi_{\text{v}}^{0}=\prod_{\nu}\Psi_{\text{v}_\nu}^{0}(q_\nu) \quad [4.37]$$

where $\Psi_{\text{v}_\nu}^{0}(q_\nu)$ (also denoted by $|\text{v}_\nu\rangle_0$) are the eigenvectors associated with the harmonic oscillator characterizing the νth vibrational mode. They are written as:

$$\Psi_{\text{v}_\nu}^{0}(q_\nu)=N_{\text{v}_\nu}e^{-q_\nu^2/2}H_{\text{v}_\nu}(q_\nu) \quad [4.38]$$

and form a complete orthonormal set.

In the above expression, $N_{\text{v}_\nu}=\left(2^{\text{v}_\nu}\,\text{v}_\nu!\sqrt{\pi}\right)^{-1/2}$ is a normalization coefficient and $H_{\text{v}_\nu}(q_\nu)$ is the Hermite polynomial of order v_ν defined by:

$$H_{\text{v}_\nu}(q_\nu)=(-1)^{\text{v}_\nu}e^{q_\nu^2}\frac{d^{\text{v}_\nu}}{dq_\nu^{\text{v}_\nu}}\left(e^{-q_\nu^2}\right) \quad [4.39]$$

The expressions of the first Hermite polynomials (v = 0, …, 4) were presented in expressions [2.25] of Chapter 2 of Volume 1 [DAH 17].

Therefore, according to the harmonic approximation, an eigenstate of the vibrational Hamiltonian of a molecule XY_3 is characterized by its energy and its wave function as:

$$\begin{aligned}E_{\text{v}}^{0}&=E_{\text{v}_1}^{0}+E_{\text{v}_2}^{0}+E_{\text{v}_{3a}}^{0}+E_{\text{v}_{3b}}^{0}+E_{\text{v}_{4a}}^{0}+E_{\text{v}_{4b}}^{0}\\&=\hbar\omega_{0_1}\left(\text{v}_1+\frac{1}{2}\right)+\hbar\omega_{0_2}\left(\text{v}_2+\frac{1}{2}\right)+\hbar\omega_{0_3}(\text{v}_{3a}+\text{v}_{3b}+1)+\hbar\omega_{0_4}(\text{v}_{4a}+\text{v}_{4b}+1) \quad [4.40]\\\Psi_{\text{v}}^{0}&=|\text{v}_1\text{v}_2\text{v}_{3a}\text{v}_{3b}\text{v}_{4a}\text{v}_{4b}\rangle_0=|\text{v}_1\rangle_0|\text{v}_2\rangle_0|\text{v}_{3a}\rangle_0|\text{v}_{3b}\rangle_0|\text{v}_{4a}\rangle_0|\text{v}_{4b}\rangle_0\end{aligned}$$

These expressions show that modes ν_1 and ν_2 are non-degenerate modes, each corresponding to a one-dimensional harmonic oscillator (see Chapter 2 of Volume 1). It can also be noted that the energy of modes ν_s (s = 3 and 4) depends only on quantum numbers $\mathrm{v}_s = \mathrm{v}_{sa} + \mathrm{v}_{sb}$ that can take the values 0, 1, 2, and so on. These modes are doubly degenerate and each of them corresponds to a two-dimensional isotropic harmonic oscillator. The immediate consequence is that if $\mathrm{v}_s \neq 0$, the eigenvalues are degenerate of degree of degeneracy $\mathrm{v}_s + 1$. The wave function associated with an energy $E^0_{\mathrm{v}_s}$ is a linear combination of products $|\mathrm{v}_{sa}\rangle_0 |\mathrm{v}_{sb}\rangle_0$ of wave functions $|\mathrm{v}_{sa}\rangle_0$ and $|\mathrm{v}_{sb}\rangle_0$.

Another way to deal with doubly degenerate modes ν_s for s = 3 and 4 is to use the polar coordinates r_s and γ_s defined by $q_{sa} = r_s \cos\gamma_s$ and $q_{sb} = r_s \sin\gamma_s$.

Solving the associated Schrödinger equation that depends on these new coordinates leads to the same eigenvalue $E^0_{\mathrm{v}_s} = \hbar\omega_{0_s}(\mathrm{v}_s + 1)$, but to eigenfunctions defined by two quantum numbers v_s and ℓ_s as follows:

$$|\mathrm{v}_s \ell_s\rangle_0 = N_{\mathrm{v}_s|\ell_s|} e^{-r_s^2/2} r_s^{|\ell_s|} L^{|\ell_s|}_{\tau_s}(r_s^2) e^{i\ell_s\gamma_s} \qquad [4.41]$$

where:

$$\tau_s = \tfrac{1}{2}(\mathrm{v}_s - |\ell_s|), \quad N_{\mathrm{v}_s|\ell_s|} = \left[\left(\frac{\mathrm{v}_s - |\ell_s|}{2}\right)!\right]^{1/2} \Bigg/ \sqrt{\pi}\left[\left(\frac{\mathrm{v}_s + |\ell_s|}{2}\right)!\right]^{3/2}$$

is the normalization coefficient and $L^{|\ell_s|}_{\tau_s}(r_s^2)$ is the associated Laguerre polynomial; the quantum numbers v_s and ℓ_s can take the values v_s = 0, 1, 2, 3, …, and $\ell_s = \mathrm{V}_s$, $\mathrm{V}_s - 2$, $\mathrm{V}_s - 4$, …, $-\mathrm{V}_s + 2$, $-\mathrm{V}_s$, therefore ($\mathrm{V}_s + 1$) values. The energy levels of these modes are conventionally denoted by $\mathrm{v}_s^{\ell_s}$.

NOTE.– In spectroscopy, the angular frequency ω (frequency ν) and the energy are commonly expressed in cm^{-1} (wave numbers). Then, $\omega(\mathrm{cm}^{-1}) = \omega(\mathrm{Hz})/c$ and $E_\mathrm{v}(\mathrm{cm}^{-1}) = E_\mathrm{v}(\mathrm{joule})/c$, where c is the speed of light expressed in cm.s^{-1}.

4.3.5. *Molecular vibrations beyond the harmonic approximation*

Beyond the harmonic approximation, the vibrational potential energy function V_{vib} expressed in terms of dimensionless normal coordinates q_v is written as:

$$V_{\text{vib}} = \frac{1}{2}\sum_{v} \hbar\omega_{0_v} q_v^2 + \left\{ \begin{array}{l} \sum\limits_{v',v'',v'''} k_{v'v''v'''} q_{v'} q_{v''} q_{v'''} \\ + \sum\limits_{v',v'',v''',v''''} k_{v'v''v'''v''''} q_{v'} q_{v''} q_{v'''} q_{v''''} + \ldots \end{array} \right\} \quad [4.42]$$

$$= V_{\text{vib}}^{\text{har}} + \Delta V_{\text{vib}}^{\text{anhar}}$$

with $v, v', v'', \ldots = 1, 2, 3a, 3b, 4a, 4b$.

It is important to recall that, because of successive changes of variables (Cartesian coordinates – internal coordinates – … – dimensionless normal coordinates), the force constants of the cross terms $q_{v'} q_{v''}$ are equal to zero; and in the harmonic approximation, the normal modes of vibrations are fully separated.

Moreover, $\Delta V_{\text{vib}}^{\text{anhar}}$ (terms between curly brackets) represent the part characterizing the anharmonicity of vibrational motions. It is generally much smaller than the harmonic part and is therefore treated as a perturbation. However, for polyatomic molecules (number of nuclei N > 3), $\Delta V_{\text{vib}}^{\text{anhar}}$ contains a large number of force constants k of orders 3, 4, …, and consequently, the analytical calculation is more complex and tedious.

A numerical calculation using the stationary perturbation theory, for example, makes it possible to determine, beyond the harmonic approximation the eigenenergy and the eigenvector associated with a vibrational state v up to the second-order perturbation expansion as:

$$E_{\text{v}} = E_{\text{v}}^0 + {}_0\langle \text{v}_1\text{v}_2\text{v}_{3a}\text{v}_{3b}\text{v}_{4a}\text{v}_{4b} | \Delta V_{\text{vib}}^{\text{anhar}} | \text{v}_1\text{v}_2\text{v}_{3a}\text{v}_{3b}\text{v}_{4a}\text{v}_{4b} \rangle_0$$
$$+ \sum_{\text{v}'\neq\text{v}} \frac{\left| {}_0\langle \text{v}_1'\text{v}_2'\text{v}_{3a}'\text{v}_{3b}'\text{v}_{4a}'\text{v}_{4b}' | \Delta V_{\text{vib}}^{\text{anhar}} | \text{v}_1\text{v}_2\text{v}_{3a}\text{v}_{3b}\text{v}_{4a}\text{v}_{4b} \rangle_0 \right|^2}{E_{\text{v}}^0 - E_{\text{v}'}^0} + \ldots$$

and

$$\left|v_1 v_2 v_{3a} v_{3b} v_{4a} v_{4b}\right\rangle = \left|v_1 v_2 v_{3a} v_{3b} v_{4a} v_{4b}\right\rangle_0 + \sum_{v' \neq v} \frac{{}_0\left\langle v_1' v_2' v_{3a}' v_{3b}' v_{4a}' v_{4b}'\right| \Delta V_{\text{vib}}^{\text{anhar}} \left|v_1 v_2 v_{3a} v_{3b} v_{4a} v_{4b}\right\rangle_0}{E_{\text{v}}^0 - E_{\text{v}'}^0} \times \left|v_1' v_2' v_{3a}' v_{3b}' v_{4a}' v_{4b}'\right\rangle_0 + \ldots \quad [4.43]$$

where the average of odd powers of the coordinate q associated with a harmonic oscillator in a state v is such that ${}_0\langle v_\nu | q_\nu^{2n+1} | v_\nu \rangle_0 = 0$, and the expressions of elements ${}_0\left\langle v_\nu' \right| q_\nu^n \left| v_\nu \right\rangle_0$ can be found in section 2.5.3 of Volume 1.

The above expressions [4.42] and [4.43] indicate that the eigenstates of the vibrational Hamiltonian are mutually contaminated through the anharmonic part of the potential energy function $\Delta V_{\text{vib}}^{\text{anhar}}$: (1) the vibrational energy levels are shifted; and (2) the associated eigenvectors are linear combinations of eigenvectors of the harmonic vibrational Hamiltonian.

Moreover, if in the anharmonic part $\Delta V_{\text{vib}}^{\text{anhar}}$, there are terms coupling the coordinates q_{3a}, q_{3b}, on the one hand, and/or q_{4a}, q_{4b}, on the other hand, of the doubly degenerate normal modes, then these terms should be introduced in the matrices associated with these modes and they should be diagonalized in order to calculate the splitting of the degenerate levels before the introduction of the remaining anharmonic terms by the stationary perturbation theory.

4.3.6. *Intrinsic inversion phenomenon of certain pyramidal molecules of type XY_3*

At the beginning of 1930s, a splitting of certain vibration bands in the IR spectrum of ammonia NH_3 in the gas phase was noted by Dennison *et al.* [DEN 32a, DEN 32b]. These authors associated this phenomenon with the existence of two equivalent, indiscernible configurations of the molecule as shown in Figure 4.7: the nitrogen N atom (X atom) can be on either side of the plane formed by the three hydrogen H atoms (Y atoms). They also predicted the splitting of the energy levels of all the vibrational modes of the molecule. However, this splitting is particularly important for the excited levels of mode ν_2, the angular deformation mode.

From the classical perspective, during this mode, the amplitude of the motion increases so that atom X gets closer to the plane formed by the Y atoms and can cross it in order to reach a symmetric equilibrium position: the molecule reverses. However, crossing the energy barrier corresponding to the plane configuration of the molecule requires sufficiently high energy.

In fact, this vibration–inversion mode related to the invariance of the internal Hamiltonian under symmetry in a plane passing through the center of mass of the molecule is purely quantum.

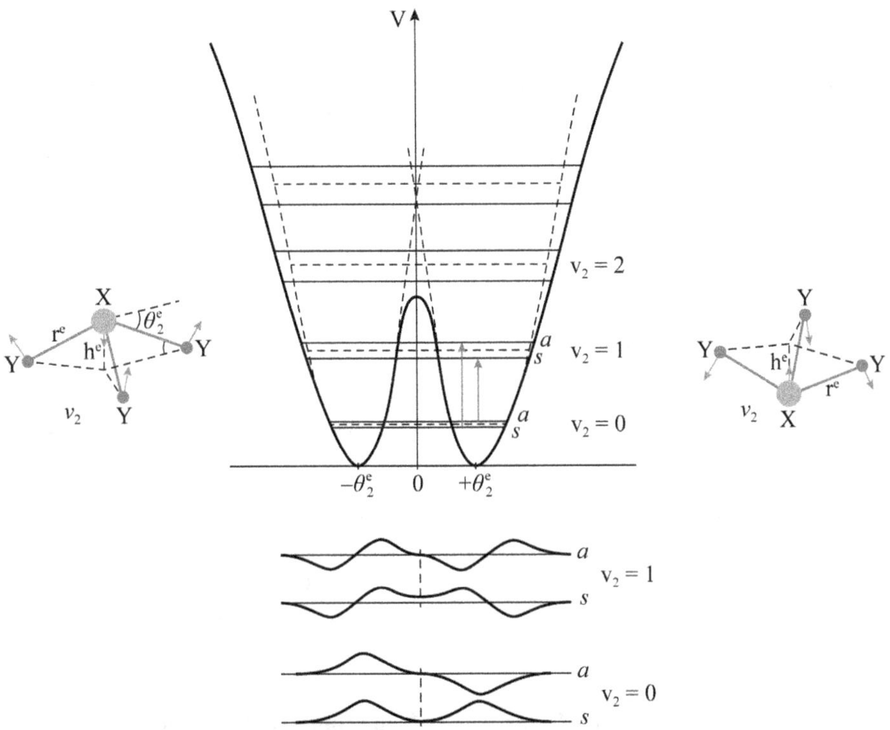

Figure 4.7. *Vibration–inversion mode of a pyramidal symmetric top molecule XY_3. Potential energy function, energy levels and wave functions. For a color version of this figure, see www.iste.co.uk/dahoo/infrared.zip*

The simplest manner to approach this phenomenon is to consider it as the motion of a fictitious particle of mass μ, equal to the reduced mass of the molecule, evolving in a potential energy function *V*.

Figure 4.7 shows the two potential energy functions V (dashed lines) as a function of θ_2, the angle between the internuclear bond X-Y and the plane of Y atoms. These functions are each associated with a vibrational motion about their respective equilibrium position $-\theta_2^e$ and $+\theta_2^e$.

These curves characterize two anharmonic oscillators whose energy levels are represented by dashed lines. However, these two oscillators are not independent.

Indeed, the two curves intersect at the plane configuration of the molecule $(\theta_2 = 0)$. The result is a total potential energy function with two symmetric wells that are separated by a finite energy barrier (solid curved line). The molecule is inverted by a "tunneling effect", the energy levels split (undergo degeneracy lifting) as indicated by the solid line in Figure 4.7. This splitting increases rapidly with the quantum number v_2 of the energy level and/or the lowering of the potential energy barrier.

The Hamiltonian associated with this vibration–inversion mode V_2 is written as:

$$H_{V_2} = -\frac{\hbar^2}{2{r^e}^2\mu}\frac{\partial^2}{\partial\theta_2^2} + V(\theta_2) \qquad [4.44]$$

where r^e is the internuclear distance X-Y, and μ is the reduced mass of the molecule.

Moreover, the potential energy function V being symmetric, it is possible to approximately represent the eigenfunctions of the vibration–inversion Hamiltonian for the first energy levels by a symmetric linear combination (s) and an antisymmetric linear combination (a) of eigenfunctions associated with the vibration motions v_2 of each configuration about their respective equilibrium positions $-\theta_2$ and $+\theta_2$.

The eigenenergies and functions of this Hamiltonian are written as: $E_{v_2\alpha}$ and $|v_2\alpha\rangle$, where $\alpha = s, a$ associated with symmetric and antisymmetric energy sub-levels. As an illustration, the eigenfunctions associated with levels $v_2 = 0$ and $v_2 = 1$ are shown in Figure 4.7.

4.3.7. *Transitions between two vibrational levels: selection rules*

When a molecule is subjected to an electromagnetic radiation, its electric dipole moment $\vec{\mu}$ interacts with the electric field $\vec{E}$ of the radiation. If the molecule has no permanent dipole moment, a vibration-induced dipole moment can be present.

The electric field can also induce a dipole moment in the molecule, which is defined by $\vec{\mu}_{\text{ind}} = \vec{\vec{\alpha}}.\vec{E}$, where $\vec{\vec{\alpha}}$ is the polarizability tensor of the molecule.

The interaction term is generally written in the form $H' = -\vec{\mu}.\vec{E}$. It is this term that, depending on the existence of the dipole moment, leads to electric dipole transitions (IR absorption or emission spectroscopy) or to Raman transitions (radiation scattering).

When the molecule absorbs or emits a photon of energy $\hbar\omega_{0_v}$, in the frequency range of a vibration mode v, its vibrational state undergoes a modification. It undergoes a transition between its initial state of energy $E^0_{\mathrm{v}_{vi}}$ and a final state of energy $E^0_{\mathrm{v}_{vf}}$. This is reflected by the presence of a line in the near infrared spectrum and by small variations of its dipole moment $\vec{\mu}$, leading to constraints on the vibrational transition, termed as "selection rules". The transition dipole moment ${}_0\left\langle \mathrm{v}_{v_f} \middle| \vec{\mu} \middle| \mathrm{v}_{v_i} \right\rangle_0$ must be *non-zero*.

In order to determine these selection rules, it is possible to write a series expansion, in terms of normal vibration coordinates $\{q_v\}$, of the dipole moment of the molecule around its equilibrium value $\vec{\mu}^e$ (a permanent dipole moment that can be equal to zero) in the mobile reference frame (G, $\vec{x}$, $\vec{y}$, $\vec{z}$):

$$\vec{\mu} = \vec{\mu}^e + \sum_v \left(\frac{\partial \vec{\mu}}{\partial q_v} \right)_0 q_v + \frac{1}{2} \sum_{v,v'} \left(\frac{\partial^2 \vec{\mu}}{\partial q_v \partial q_{v'}} \right)_0 q_v q_{v'} + \ldots \qquad [4.45]$$

where the various terms (except $\vec{\mu}^e$) of the expansion correspond to the vibration-induced dipole moment. This expansion is generally truncated at second order, the variations being assumed to be small.

In practice, the electric field of the electromagnetic radiation is directed along the fixed OZ axis of the laboratory reference frame $(\mathrm{O},\vec{\mathrm{X}},\vec{\mathrm{Y}},\vec{\mathrm{Z}})$. The electric dipole moment $\vec{\mu}$ must therefore be expressed in the reference frame $(\mathrm{G},\vec{\mathrm{X}},\vec{\mathrm{Y}},\vec{\mathrm{Z}})$ attached to the molecule: $\vec{\mu}_A$ The two moments are connected by the following relation:

$$\vec{\mu}_A = \mathbf{M}^{-1}(\phi,\theta,\chi)\vec{\mu} \qquad [4.46]$$

where $\mathbf{M}^{-1}(\varphi,\theta,\chi)$ is the inverse matrix of the unitary transformation matrix $\mathbf{M}(\varphi,\theta,\chi)$, making it possible to define the orientation (φ for precession angle, θ for nutation angle and χ for intrinsic rotation, called Euler angles) of the reference frame $(\mathrm{G},\vec{x},\vec{y},\vec{z})$ with respect to the reference frame $(\mathrm{G},\vec{\mathrm{X}},\vec{\mathrm{Y}},\vec{\mathrm{Z}})$ (see section 4.7.1).

According to the harmonic approximation, the pure vibrational transition moment leading to an absorption or emission of a photon of vibration mode v is written as

$${}_0\left\langle \mathrm{v}_{V_f}\middle|\vec{\mu}_A\middle|\mathrm{v}_{V_i}\right\rangle_0 = \int \Psi^{0*}_{\mathrm{v}_{V_f}}\vec{\mu}_A\Psi^{0}_{\mathrm{v}_{V_i}}\,dq_V \neq 0 \qquad [4.47]$$

and if $\vec{\mu}_A$ is replaced by its expression [4.44]:

$$\begin{aligned}{}_0\left\langle \mathrm{v}_{V_f}\middle|\vec{\mu}_A\middle|\mathrm{v}_{V_i}\right\rangle_0 &= \vec{\mu}^e_A\,{}_0\left\langle \mathrm{v}_{V_f}\middle|\mathrm{v}_{V_i}\right\rangle_0 + \left(\frac{\partial\vec{\mu}_A}{\partial q_V}\right)_0{}_0\left\langle \mathrm{v}_{V_f}\middle|q_V\middle|\mathrm{v}_{V_i}\right\rangle_0 + \ldots \\ &= \vec{\mu}^e_A\delta_{\mathrm{v}_{V_i},\mathrm{v}_{V_f}} + \left(\frac{\partial\vec{\mu}_A}{\partial q_V}\right)_0\sqrt{\frac{2^{\mathrm{v}_{V_f}}\,\mathrm{v}_{V_f}!}{2^{\mathrm{v}_{V_i}}\,\mathrm{v}_{V_i}!}}\left(\frac{1}{2}\delta_{\mathrm{v}_{V_i}+1,\mathrm{v}_{V_f}} + \mathrm{v}_{V_i}\delta_{\mathrm{v}_{V_i}-1,\mathrm{v}_{V_f}}\right) + \ldots\end{aligned} \qquad [4.48]$$

where δ is the Kronecker symbol.

In this expression, only one term can be non-zero:

– if $\delta_{\mathrm{v}_{V_i},\mathrm{v}_{V_f}} = 1$, then $\mathrm{v}_{V_i} = \mathrm{v}_{V_f}$, or $\Delta\mathrm{v} = 0$, no transition occurs;

– if $\delta_{\mathrm{v}_{V_i}+1,\mathrm{v}_{V_f}} = 1$, then $\mathrm{v}_{V_i} + 1 = \mathrm{v}_{V_f}$, or $\Delta\mathrm{v} = +1$, which corresponds to an absorption transition;

– if $\delta_{\mathrm{v}_{V_i}-1,\mathrm{v}_{V_f}}=1$, then $\mathrm{v}_{V_i}-1=\mathrm{v}_{V_f}$, or $\Delta\mathrm{v}=-1$, which corresponds to an emission transition.

NOTE.– If second-order terms are also taken into account in expression [4.47], two photon transitions (of the same mode or combinations of distinct modes) may occur.

Vibration–inversion mode ν_2: for reasons related to reflection symmetry of the dipole moment with respect to the origin, two additional selection rules must be taken into account during the transitions, $s \rightarrow a$ and $a \rightarrow s$ (see Figure 4.7).

4.4. Rotational motion of an isolated rigid symmetric top molecule

4.4.1. *Rotational kinetic Hamiltonian and energy level scheme*

As mentioned in Chapter 1, a symmetric top molecule can be:

– oblate, when its rotational constants are such that $A^e = B^e > C^e$ (NH_3, ND_3, CHD_3);

– prolate, when its rotational constants are such that $A^e > B^e = C^e$ (CH_3F, CH_3D);

where A^e, B^e and C^e are, respectively, determined with respect to Gx, Gy and Gz axes. For further details, the reader may refer to Chapter 2 of Volume 2 [DAH 19].

The zeroth-order rotational Hamiltonian of an oblate symmetric top molecule is written as:

$$H_{\mathrm{rot}}=T_{\mathrm{rot}}=-B^e\left\{\frac{\partial^2}{\partial\theta^2}+\cot\theta\frac{\partial}{\partial\theta}+\frac{1}{\sin^2\theta}\frac{\partial^2}{\partial\phi^2}+\left(\cot^2\theta+\frac{C^e}{B^e}\right)\frac{\partial^2}{\partial\chi^2}-2\frac{\cos\theta}{\sin^2\theta}\frac{\partial^2}{\partial\phi\partial\chi}\right\} \quad [4.49]$$

For a prolate molecule, the rotational constant C^e is replaced by A^e in expression [4.49].

The rotational eigenstates are characterized by three rotational quantum numbers: the angular momentum J, its projection K on Gz axis of the molecule and its projection M on the fixed axis GZ. An eigenstate is then defined by its eigenenergy E_{rot} and its eigenfunction Ψ_{rot}, solutions of Schrödinger's equation associated with the Hamiltonian H_{rot}. They are written as:

$$E_{\text{rot}} = E_{JK} = B^e J(J+1) + (C^e - B^e)K^2 \qquad [4.50a]$$

for the oblate symmetric top, and

$$E_{\text{rot}} = E_{JK} = B^e J(J+1) + (A^e - B^e)K^2 \qquad [4.50b]$$

for the prolate symmetric top, and

$$\Psi_{\text{rot}} = |JKM\rangle = N_{JKM}\Theta_{JKM}(\theta)e^{iM\varphi}e^{iK\chi} \qquad [4.51]$$

where:

N_{JKM} is the normalization constant;

$\Theta_{JKM}(\theta)$ is the function of the nutation angular coordinate θ.

As a qualitative example, Figure 4.8 presents the rotational level scheme of an oblate symmetric top.

In both cases, the quantum numbers *J*, *K* and *M* are good quantum numbers verifying the relations: $J \geq 0$, $-J \leq M \leq J$ and $-J \leq K \leq J$.

The quantum number *M* does not appear in the expression of the eigenenergy of the Hamiltonian; each energy level of quantum number *J* is associated with $2J+1$ eigenfunctions, such that it is $2J+1$ times degenerate. Moreover, since the quantum number *K* is squared in expressions [4.50], the energy levels are doubly degenerate, except for $K = 0$ [LAN 88, SUG 65, TOW 75].

The eigenfunctions can also be written in terms of elements of the rotational matrix [ROS 67], which was defined in section 2.7.5 of Volume 2 with a calculation subroutine in FORTRAN:

$$|JKM\rangle = (-1)^{M-K}\sqrt{\frac{2J+1}{8\pi^2}}D^J_{-M,-K}(\phi,\theta,\chi) \qquad [4.52]$$

φ, θ and χ being the Euler angles associated, at each instant, with the orientation of the equilibrium configuration reference frame $(\mathrm{G},\vec{x},\vec{y},\vec{z})$ of the molecule, with respect to the reference frame $(\mathrm{G},\vec{\mathrm{X}},\vec{\mathrm{Y}},\vec{\mathrm{Z}})$.

Centrifugal distortion: this distortion stems from the non-rigidity of the molecule during its rotational motion. Indeed, the nuclei tend to move apart and the moments of inertia tend to increase. Additional terms must be introduced in the rotational Hamiltonian. There is a shift in the energy levels, which remains a very small one. The corrective term to expressions [4.49] is written as:

$$\Delta E_{JK} = -D_{JK}^{e} J(J+1)K^{2} - D_{K}^{e} K^{4} - D_{J}^{e} \left[J(J+1)\right]^{2} + \ldots \qquad [4.53]$$

where D^e is the first-order centrifugal distortion constants, and, as the higher-order terms are very small, they were truncated.

4.4.2. *Transitions between two rotational levels: selection rules*

When a rigid symmetric top molecule undergoes rotational transition between an initial state $\left|J_i K_i M_i\right\rangle$ of energy $E_{J_i K_i}$ and a final state $\left|J_f K_f M_f\right\rangle$ of energy $E_{J_f K_f}$, it absorbs or emits a photon of energy $\Delta E_{J_i K_i, J_f K_f}$ depending on whether $E_{J_i K_i} \langle E_{J_f K_f}$ or $E_{J_i K_i} \rangle E_{J_f K_f}$. This is reflected by the presence of a line in the IR spectrum in the microwave range. Its dipole moment $\vec{\mu}$ coincides with its permanent moment $\vec{\mu}^e$ of components: $\mu_x^e = 0$, $\mu_y^e = 0$ and $\mu_z^e = \mu^e$ with respect to the equilibrium configuration reference frame $(\mathrm{G},\vec{x},\vec{y},\vec{z})$. The dipole moment $\vec{\mu}_A^e$, defined with respect to the reference frame $(\mathrm{G},\vec{\mathrm{X}},\vec{\mathrm{Y}},\vec{\mathrm{Z}})$ attached to the molecule, of components μ_{X}^e, μ_{Y}^e and μ_{Z}^e, is given by

$$\vec{\mu}_A^e = \begin{pmatrix} \mu_{\mathrm{X}}^e \\ \mu_{\mathrm{Y}}^e \\ \mu_{\mathrm{Z}}^e \end{pmatrix} = \mathbf{M}^{-1}(\varphi,\theta,\chi) \begin{pmatrix} 0 \\ 0 \\ \mu^e \end{pmatrix} = \mu^e \begin{pmatrix} \sin\theta\cos\varphi \\ \sin\theta\sin\varphi \\ \cos\theta \end{pmatrix} \qquad [4.54]$$

The rotational transition does not occur unless the dipole transition moment $\left\langle J_f K_f M_f \middle| \vec{\mu}_A^e \middle| J_i K_i M_i \right\rangle$ is *non-zero*, hence:

$$\langle J_f K_f M_f | \vec{\mu}_A^e | J_i K_i M_i \rangle = (-1)^{M_f - K_f} (-1)^{M_i - K_i} \sqrt{\frac{2J_f + 1}{8\pi^2}} \sqrt{\frac{2J_i + 1}{8\pi^2}}$$
$$\times \int_0^{2\pi} \int_0^{\pi} \int_0^{2\pi} D_{-M_f, -K_f}^{J_f *} (\varphi, \theta, \chi) \vec{\mu}_A^e D_{-M_i, -K_i}^{J_i} (\varphi, \theta, \chi) d\varphi \sin\theta d\theta d\chi \neq 0 \quad [4.55]$$

The steps followed in the calculation of these transition moments are presented in Appendix D, section 4.7.3. The rotational selection rules are as follows:

$$\Delta J = \pm 1 \qquad \Delta M = 0, \pm 1 \qquad \text{and} \qquad \Delta K = 0$$

completed by the $s \rightarrow a$ and $a \rightarrow s$ rules, due to the inversion motion at the fundamental vibrational level [TOW 75]. It is important to note that in the rotational spectrum, the transitions leading to the quantum number M do not appear, as the energy levels J are $(2J + 1)$ times degenerate in M. This degeneracy is schematically presented in Figure 4.8 by the thickness of the line associated with each level.

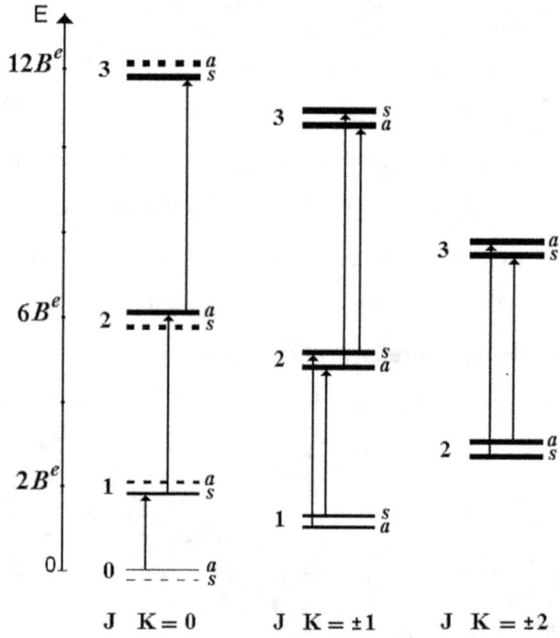

Figure 4.8. *Rotational levels scheme of an oblate symmetric top of type XY_3. The degeneracy in M of each level is schematically represented by the thickness of its line*

4.5. Rovibrational energy levels of an isolated symmetric top and selection rules

Until now, the rotational and vibrational motions were considered to be independent. In fact, their simultaneous existence leads to the appearance of a specific interaction.

To determine the vibration–rotation energy levels of an isolated molecule, it is necessary to include the vibration–rotation coupling terms as a perturbation term in the Hamiltonian. Indeed, the molecule is deformed during its rotational motion. Schrödinger's equation associated with rovibrational motions must be solved beyond the zeroth-order Born–Oppenheimer approximation.

However, given the difference in energies between vibrations and rotation, it is possible to use a perturbation calculation in order to determine the corrections to the previous results.

For a symmetric top molecule, for a vibration mode v, this leads to the eigenenergy:

$$E_{\mathrm{v}_v,JK} = E_{\mathrm{v}_v} + E_{JK}^{\mathrm{v}_v} \qquad [4.56]$$

where E_{v_v} is the vibrational eigenenergy associated with the anharmonic oscillator in state v_v, and $E_{JK}^{\mathrm{v}_v}$ is the rotational eigenenergy associated with state JKM of the molecule in the vibrational state v_v. The latter is expressed in terms of rotational constants of the molecule in the vibrational state v_v.

The corresponding eigenfunctions are expressed as:

$$\Psi_{\mathrm{v}_v,JKM} = \left|\mathrm{v}_v,JKM\right\rangle = \left|\mathrm{v}_v\right\rangle \otimes \left|JKM\right\rangle_{\mathrm{v}_v} \qquad [4.57]$$

where $\left|\mathrm{v}_v\right\rangle$ is the eigenfunction of the vibrational state v_v of the anharmonic oscillator associated with the mode v, which was previously determined (see equation [4.43]), and $\left|JKM\right\rangle_{\mathrm{v}_v}$ is the eigenfunction of the rigid rotor depending on the vibrational state of the molecule. This is reflected by a distribution of rotational states that is slightly different from one vibrational state to another.

In the case of an oblate symmetric top molecule, the rotational eigenenergy is typically written as:

$$
\begin{aligned}
E_{JK}^{\mathrm{v}_v} &= B^{\mathrm{v}_v} J(J+1) + (C^{\mathrm{v}_v} - B^{\mathrm{v}_v})K^2 - D_J^{\mathrm{v}_v}\left[J(J+1)\right]^2 \\
&\qquad - D_{JK}^{\mathrm{v}_v} J(J+1)K^2 - D_K^{\mathrm{v}_v} K^4 + \ldots \\
B^{\mathrm{v}_v} &= B^e - \alpha^e(\mathrm{v}_v + \tfrac{g_v}{2}), \\
C^{\mathrm{v}_v} &= C^e - \beta^e(\mathrm{v}_v + \tfrac{g_v}{2}), \\
D^{\mathrm{v}_v} &= D^e - \gamma^e(\mathrm{v}_v + \tfrac{g_v}{2}), \ldots
\end{aligned}
\qquad [4.58]
$$

where $\alpha, \beta, \gamma, \ldots$ are the vibration–rotation interaction constants, and g_v is the degeneracy of the mode of vibration v.

The condition for a rovibrational transition $\left|\mathrm{v}_{v_i}, J_i K_i M_i\right\rangle \rightarrow \left|\mathrm{v}_{v_f}, J_f K_f M_f\right\rangle$ to take place is that the transition dipole moment is *non-zero*. This requires the rotational and vibrational transition moments to be simultaneously *non-zero*. If the series expansion of the dipole moment of the molecule is limited to the first order, then, in the framework of the harmonic approximation, the transition moment is written as

$$
\begin{aligned}
\left\langle \mathrm{v}_{v_f}, J_f K_f M_f \left| \vec{\mu}_A \right| \mathrm{v}_{v_i}, J_i K_i M_i \right\rangle &= {}_0\left\langle \mathrm{v}_{v_f} \left| q_v \right| \mathrm{v}_{v_i} \right\rangle_0 \\
&\times {}_{\mathrm{v}_{v_f}}\left\langle J_f K_f M_f \right| \left(\frac{\partial \vec{\mu}_A}{\partial q_v} \right)_0 \left| J_i K_i M_i \right\rangle_{\mathrm{v}_{v_i}} \neq 0
\end{aligned}
\qquad [4.59]
$$

The next step is then to express the vector $\left(\frac{\partial \vec{\mu}_A}{\partial q_v} \right)_0$ in the reference frame $(\mathrm{G}, \vec{\mathrm{X}}, \vec{\mathrm{Y}}, \vec{\mathrm{Z}})$ in the same way as in equation [4.55]. Hence:

$$
\left(\frac{\partial \vec{\mu}_A}{\partial q_v} \right)_0 = \begin{pmatrix} \left(\frac{\partial \mu_X}{\partial q_v} \right)_0 \\ \left(\frac{\partial \mu_Y}{\partial q_v} \right)_0 \\ \left(\frac{\partial \mu_Z}{\partial q_v} \right)_0 \end{pmatrix} = \mathbf{M}^{-1}(\phi, \theta, \chi) \begin{pmatrix} \left(\frac{\partial \mu_x}{\partial q_v} \right)_0 \\ \left(\frac{\partial \mu_y}{\partial q_v} \right)_0 \\ \left(\frac{\partial \mu_z}{\partial q_v} \right)_0 \end{pmatrix} \qquad [4.60]
$$

Equations [4.59] and [4.60] can be used to determine the rovibrational selection rules. However, determining the rotational selection rules requires knowledge on the given mode of vibration.

4.6. Application to the ammonia NH_3 molecule

4.6.1. *Geometric, rotational and vibrational characteristics*

NH_3 is a symmetric top molecule in which the nitrogen N nucleus is at the top of the pyramidal structure and the hydrogen nuclei form the base plane of the structure. Its geometric characteristics (defined in Figure 4.1), its rotational constants, and its dipole and quadrupole moments in its equilibrium configuration are given in Table 4.2.

r^e (Å)	h^e (Å)	β^e (°)	α^e (°)	$A^e=B^e$ (cm^{-1})	C^e (cm^{-1})	μ^e (D)	Θ^e (DÅ)
1.0156	0.3811	67.96	106.78	9.941	6.309	1.476	– 2.930

Table 4.2. *Geometric characteristics, rotational constants, and dipole and quadrupole moments of the ammonia NH_3 molecule in its equilibrium configuration*

As can be noted, the values of the rotational constants are such that $A^e = B^e > C^e$. Consequently the molecule has the form of an oblate symmetric top.

The Cartesian coordinates of the nuclei, with respect to the equilibrium configuration reference frame of the rigid molecule, are presented in Table 4.3.

Nucleus coordinates	$\eta^e_{i_x}$ (Å)	$\eta^e_{i_y}$ (Å)	$\eta^e_{i_z}$ (Å)
N (i = 1)	0	0	– 0.0676
H (i = 2)	0.9414	0	0.3133
H (i = 3)	– 0.4707	0.8153	0.3133
H (i = 4)	– 0.4707	– 0.8153	0.3133

Table 4.3. *Cartesian coordinates of the nuclei of the ammonia NH_3 molecule in the reference frame $(G,\vec{x},\vec{y},\vec{z})$ attached to its equilibrium configuration. m_N and m_H are, respectively, the mass of the nitrogen nucleus and that of the hydrogen nucleus*

The experimental frequencies of the vibrational modes of the ammonia molecule in the gas phase are as follows:

– $\omega_1 = 3{,}337\ \text{cm}^{-1}$, symmetric mode of species A_1, of degeneracy $g_1 = 1$;

– $\omega_2 = 950\ \text{cm}^{-1}$, symmetric mode of species A_1, of degeneracy $g_2 = 1$;

– $\omega_3 = 3{,}448\ \text{cm}^{-1}$, antisymmetric mode of species E, of degeneracy $g_3 = 1$;

– $\omega_4 = 1{,}627\ \text{cm}^{-1}$, antisymmetric mode of species E, of degeneracy $g_4 = 1$.

In order to apply the methods presented in section 4.3, the expression used here is that of the vibrational potential energy function obtained by *ab initio* methods by Martin *et al.* [MAR 92], according to the *hypothesis of valence forces.*

4.6.2. *Vibrational motions in the harmonic approximation*

In expression [4.15], the harmonic potential energy function $V_{\text{vib}}^{\text{har}}$ is expressed in terms of the internal force constants f', which are presented in Table 4.4 [MAR 92].

f'	Units: attojoule, Å, rad.	Units: cm^{-1}, Å, rad.
f_1'	7.033675	354082.9
f_2'	0.644165	32428.0
f_3'	– 0.036958	– 1860.5
f_4'	– 0.050515	– 2543.0
f_5'	0.040950	2061.5
f_6'	0.241401	12152.4

Table 4.4. *Internal harmonic force constants of the ammonia NH_3 molecule. Energy unit conversion: 1 attojoule = 50,341.1 cm^{-1}*

The replacement of the internal bond coordinates q_{1j} and angular coordinates $\Delta\alpha_{ij}$ as a function of symmetry coordinates $\{S_v\}$, given by expressions [4.19], as well as the force constants f' in expressions [4.15] of the potential energy $V_{\text{vib}}^{\text{har}}$ and

[4.16] of the kinetic energy T_{vib}, yields expression [4.20] for T_{vib} and [4.22] for V_{vib}^{har}. The respective mass constants $d_{vv'}$ and force constants $k'_{vv'}$ (see section 4.7.2) are given in Table 4.5 for species A_1 and Table 4.6 for species E.

Moreover, it is important to keep in mind that the normal modes of vibration of species A_1 and those of species E are totally separated. The frequencies and symmetry coordinates functions associated with the vibration modes ν_1 and ν_2, on the one hand, and ν_3 and ν_4 (doubly degenerate), on the other hand, are obtained by the diagonalization of 2×2 matrices $\left[\boldsymbol{d}^{-1}\boldsymbol{k}'\right]_{A_1}$ and $\left[\boldsymbol{d}^{-1}\boldsymbol{k}'\right]_{E}$, respectively.

Matrices	**Species A₁**	
Matrix [d] (g)	$5.1741 \cdot 10^{-24}$ $-0.0377 \cdot 10^{-24}$	$-0.3770 \cdot 10^{-24}$ $5.9532 \cdot 10^{-24}$
Inverse matrix [d^{-1}] (g^{-1})	$19.4164 \cdot 10^{+22}$ $1.2295 \cdot 10^{+22}$	$1.2295 \cdot 10^{+22}$ $16.8756 \cdot 10^{+22}$
Matrix[k'] (erg/cm^2)	2,191,696.25 28,684.32	28,684.32 245,378.43
Matrix [$d^{-1}.k'$] (s^{-2})	$425,902.06 \cdot 10^{+24}$ $31,787.33 \cdot 10^{+24}$	$8,586.38 \cdot 10^{+24}$ $41,761.70 \cdot 10^{+24}$

Table 4.5. *Mass and force constant matrix elements and elements of the matrix to be diagonalized associated with species A_1. Force constant unit conversion: 1erg/cm^2 = 0.503411 cm^{-1}/Å^2*

Matrices	**Species E**	
Matrix [d] (g)	$10.9748 \cdot 10^{-24}$ $-4.6549 \cdot 10^{-24}$	$-4.6549 \cdot 10^{-24}$ $5.0216 \cdot 10^{-24}$
Inverse matrix [d^{-1}] (g^{-1})	$15.0157 \cdot 10^{+22}$ $13.9194 \cdot 10^{+22}$	$13.9194 \cdot 10^{+22}$ $32.8171 \cdot 10^{+22}$
Matrix [k'] (erg/cm^2)	5,088,406.89 −2,056,729.00	−2,056,729.00 1,143,244.78
Matrix [$d^{-1}.k'$] (s^{-2})	$477,776.40 \cdot 10^{+24}$ $33,314.86 \cdot 10^{+24}$	$-149,699.90 \cdot 10^{+24}$ $88,896.27 \cdot 10^{+24}$

Table 4.6. *Mass and force constant matrix elements and elements of the matrix to be diagonalized associated with species E. Force constant unit conversion: 1erg/cm^2 = 0.503411 cm^{-1}/Å^2*

Finally, after the diagonalization of the two matrices, the matrices of the eigenvalues (in s^{-2}) and their corresponding eigenvectors are obtained as follows:

– Species A_1 (simple degenerate)

$$(\lambda_0)_{A_1} = \begin{pmatrix} \lambda_{0_1} = 42.66110 \cdot 10^{+28} & 0 \\ 0 & \lambda_{0_2} = 4.10525 \cdot 10^{+28} \end{pmatrix}$$

$S_1 = 0.996606 \times Q_1 - 0.022305 \times Q_2$ associated with λ_{0_1}

$S_2 = 0.082317 \times Q_1 + 0.999751 \times Q_2$ associated with λ_{0_2} ;

– Species E (doubly degenerate)

$$(\lambda_0)_E = \begin{pmatrix} \lambda_{0_{3a}} = \lambda_{0_{3b}} = 46.44980 \cdot 10^{+28} & 0 \\ 0 & \lambda_{0_{4a}} = \lambda_{0_{4b}} = 10.21740 \cdot 10^{+28} \end{pmatrix}$$

$$\left.\begin{aligned} S_{3a} &= 0.996089 \times Q_{3a} + 0.370237 \times Q_{4a} \\ S_{3b} &= 0.996089 \times Q_{3b} + 0.370237 \times Q_{4b} \end{aligned}\right\} \text{associated with } \lambda_{0_3};$$

$$\left.\begin{aligned} S_{4a} &= 0.088350 \times Q_{3a} + 0.928937 \times Q_{4a} \\ S_{4b} &= 0.088350 \times Q_{3b} + 0.928937 \times Q_{4b} \end{aligned}\right\} \text{associated with } \lambda_{0_4};$$

where Q_v are the vibrational normal coordinates, defined in section 4.3.4 and representing each harmonic oscillator (expression [4.32]).

ω_{0_v} (cm^{-1})	**Application**	**[MAR 92]**	**Observation**
ω_{0_1}	3,467	3,480	3,337
ω_{0_2}	1,076	1,084	950
$\omega_{0_{3a}} = \omega_{0_{3b}}$	3,618	3,609	3,448
$\omega_{0_{4a}} = \omega_{0_{4b}}$	1,697	1,680	1,627

Table 4.7. *Harmonic frequencies of the vibration modes of the NH_3 molecule*

Table 4.7 presents the values obtained for the harmonic frequencies of various vibration modes, in comparison with the values obtained by Martin *et al.* [MAR 92] and the observed values.

Finally, the use of expression [4.33] yields the dimensionless normal coordinates q_v as a function of normal coordinates Q_v :

$$q_1 = 17.902 \times Q_1$$
$$q_2 = 10.695 \times Q_2$$
$$q_{3a} = 26.633 \times Q_{3a}$$
$$q_{3b} = 26.633 \times Q_{3b}$$
$$q_{4a} = 12.338 \times Q_{4a}$$
$$q_{4b} = 12.338 \times Q_{4b}$$

4.6.3. *Vibrational motions beyond the harmonic approximation*

As previously mentioned, the anharmonic part $\Delta V_{\text{vib}}^{\text{anhar}}$ (limited at orders 3 and 4) of the vibrational potential energy function V_{vib} determined by Martin *et al.* [MAR 92] contains a large number of force constant terms $f_{ijk}^{'}$, $f_{ijkl}^{'}$. The values of these force constants are not presented in this chapter. Moreover, the switch from expression $\Delta V_{\text{vib}}^{\text{anhar}}$, in terms of internal coordinates q_{1j} and $\Delta\alpha_{i'j'}$, to that in terms of dimensionless normal coordinates q_v [4.41], is quite tedious. However, it can be achieved by a formal calculation software (open-source Xcas, mathematica, etc.). The values of third-order $k_{v'v''v'''}$ and fourth-order $k_{v'v''v'''v''''}$ force constants in expression [4.41] are presented in Box 4.1 in Appendix D, section 4.7.4. It can be noted that there are no terms coupling the coordinates q_{sa} and q_{sb} of the doubly degenerate modes of vibration.

The introduction of $\Delta V_{\text{vib}}^{\text{anhar}}$ in expressions [4.42] makes it possible to determine the corrected eigenenergies E_v and the corrected eigenvectors $\left|v_1 v_2 v_{3a} v_{3b} v_{4a} v_{4b}\right\rangle$, of the fundamental level and the first excited levels (with the normalization coefficients taken into account).

The corrected frequencies of the vibrational modes of the molecule are summarized in Table 4.8. The eigenvectors are presented in Appendix D, section 4.7.4. These are used for calculating the vibrational frequency shift when the molecule is trapped in a nanocage or in the vicinity of a surface.

ω_v (cm^{-1})	**Application**	**[MAR 92]**	**Observation**
ω_1	3,309	3,330	3,337
ω_2	1,070	1,012	950
$\omega_{3a} = \omega_{3b}$	3,453	3,432	3,448
$\omega_{4a} = \omega_{4b}$	1,689	1,630	1,627

Table 4.8. *Frequencies beyond the harmonic approximation of the vibrational modes of NH_3 molecule*

4.6.4. *Vibration-inversion mode*

As previously noted, this mode was highlighted by the splitting of certain vibrational bands of the NH_3 molecule [DEN 32a, DEN 32b].

Various functions were used to model the double-well potential profile [MAN 35, NEW 48] of the NH_3 molecule and its isotopologues. However, the relative differences between the calculated values and the experimental values of the level splitting may reach 20%.

Swalen and Ibers [SWA 62] proposed a better function to describe the double-well potential: superimposing on a well with a harmonic profile of a minimum at $\theta_2 = 0$ and of force constant k_0, an exponential (Gaussian) curve of a maximum at $\theta_2 = 0$ and characterized by two constants A_0 and a_0, such that:

$$V(\theta_2) = \frac{1}{2} k_0 \theta_2^2 + A_0 \left(e^{-a_0 \theta_2^2} - 1 \right) \qquad [4.61]$$

Moreover, if the internuclear N-H distance is assumed to be constant during the inversion motion, the reduced mass of the molecule is written as [SWA 62]

$$\mu = \frac{3m_H (m_N + 3m_H \sin^2 \theta_2^e)}{(m_N + 3m_H)}$$

and therefore: $\mu = 2.563$ g.mole^{-1}.

Various numerical methods can be used to solve the eigenvalue equation associated with the Hamiltonian H_{v_2} of expression [4.43]. This application uses the

Discrete Variable Representation (DVR) method [LIG 85]. The constants k_0, A_0 and a_0 of the potential function [4.61] are: k_0 = 60,137 $cm^{-1}.rad^{-2}$, A_0 = 12,469 cm^{-1}, a_0 = 4.974 rad^{-2}. The height of the barrier between the two wells is then equal to 2,047 cm^{-1}.

Figure 4.9 presents the vibration–inversion potential function $V(\theta_2)$, the energy levels and their associated wave function and the transitions between these levels in the near IR.

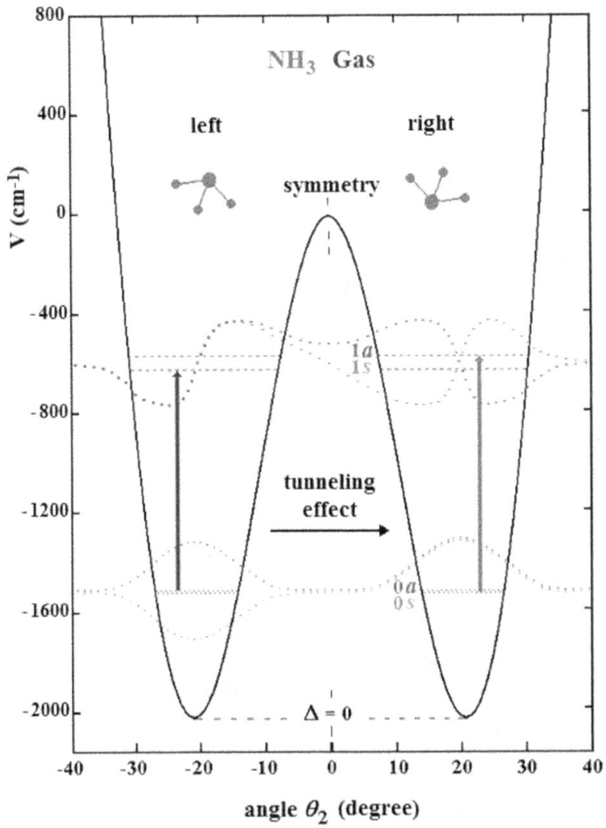

Figure 4.9. *Potential energy of the vibration–inversion mode ν_2 of an ammonia NH_3 molecule. Δ represents the energy gap between the potential wells, $\Delta = 0$ for the isolated NH_3 molecule. For a color version of this figure, see www.iste.co.uk/dahoo/infrared.zip*

The calculated and observed values [BEN 57] of level splitting $\delta(v_2)$ and of transition frequencies $\omega_{0s\to 1a}$, $\omega_{0a\to 1s}$ are summarized in Table 4.9.

	$\delta(v_2 = 0)$	$\delta(v_2 = 1)$	$\omega_{0s\rightarrow 1a}$	$\omega_{0a\rightarrow 1s}$
Calculated (cm^{-1})	0.92	35.99	970.16	933.34
Observed[1] (cm^{-1})	0.79	35.81	968.32	932.51

Table 4.9. *Level splitting and allowed transition frequencies of the vibration–inversion mode of the NH_3 molecule in the gas phase*

4.6.5. *Dipole moment as a function of normal coordinates*

At the first order of the expansion as a function of normal coordinates (expression [4.44]), the dipole moment of a molecule XY_3 is written as:

$$\vec{\mu} = \mu^e \vec{e}_z + \left(\left(\frac{\partial \mu_z}{\partial q_1}\right)_0 q_1 + \left(\frac{\partial \mu_z}{\partial q_2}\right)_0 q_2\right)\vec{e}_z + \left(\left(\frac{\partial \mu_x}{\partial q_{3a}}\right)_0 q_{3a} + \left(\frac{\partial \mu_x}{\partial q_{4a}}\right)_0 q_{4a}\right)\vec{e}_x + \left(\left(\frac{\partial \mu_y}{\partial q_{3b}}\right)_0 q_{3b} + \left(\frac{\partial \mu_y}{\partial q_{4b}}\right)_0 q_{4b}\right)\vec{e}_y \quad [4.62]$$

For the NH_3 molecule, it is written as:

$$\left(\frac{\partial \mu_z}{\partial q_1}\right)_0 = 7.83\cdot 10^{-2}\,\mathrm{D},\quad \left(\frac{\partial \mu_z}{\partial q_2}\right)_0 = -\,41.17\cdot 10^{-2}\,\mathrm{D}$$

$$\left(\frac{\partial \mu_x}{\partial q_{3a}}\right)_0 = -\,15.04\cdot 10^{-2}\,\mathrm{D},\quad \left(\frac{\partial \mu_x}{\partial q_{4a}}\right)_0 = 16.06\cdot 10^{-2}\,\mathrm{D}$$

$$\left(\frac{\partial \mu_y}{\partial q_{3b}}\right)_0 = 9.24\cdot 10^{-2}\,\mathrm{D},\quad \left(\frac{\partial \mu_y}{\partial q_{4b}}\right)_0 = -\,20.71\cdot 10^{-2}\,\mathrm{D}$$

4.7. Appendices

4.7.1. *Appendix A: Rotation matrix*

Consider $(O, \vec{X}, \vec{Y}, \vec{Z})$ and two direct orthonormal reference frames. At each instant, the orientation of the reference frame $(O, \vec{x}, \vec{y}, \vec{z})$ with respect to the

1 Source: [BEN 57].

reference frame $(O,\vec{X},\vec{Y},\vec{Z})$ can be decomposed into three successive elementary rotations: φ for precession, θ for nutation and χ for intrinsic rotation (Euler angles). The unitary transformation matrix or the rotation matrix is given by (see section 2.7.1, Volume 2 [DAH 19]) [ROS 67]:

$$\mathbf{M}(\varphi,\theta,\chi)=\begin{pmatrix} \cos\varphi\cos\theta\cos\chi - \sin\varphi\sin\chi & \sin\varphi\cos\theta\cos\chi + \cos\varphi\sin\chi & -\sin\theta\cos\chi \\ -\cos\varphi\cos\theta\sin\chi - \sin\varphi\cos\chi & -\sin\varphi\cos\theta\sin\chi + \cos\varphi\cos\chi & \sin\theta\sin\chi \\ \cos\varphi\sin\theta & \sin\varphi\sin\theta & \cos\theta \end{pmatrix}$$

NOTE.– Being a unitary matrix, its inverse is equal to its transpose.

4.7.2. Appendix B: Expressions of force constants

The force constants k and k', involved in expressions [4.20] and [4.21] of the vibrational potential energy, are as follows:

– Central force hypothesis

$$k_{11}=3(f_1+2f_3)p^2+9(f_2+2f_4)\sin^2\beta^e+6\sqrt{3}(f_5+2f_6)p\sin\beta^e$$

$$k_{22}=3(f_1+2f_3)(p-1)(u-1)+9(f_2+2f_4)\cos^2\beta^e -6\sqrt{3}(f_5+2f_6)(p-1)\sin\beta^e$$

$$k_{12}=-3(f_1+2f_3)p\sqrt{(p-1)(u-1)}+9(f_2+2f_4)\sin\beta^e\cos\beta^e +3\sqrt{3}(f_5+2f_6)(p-u+1)\cos\beta^e$$

$$k_{3a3a}=k_{3b3b}=3(f_1-f_3)(u+1)^2+\frac{9}{2}(f_2-f_4)\sin^2\beta^e-3\sqrt{3}(f_5-f_6) \times(u+1)\sin\beta^e$$

$$k_{4a4a}=k_{4b4b}=\frac{3}{2}(f_1-f_3)\sin^2\beta^e+\frac{9}{2}(f_2-f_4)-3\sqrt{3}(f_5-f_6)\sin\beta^e$$

$$k_{3a4a}=k_{3b4b}=-\frac{3}{2}(f_1-f_3)(u+1)\sin\beta^e-\frac{9}{2}(f_2-f_4)\sin\beta^e+\frac{3}{2}\sqrt{3}(f_5-f_6) \times(u+1+\sin^2\beta^e)$$

– valence force hypothesis

$$k'_{11} = 3\left(f'_1 + 2f'_3\right)p^2 + 9\left[f'_2 + \left(f'_2 + 2f'_4\right)(3p-1)(p-1)\right]\frac{\sin^4\beta^e}{r^{e2}\sin^2\alpha^e}$$
$$-6\left(f'_5 + 2f'_6\right)p(3p-2)\frac{\sin^2\beta^e}{r^e\sin\alpha^e}$$

$$k'_{22} = 3\left(f'_1 + 2f'_3\right)(p-1)(u-1) + 18\left(f'_2 + f'_4\right)\frac{\sin^2\beta^e\cos^2\beta^e}{r^{e2}\sin^2\alpha^e}$$
$$+9\left(f'_2 + 2f'_4\right)(p-1)(3u+1)\frac{\sin^4\beta^e}{r^{e2}\sin^2\alpha^e}$$
$$-6\left(f'_5 + 2f'_6\right)(p-1)(3u-1)\frac{\sin^2\beta^e}{r^e\sin\alpha^e}$$

$$k'_{3a3a} = \frac{3}{2}\left(f'_1 - f'_3\right)(u+1)^2 + \frac{9}{8}\left[4\left(f'_2 + f'_4\right) + \left(f'_2 - f'_4\right)(3u-1)(u+1)\right]\frac{\sin^4\beta^e}{r^{e2}\sin^2\alpha^e}$$
$$+\frac{3}{2}\left(f'_5 - f'_6\right)(3u+1)(u+1)\frac{\sin^2\beta^e}{r^e\sin\alpha^e}$$

$$k'_{3b3b} = \frac{3}{2}\left(f'_1 - f'_3\right)(u+1)^2 + \frac{9}{8}\left[8f'_2 + 3\left(f'_2 - f'_4\right)(u-1)^2\right]\frac{\sin^4\beta^e}{r^{e2}\sin^2\alpha^e}$$
$$+\frac{9}{2}\left(f'_5 - f'_6\right)(u+1)(u-1)\frac{\sin^2\beta^e}{r^e\sin\alpha^e}$$

$$k'_{4a4a} = \frac{3}{2}\left(f'_1 - f'_3\right)\sin^2\beta^e + \frac{9}{2}\left(f'_2 + f'_4\right)\frac{\sin^2\beta^e}{r^{e2}\sin^2\alpha^e}$$
$$-\frac{9}{2}\left(f'_2 - f'_4\right)\left[1 - \frac{3}{4}\sin^2\beta^e\right]\frac{\sin^4\beta^e}{r^{e2}\sin^2\alpha^e}$$
$$-3\left(f'_5 - f'_6\right)\left[1 - \frac{3}{2}\sin^2\beta^e\right]\frac{\sin^2\beta^e}{r^e\sin\alpha^e}$$

$$k'_{4b4b} = \frac{3}{2}\left(f'_1 - f'_3\right)\sin^2\beta^e + \frac{9}{2}\left(5f'_2 - 3f'_4\right)\frac{\sin^2\beta^e}{r^{e2}\sin^2\alpha^e}$$
$$-\frac{27}{2}\left(f'_2 - f'_4\right)\left[1 - \frac{1}{4}\sin^2\beta^e\right]\frac{\sin^4\beta^e}{r^{e2}\sin^2\alpha^e}$$
$$-9\left(f'_5 - f'_6\right)\left[1 - \frac{1}{2}\sin^2\beta^e\right]\frac{\sin^2\beta^e}{r^e\sin\alpha^e}$$

$$k'_{12} = -3\left(f'_1 + 2f'_3\right)p\sqrt{(p-1)(u-1)}$$
$$+9\left[f'_2 - \left(f'_2 + 2f'_4\right)(3pu - 2u - p + 1)\right]\frac{\sin^3\beta^e\cos\beta^e}{r^{e2}\sin^2\alpha^e}$$
$$+6\left(f'_5 + 2f'_6\right)(3pu - u - 2p + 1)\frac{\sin\beta^e\cos\beta^e}{r^e\sin\alpha^e}$$

$$k'_{13a} = \frac{9}{2}\left[2\left(f'_2 + 2f'_4\right)p + \left(f'_2 - f'_4\right)u - \left(f'_2 + 3f'_4\right)\right]\frac{\sin^4\beta^e}{r^{e2}\sin^2\alpha^e}$$
$$-3\left[\left(f'_5 + 2f'_6\right)p - \left(f'_5 - f'_6\right)(u+1)\right]\frac{\sin^2\beta^e}{r^e\sin\alpha^e}$$

$$k'_{13b} = 9\left[\left(f'_2 + 2f'_4\right)(p-1) - f'_2\right]\frac{\sin^4\beta^e}{r^{e2}\sin^2\alpha^e} + 3\left(f'_5 + 2f'_6\right)p\frac{\sin^2\beta^e}{r^e\sin\alpha^e}$$

$$k'_{14a} = \frac{9}{2}\left[f'_4 - \left(f'_2 + 2f'_4\right)(2p-1)\right]\frac{\sin^3\beta^e}{r^{e2}\sin^2\alpha^e}$$
$$+3\left(f'_5 + 2f'_6\right)p\frac{\sin\beta^e}{r^e\sin\alpha^e} - 3\left(f'_5 - f'_6\right)\frac{\sin^3\beta^e}{r^e\sin\alpha^e}$$

$$k'_{14b} = 9\left[f'_2 + \left(f'_2 + 2f'_4\right)(p-1)\right]\frac{\sin^3\beta^e}{r^{e2}\sin^2\alpha^e} - 3\left(f'_5 + 2f'_6\right)p\frac{\sin\beta^e}{r^e\sin\alpha^e}$$

$$k'_{23a} = \frac{9}{2}\left[f'_2(u+1) + f'_4(5u+1)\right]\frac{\sin^3\beta^e\cos\beta^e}{r^{e2}\sin^2\alpha^e}$$
$$+3\left[2f'_5u + f'_6(u-3)\right]\frac{\sin\beta^e\cos\beta^e}{r^e\sin\alpha^e}$$

$$k'_{23b} = 3\left[2\left(f'_2 + 3f'_4\right)u - 3f'_2\right]\frac{\sin^3\beta^e\cos\beta^e}{r^{e2}\sin^2\alpha^e} - 3\left(f'_5 + 2f'_6\right)(u-1)\frac{\sin\beta^e\cos\beta^e}{r^e\sin\alpha^e}$$

$$k'_{24a} = 9\left[\left(f'_2 + 2f'_4\right)u - f'_4\right]\frac{\sin^2\beta^e\cos\beta^e}{r^{e2}\sin^2\alpha^e} - \frac{9}{2}\left(f'_2 - f'_4\right)\frac{\sin^4\beta^e\cos\beta^e}{r^{e2}\sin^2\alpha^e}$$
$$-3\left(f'_5 + 2f'_6\right)(u-1)\frac{\cos\beta^e}{r^e\sin\alpha^e} - 3\left(f'_5 - f'_6\right)\frac{\sin^2\beta^e\cos\beta^e}{r^e\sin\alpha^e}$$

$$k'_{24b} = 9\left[f'_2 - \left(f'_2 + 2f'_4\right)u\right]\frac{\sin^2\beta^e\cos\beta^e}{r^{e2}\sin^2\alpha^e} + 3\left(f'_5 + 2f'_6\right)(u-1)\frac{\cos\beta^e}{r^e\sin\alpha^e}$$

$$k'_{3a3b} = -\frac{9}{2}\left[\left(f'_2 - f'_4\right)u + \left(f'_2 + f'_4\right)\right]\frac{\sin^4\beta^e}{r^{e2}\sin^2\alpha^e} - 3\left(f'_5 - f'_6\right)(u+1)\frac{\sin^2\beta^e}{r^e\sin\alpha^e}$$

$$k'_{3a4a} = -\frac{3}{2}\left(f'_1 - f'_3\right)(u+1)\sin\beta^e + \frac{9}{4}\left[\left(f'_2 - f'_4\right)u - \left(f'_2 + 3f'_4\right)\right]\frac{\sin^3\beta^e}{r^{e2}\sin^2\alpha^e}$$
$$-\frac{9}{8}\left(f'_2 - f'_4\right)(3u-1)\frac{\sin^5\beta^e}{r^{e2}\sin^2\alpha^e} - \frac{3}{2}\left(f'_5 - f'_6\right)(u-3)\frac{\sin\beta^e}{r^e\sin\alpha^e}$$
$$+\frac{3}{2}\left(f'_5 - f'_6\right)(3u-8)\frac{\sin^3\beta^e}{r^e\sin\alpha^e}$$

$$k'_{3a4b} = \frac{9}{2}\left[\left(f'_2 - f'_4\right)u + \left(f'_2 + f'_4\right)\right]\frac{\sin^3\beta^e}{r^{e2}\sin^2\alpha^e}$$
$$+6\left(f'_5 - f'_6\right)\frac{\sin\beta^e}{r^e\sin\alpha^e} + 3\left(f'_5 - f'_6\right)(u-1)\frac{\sin^3\beta^e}{r^e\sin\alpha^e}$$

$$k'_{3b4a} = 9f'_4\frac{\sin^3\beta^e}{r^{e2}\sin^2\alpha^e} + \frac{9}{2}\left(f'_2 - f'_4\right)\frac{\sin^5\beta^e}{r^{e2}\sin^2\alpha^e} + 3\left(f'_5 - f'_6\right)\frac{\sin^3\beta^e}{r^e\sin\alpha^e}$$

$$k'_{3b4b} = -\frac{3}{2}\left(f'_1 - f'_3\right)(u+1)\sin\beta^e + \frac{9}{4}\left[3\left(f'_2 - f'_4\right)(u-1) - 4f'_2\right]\frac{\sin^3\beta^e}{r^{e2}\sin^2\alpha^e}$$
$$-\frac{27}{4}\left(f'_2 - f'_4\right)(u-1)\frac{\sin^5\beta^e}{r^{e2}\sin^2\alpha^e} + \frac{9}{2}\left(f'_5 - f'_6\right)(u+1)\frac{\sin\beta^e}{r^e\sin\alpha^e}$$
$$-\frac{9}{2}\left(f'_5 - f'_6\right)u\frac{\sin^3\beta^e}{r^e\sin\alpha^e}$$
$$k'_{4a4b} = -9f'_4\frac{\sin^2\beta^e}{r^{e2}\sin^2\alpha^e} - \frac{9}{2}\left(f'_2 - f'_4\right)\frac{\sin^4\beta^e}{r^{e2}\sin^2\alpha^e} - 3\left(f'_5 - f'_6\right)\frac{\sin^2\beta^e}{r^e\sin\alpha^e}$$

where

$$p = 1 + \frac{3m_Y}{m_X}\cos^2\beta^e \text{ and } u = 1 + \frac{3m_Y}{m_X}\sin^2\beta^e.$$

4.7.3. *Appendix C: Rotational moments of transition*

The rotational moments of transition of the components of the dipole moment are:

$$\langle J_f K_f M_f|\sin\theta\cos\varphi|J_i K_i M_i\rangle = \int_0^{2\pi}\int_0^{\pi}\int_0^{2\pi}\left\{\begin{matrix} D^{J_f *}_{-M_f,-K_f}(\varphi,\theta,\chi)\sin\theta\cos\varphi \\ D^{J_i}_{-M_i,-K_i}(\varphi,\theta,\chi)d\varphi\sin\theta d\theta d\chi\end{matrix}\right\}$$

$$\langle J_f K_f M_f|\sin\theta\sin\varphi|J_i K_i M_i\rangle = \int_0^{2\pi}\int_0^{\pi}\int_0^{2\pi}\left\{\begin{matrix} D^{J_f *}_{-M_f,-K_f}(\varphi,\theta,\chi)\sin\theta\sin\varphi \\ D^{J_i}_{-M_i,-K_i}(\varphi,\theta,\chi)d\varphi\sin\theta d\theta d\chi\end{matrix}\right\}$$

$$\langle J_f K_f M_f|\cos\theta|J_i K_i M_i\rangle = \int_0^{2\pi}\int_0^{\pi}\int_0^{2\pi}\left\{\begin{matrix} D^{J_f *}_{-M_f,-K_f}(\varphi,\theta,\chi)\cos\theta \\ D^{J_i}_{-M_i,-K_i}(\varphi,\theta,\chi)d\varphi\sin\theta d\theta d\chi\end{matrix}\right\}$$

Determining these elements requires, first, expressing the terms $\sin\theta\cos\varphi$, $\sin\theta\sin\varphi$ and $\cos\theta$ as a function of the elements of the rotation matrix $D^1_{+1,0}(\varphi,\theta,\chi)$, $D^1_{-1,0}(\varphi,\theta,\chi)$ and $D^1_{0,0}(\varphi,\theta,\chi)$, hence

$$\sin\theta\cos\varphi = \frac{1}{\sqrt{2}}\left(D^1_{-1,0}(\varphi,\theta,\chi) - D^1_{+1,0}(\varphi,\theta,\chi)\right)$$

$$\sin\theta\sin\varphi = \frac{i}{\sqrt{2}}\cdot\left(D^1_{-1,0}(\varphi,\theta,\chi) + D^1_{+1,0}(\varphi,\theta,\chi)\right)$$

$$\cos\theta = D^1_{0,0}(\varphi,\theta,\chi)$$

where i is the purely imaginary complex number.

The reader is invited to calculate the previous rotational transition moments using the general relation [ROS 67]:

$$\int_0^{2\pi}\int_0^{\pi}\int_0^{2\pi} D^{J*}_{-M,-K}(\varphi,\theta,\chi)D^{l}_{p,q}(\varphi,\theta,\chi)D^{J'}_{-M',-K'}(\varphi,\theta,\chi)d\phi\sin\theta d\theta d\chi$$

$$= \frac{8\pi^2}{(2J+1)}C(J'lJ;-M',p,-M)C(J'lJ;-K',q,-K)\delta_{p-M',-M}\delta_{q-K',-K}$$

where δ is the Kronecker symbol and C the Clebsch–Gordan coefficients, whose expression can be found in the reference [ROS 67]. A FORTRAN subroutine that can be used for their calculation is presented as follows:

```
C*******************************************************************
C-CALCUL DES COEFFICIENTS DE CLEBCH-GORDAN
C*******************************************************************
      WRITE(0,500)
 500  FORMAT(4X,'J1',4X,'J2',4X,'J3',4X,'M1',4X,'M2',
     &3X,'M12',10X,'COEF. CLEB.')
      DO 1 JF1=1,7
      J1=JF1-1
      MJ1=2*J1+1
      DO 2 MF1=1,MJ1
      M1=MF1-J1-1
      DO 5 JF2=1,7
      J2=JF2-1
      MJ2=2*J2+1
      DO 6 MF2=1,MJ2
      M2=MF2-J2-1
      IS3=J1+J2
      IR3=J1-J2
      IAR3=IABS(IR3)
      J3MIN=IAR3+1
      J3MAX=IS3+1
      DO 3 JF3=J3MIN,J3MAX
      J3=JF3-1
      MJ3=2*J3+1
      DO 4 MF12=1,MJ3
      M12=MF12-J3-1
```

```
      IF(M12.NE.(M1+M2))GO TO 4
      MA12=IABS(M12)
      IF(J3-IS3)100,100,101
 100  IF(J3-IAR3)101,102,102
 102  IF(J3-MA12)101,103,103
 103  CC=((2*J3+1)*FACT(J1+J2-J3)*FACT(J3+J1-J2)/
     &FACT(J1+J2+J3+1))*FACT(J1+M1)*FACT(J1-M1)*
     &FACT(J2+M2)*FACT(J2-M2)*FACT(J3+M12)*
     &FACT(J3-M12)*FACT(J3+J2-J1)
      CC=SQRT(CC)
      JP1=2*J1
      JP2=2*J2
      IF(JP1-JP2)104,104,105
 104  NUBO=JP1+1
      GO TO 106
 105  NUBO=JP2+1
 106  CA=0.
      DO 107 NU1=1,NUBO
      NU=NU1-1
      DNU=FACT(J1+J2-J3-NU)*FACT(NU)*FACT(J1-M1-NU)*
     &FACT(J2+M2-NU)*FACT(J3-J2+M1+NU)*FACT(J3-J1-M2+NU)
      IF(DNU)108,108,109
 108  CNU=0.
      GO TO 107
 109  CNU=(-1.)**NU/DNU
      CA=CA+CNU
 107  CONTINUE
      C=CA*CC
      GO TO 110
 101  C=0.
 110  WRITE(0,400)J1,J2,J3,M1,M2,M12,C
 400  FORMAT(6(3X,I3),10X,'C=',F9.6)
   4  CONTINUE
   3  CONTINUE
   6  CONTINUE
   5  CONTINUE
   2  CONTINUE
   1  CONTINUE
      STOP
      END
C
C*******************************************************************
C-CALCUL DE LA FONCTION FACTORIEL
C*******************************************************************
      FUNCTION FACT(NF)
      IF(NF)121,122,123
 123  X=1.
      DO 125 IF=1,NF
 125  X=X*IF
      GO TO 124
 122  X=1.
      GO TO 124
 121  X=0.
 124  FACT=X
      END
C*******************************************************************
```

Finally, the expression of the elements of the rotation matrix $D^{l}_{p,q}(\phi,\theta,\chi)$ as a function of Euler angles, and also the FORTRAN program for their calculation, were presented in section 2.7.5 of Volume 2 [DAH 19].

4.7.4. *Appendix D: Values of non-zero anharmonic vibrational force constants and corrected eigenvectors*

The non-negligible values of anharmonic force constants of orders 3 and 4 (unit: cm^{-1}), given by expression [4.41], are indicated in Box 4.1:

$k_{111} = -211.4$	$k_{223a} = k_{223b} = 0.6$
$k_{222} = -12.1$	$k_{23a3a} = k_{23b3b} = 96.5$
$k_{3a3a3a} = k_{3b3b3b} = 137.5$	$k_{24a4a} = k_{24b4b} = -66.9$
$k_{4a4a4a} = k_{4b4b4b} = 1.4$	$k_{3a3a4a} = k_{3b3b4b} = -30.0$
$k_{112} = 93.4$	$k_{3a4a4a} = k_{3b4b4b} = 4.7$
$k_{122} = -6.1$	$k_{123a} = k_{123b} = -3.9$
$k_{13a3a} = k_{13b3b} = -601.4$	$k_{13a4a} = k_{13b4b} = 47.5$
$k_{14a4a} = k_{14b4b} = -23.9$	$k_{23a4a} = k_{23b4b} = -38.2$
$k_{1111} = 16.5$	$k_{24a4a4a} = k_{24b4b4b} = 3.4$
$k_{2222} = -1.5$	$k_{3a3a3a4a} = k_{3b3b3b4b} = -5.9$
$k_{3a3a3a3a} = k_{3b3b3b3b} = 23.5$	$k_{3a3a4a4a} = k_{3b3b4b4b} = -1.3$
$k_{4a4a4a4a} = k_{4b4b4b4b} = 0.2$	$k_{3a4a4a4a} = k_{3b4b4b4b} = 0.8$
$k_{1112} = -12.2$	$k_{113a4a} = k_{113b4b} = -11.3$
$k_{1122} = 8.1$	$k_{13a3a4a} = k_{13b3b4b} = 7.9$
$k_{113a3a} = k_{113b3b} = 97.4$	$k_{123a3a} = k_{123b3b} = -19.2$
$k_{1222} = 3.3$	$k_{124a4a} = k_{124b4b} = 2.5$
$k_{13a3a3a} = k_{13b3b3b} = -45.2$	$k_{13a4a4a} = k_{13b4b4b} = -0.4$
$k_{14a4a4a} = k_{14b4b4b} = -0.3$	$k_{223a4a} = k_{223b4b} = 12.3$
$k_{223a3a} = k_{223b3b} = 3.1$	$k_{23a3a4a} = k_{23b3b4b} = -0.5$
$k_{224a4a} = k_{224b4b} = -3.3$	$k_{23a4a4a} = k_{23b4b4b} = 1.0$
$k_{23a3a3a} = k_{23b3b3b} = 8.3$	$k_{123a4a} = k_{123b4b} = 5.6$

Box 4.1. *Values of anharmonic vibrational force constants*

The corrected eigenvectors of the fundamental level and the first excited levels expressed in terms of harmonic eigenvectors are written as:

$$\begin{aligned}|000000\rangle = {} & 0.978\cdot|000000\rangle_0 + 0.189\cdot|100000\rangle_0 - 0.040\cdot|001000\rangle_0 \\ & -0.040\cdot|000100\rangle_0 - 0.037\cdot|010000\rangle_0 + 0.027\cdot|102000\rangle_0 \\ & +0.027\cdot|100200\rangle_0 + 0.018\cdot|300000\rangle_0 - 0.015\cdot|200000\rangle_0 \\ & -0.012\cdot|002000\rangle_0 - 0.012\cdot|000200\rangle_0 - 0.011\cdot|003000\rangle_0 \\ & -0.011\cdot|000300\rangle_0\end{aligned}$$

$$\begin{aligned}|100000\rangle = {} & 0.858\cdot|100000\rangle_0 + 0.312\cdot|200000\rangle_0 + 0.194\cdot|001000\rangle_0 \\ & +0.194\cdot|000100\rangle_0 - 0.166\cdot|000000\rangle_0 - 0.086\cdot|110000\rangle_0 \\ & +0.069\cdot|002000\rangle_0 + 0.069\cdot|000200\rangle_0 - 0.035\cdot|101000\rangle_0 \\ & -0.035\cdot|100100\rangle_0 + 0.034\cdot|202000\rangle_0 + 0.034\cdot|200200\rangle_0 \\ & +0.030\cdot|400000\rangle_0 - 0.028\cdot|300000\rangle_0 - 0.018\cdot|102000\rangle_0 \\ & -0.018\cdot|100200\rangle_0 - 0.009\cdot|103000\rangle_0 - 0.009\cdot|100300\rangle_0\end{aligned}$$

$$\begin{aligned}|010000\rangle = {} & 0.977\cdot|010000\rangle_0 + 0.190\cdot|110000\rangle_0 - 0.040\cdot|011000\rangle_0 \\ & -0.040\cdot|010100\rangle_0 + 0.037\cdot|000000\rangle_0 - 0.036\cdot|020000\rangle_0 \\ & +0.027\cdot|112000\rangle_0 + 0.027\cdot|110200\rangle_0 + 0.018\cdot|310000\rangle_0 \\ & -0.016\cdot|210000\rangle_0 + 0.014\cdot|000020\rangle_0 + 0.014\cdot|000002\rangle_0 \\ & -0.012\cdot|012000\rangle_0 - 0.012\cdot|010200\rangle_0 - 0.011\cdot|013000\rangle_0 \\ & -0.011\cdot|010300\rangle_0 + 0.011\cdot|020020\rangle_0 + 0.011\cdot|020002\rangle_0\end{aligned}$$

$$\begin{aligned}|001000\rangle = {} & 0.919\cdot|001000\rangle_0 + 0.290\cdot|101000\rangle_0 - 0.208\cdot|100000\rangle_0 \\ & -0.106\cdot|002000\rangle_0 - 0.094\cdot|011000\rangle_0 + 0.045\cdot|103000\rangle_0 \\ & +0.038\cdot|000000\rangle_0 - 0.038\cdot|001100\rangle_0 - 0.026\cdot|003000\rangle_0 \\ & +0.026\cdot|101200\rangle_0 - 0.023\cdot|201000\rangle_0 - 0.020\cdot|004000\rangle_0 \\ & +0.017\cdot|001010\rangle_0 + 0.017\cdot|301000\rangle_0 - 0.015\cdot|010010\rangle_0 \\ & +0.013\cdot|102000\rangle_0 - 0.011\cdot|001200\rangle_0 + 0.010\cdot|000020\rangle_0 \\ & -0.010\cdot|100010\rangle_0 - 0.010\cdot|001300\rangle_0 - 0.009\cdot|013000\rangle_0\end{aligned}$$

$$
\begin{aligned}
|000100\rangle = {} & 0.919\cdot|000100\rangle_0 + 0.290\cdot|100100\rangle_0 - 0.208\cdot|100000\rangle_0 \\
& - 0.106\cdot|000200\rangle_0 - 0.094\cdot|010100\rangle_0 + 0.045\cdot|100300\rangle_0 \\
& + 0.038\cdot|000000\rangle_0 - 0.038\cdot|001100\rangle_0 - 0.026\cdot|000300\rangle_0 \\
& + 0.026\cdot|102100\rangle_0 - 0.023\cdot|200100\rangle_0 - 0.020\cdot|000400\rangle_0 \\
& + 0.017\cdot|000101\rangle_0 + 0.017\cdot|300100\rangle_0 - 0.015\cdot|010001\rangle_0 \\
& + 0.013\cdot|100200\rangle_0 - 0.011\cdot|002100\rangle_0 + 0.010\cdot|000002\rangle_0 \\
& - 0.010\cdot|100001\rangle_0 - 0.010\cdot|003100\rangle_0 - 0.009\cdot|010300\rangle_0
\end{aligned}
$$

$$
\begin{aligned}
|000010\rangle = {} & 0.978\cdot|000010\rangle_0 + 0.194\cdot|100010\rangle_0 - 0.041\cdot|001010\rangle_0 \\
& - 0.040\cdot|000110\rangle_0 + 0.027\cdot|102010\rangle_0 + 0.027\cdot|100210\rangle_0 \\
& + 0.018\cdot|300010\rangle_0 - 0.015\cdot|200010\rangle_0 + 0.013\cdot|010030\rangle_0 \\
& - 0.012\cdot|002010\rangle_0 - 0.012\cdot|000210\rangle_0 - 0.011\cdot|003010\rangle_0 \\
& - 0.011\cdot|000310\rangle_0
\end{aligned}
$$

$$
\begin{aligned}
|000001\rangle = {} & 0.978\cdot|000001\rangle_0 + 0.194\cdot|100001\rangle_0 - 0.041\cdot|000101\rangle_0 \\
& - 0.040\cdot|001001\rangle_0 + 0.027\cdot|102001\rangle_0 + 0.027\cdot|100201\rangle_0 \\
& + 0.018\cdot|300001\rangle_0 - 0.015\cdot|200001\rangle_0 + 0.013\cdot|010003\rangle_0 \\
& - 0.012\cdot|002001\rangle_0 - 0.012\cdot|000201\rangle_0 - 0.011\cdot|003001\rangle_0 \\
& - 0.011\cdot|000301\rangle_0
\end{aligned}
$$

5

Spherical Top CH_4

The theoretical study of methane CH_4, a spherical top molecule, is not easy. The most complete theoretical model relies on the group theory methods, which use the tensor formalism developed by teams of researchers in Dijon, France. The subject of spherical tops would require more than one chapter to cover the various theoretical models elaborated for the calculation of the energy levels and of the transitions between these levels. The approaches described here focus, in particular, on the rotational Hamiltonian proposed by Watson, and gives an overview of the Dijon model for gas phase methane IR spectroscopy. The procedures to find the selection rules within the tensor formalism are presented. These descriptions are addressed to non-specialists in spectroscopy who devise specific apparatus and use adequate connected technologies for space observation and exploration. Based on the theoretical model developed for gas phase study, applied to the lowest order in the expansion of the vibration–rotation Hamiltonian, the effect of a low temperature environment on the spectroscopic characteristics of methane CH_4 in various media can be studied. From the symmetry considerations presented in the first three chapters, the transitions from the fundamental level at low temperature can be studied and the probable degeneracy lifting can be predicted. Group theory methods are used to determine the effects on the IR spectroscopy of the molecule when it evolves in a medium characterized by a non-spherical symmetry, in contrast to the spherical environment of an isolated molecule in gas phase.

5.1. Introduction

Chapter 2 of Volume 2 [DAH 19] described the methods and the approximations applied for the study of infrared spectroscopy of polyatomic molecules. The vibration–rotation energies are calculated from an effective Hamiltonian by applying a series of Van Vleck contact transformations (section 2.2.6, Volume 1 [DAH 17]) to the Hamiltonian proposed by Darling and Dennison and simplified by Watson (see equation [2.7], Volume 2 [DAH 19]):

$$H_N = \frac{1}{2}\sum (J_\alpha - p_\alpha)\mu_{\alpha\beta}(J_\beta - p_\beta) + T_{vib} + V_{vib} + U_{Watson}$$

This method uses the eigenfunctions of the zeroth-order Hamiltonian $H^0_{\text{vib}} + H^0_{\text{rot}}$ to calculate the corrections that need to be applied to the eigenvalues of this Hamiltonian when including the higher-order transformed terms to calculate the eigenenergies of the effective Hamiltonian (see equation [2.67], Volume 1):

$$H^{(n)} = H_0^{(0)} + \lambda H_1^{(1)} + \lambda^2 H_2^{(2)} + \ldots + \lambda^n H_n^{(n)}$$

As discussed in Chapter 2, this method can be used to cast the Hamiltonian into interacting polyads by means of unitary transformations, which leads to treating the Hamiltonian in a block matrix form, which is easier to diagonalize for the calculation of energies. The transformation of the initial Hamiltonian into interacting blocks allows the connection of experimental observations to the interacting vibrational levels. According to the calculation conducted by Amat, Nielsen and Goldsmith [AMA 57a, AMA 57b, AMA 58], the Hamiltonian can be expressed as a function of dimensionless normal coordinates up to the fourth order, as recalled in the Appendix.

The central hypothesis is to consider the non-diagonal elements of ΔH_{VR} of the matrix $H_{VR} = H_{Vib} + H_{Rot} + \Delta H_{VR}$ to be small enough so that it may be diagonalized by successive approximations by applying, for example, Jacobi or Gauss–Seidel methods. Assuming that the rotational energy levels of the same non-degenerate state are quasi-degenerate, with respect to the energy gaps between the vibrational levels, a subspace of quasi-degeneracy is assigned to each vibrational state. In this case, the non-diagonal elements couple these spaces that are in quasi-degeneracy. When several vibrational modes are present, the subspaces composed of resonant vibrations should also be considered. The set of these vibrational modes form a quasi-degenerate subspace, termed as a resonance block.

Moreover, in the case of polyatomic molecules such as methane, constituted of two triply degenerate vibrational modes, the coupling between the internal vibrational angular momentum and the overall rotational angular momentum of the molecule and also between the internal vibrational angular momenta should be considered. The first formalism that takes the coupling between the vibrational angular momentum and the rotation into consideration dates back to the works of Jahn and Childs [CHI 38, JAH 38a, JAH 38b], who addressed the problem of orientation of the bases of the two representations, the one related to the vibration and the one related to the rotation. Schaffer *et al.* [SCH 39a, SCH 39b] consequently developed a complete expression of the vibration–rotation Hamiltonian. Nevertheless, in order to be used, this expression had to be simplified, and this led to the development of tensor formalism.

The spectroscopic characteristics of the CH_4 molecule can be found in various works such as that of Herzberg [HER 45], which is one of the oldest, or the more recent work by Bunker and Jensen [BUN 98], which considers the spectroscopic study within the framework of the permutation-inversion group. Among the works focusing on this molecule, it is worth mentioning those of Hougen [HOU 76, HOU 86] on the classifications of energy levels [HOU 63] and on the ortho–para transitions in the methane [HOU 71], and also those of Watson and collaborators [ALI 76, ALI 86, DOR 72, KIR 73, WAT 67, WAT 71, WAT 75] on the rotational Hamiltonian. The works of Watson were described in Chapter 3 (see equation [3.67]) in relation to the rotation of polyatomic molecules. Watson showed that the rotational Hamiltonian could be built from operators J_x, J_y and J_z using a purely algebraic approach relying on the commutation relations. It provides the expected form of the rotational Hamiltonian at various orders of power expansion of these operators, taking into account the properties of the Hamiltonian (invariance under the symmetry operations of the overlapping group of the molecule, hermiticity and invariance under a time-reversal operation). It is important to recall that the operators J_x, J_y and J_z verify the commutation relations, which are such that two of the operators generate a third operator (the rules of commutation of angular moment operators that are generators of infinitesimal rotations). The set of these three operators constitutes a first rank Lie algebra, as there is only one operator J^2 that commutes with each of the operators J_i $(i = x, y, z)$ and constitutes the Casimir operator. If J_z is chosen as one of the three operators that commute with J^2, a complete set of observables that commute can be formed. Their eigenfunctions are common to the two operators J^2 and J_z and are generally denoted by $|JM\rangle$. Chapter 8 (sections 8.8–8.14) of Hamermesch's book [HAM 62] describes the construction of irreducible representations of Lie groups and Lie algebras. For a better understanding of the use of Lie groups and Lie algebras, the reader can refer to specialist literature. For example, the work of Lipkin [LIP 65] is very accessible as an introduction and the work of Gilmore [GIL 74] is more specialized, offering a more in-depth approach.

The tensor formalism [CHA 92] relies on an algebraic approach for the setting up of the vibration–rotation Hamiltonian of the CH_4 molecule. It is the most comprehensive approach to studying this molecule in gas phase and for potentially very high temperatures. This formalism was developed by teams of researchers in Dijon, France [CHA 92], and originates from the initial works conducted by Hecht [HEC 60a] and Moret-Bailly [MOR 61]. It will be described in section 5.3.

It is important to note that a formalization of this approach can be found in the works of Iachello and Levine [IAC 81, IAC 82, IAC 93, IAC 94] on the vibration–rotation spectroscopy, in particular, for the diatomic molecule, or the work of Van Rooasmalen [ROO 83] on the triatomic molecule. In this case, the Hamiltonian H is expressed in the form of operators that are the elements of an algebra G termed as spectrum generating algebra (SGA), the Hamiltonian H being an enveloping algebra of G, such that:

$$H = E_0 + \sum_{\alpha} \lambda_\alpha G_\alpha + \sum_{\alpha\beta} \mu_{\alpha\beta} G_{\alpha\beta} + \ldots$$

where $G_\alpha \in G$ and $\lambda_\alpha, \mu_{\alpha\beta}, \ldots$ are the algebraic parameters and $G_\alpha, G_{\alpha\beta}, \ldots$, are the algebraic operators. In certain situations, H only contains a certain number of elements of G, the Casimir operators that are invariants of G and one of its sub-algebras belonging to a chain such that $G \supset G' \supset G'' \supset \ldots$, H only contains a certain number of elements G. In algebraic theory, the eigenvalues of H are obtained in an analytical form in terms of quantum numbers characterizing the representations $G \supset G' \supset \ldots$, the eigenvalues of H. According to the works of Iachello *et al.*, the boson model [ARI 75] and the vibron model [IAC 81], the quantum mechanics problems in a d-dimensional space can be solved in a $d+1$ -dimensional space algebra $U(d+1)$, with its states assigned to the fully symmetric representation of $U(d+1)$ [IAC 93]. It can be noted that the vibron model is used by Choudhury [CHO 13] to study the stretching vibration spectrum of certain tetrahedral molecules XY_4. The theoretical model relies on the isomorphism of the Lie algebra U(2) and the one-dimensional Morse oscillator, whose eigenstates can be associated with states $U(2) \supset O(2)$ to describe the stretching modes A–B.

It is important to note that for the observations of methane spectroscopy at high temperature, in atmospheres spanning from that of Jupiter, from exoplanets to brown dwarfs [BEA 11, DIN 97, GEB 96, SWA 08, YUR 13], collaboration between research groups at national or international level is required in order to master all the aspects, ranging from the observation to the interpretation of spectra. References concerning methane spectroscopy can be found in the excellent synthesis on methane spectroscopy and its applications to the atmosphere of the planets, including that of the Earth, by Coustenis and Boudon in [COU 13].

The following section provides a brief review of the characteristics of the CH_4 molecule, followed by an overview of the tensor formalism applied to the study of methane, after giving the essentials allowing a non-specialist to acquire knowledge on the tensor approach concept. An example of the treatment of the interaction between the vibrational and rotational degrees of freedom of the molecule is provided and the procedure followed to build the vibration–rotation Hamiltonian is described. The applications will be the focus of Part 2 of Volume 4.

5.2. Characteristics of the CH_4 molecule in gas phase

The study of methane presents an interest in astrophysics for the analysis of the planetary atmosphere, earth pollution and climate heating. Methane belongs to the symmetry group T_d, as shown in Chapter 2, and has 15 degrees of freedom for the vibrational, translational and rotational motions. It is characterized by nine vibrational coordinates, as shown in Chapter 2, but only four different fundamental frequencies: a non-degenerate frequency of type A_1 denoted by ν_1 (stretching mode, 2916.50 cm^{-1}), a doubly degenerate frequency of type E denoted by ν_2 (bending mode, 1533.33 cm^{-1}) and two triply degenerate frequencies, both of symmetry type T_2 denoted by ν_3 (stretching mode, 3019.49 cm^{-1}) and ν_4 (bending mode, 1310.76 cm^{-1}) (see Figure 5.1 a, b, c and d).

The three translational degrees of freedom through the three components of the translational vector constitute a basis of representation of third degree and of symmetry type T_2. The rotational degrees of freedom through the three components of the rotational vector constitute a basis of another representation of third degree and of symmetry type T_1. Table 5.1 summarizes these symmetry properties of the methane molecule.

Vector components or normal coordinates	Symmetry types
(T_x, T_y, T_z)	T_2
(R_x, R_y, R_z)	T_1
q_1	A_1
$q_{2\sigma}$ $(\sigma = 1, 2)$	E
$q_{3\sigma}$ $(\sigma = 1, 2, 3)$	T_2
$q_{4\sigma}$ $(\sigma = 1, 2, 3)$	T_2

Table 5.1. *Symmetry types of the normal vibration coordinates of the translational polar vector and of the rotational axial vector in group T_d*

The fundamental vibrational frequencies of CH_4 approximately verify the relation $\nu_1 \cong 2\nu_2 \cong \nu_3 \cong 2\nu_4$. Consequently, the vibrational states of methane can be grouped in packets n termed as polyads P_n. Each polyad P_n is identified by the integer n given by: $n = 2(v_1 + v_3) + v_2 + v_4$, where $v_i = 0, 1, 2\ldots$, $i = 1$ to 4 are the vibrational quantum numbers. P_0 is the fundamental state or a monad. Each set (v_1, v_2, v_3, v_4) defines a vibrational level. The fundamental level is characterized by $v_i = 0$ for $i = 1$ to 4, which leads to $n = 0$, hence the use of P_0 to identify the fundamental state also termed as the "ground state" (GS). Thus, for $v_2 = 1$, with the other v_i being set to zero ($i \neq 2$) or $v_4 = 1$, and the other v_i being set to zero ($i \neq 4$), there is the dyad (ν_2, ν_4) denoted by P_1. The pentad is identified by P_2 ($2\nu_4$, $\nu_2 + \nu_4$, ν_1, $2\nu_2$, ν_3) when $v_1 = 1$,

$v_2 = 2$, $v_3 = 1$, $v_4 = 2$, and as a combination, $v_2 = 1$ and $v_4 = 1$, with the other v_i being set to zero. The octad is denoted by P_3 ($3v_4$, $v_2 + 2v_4$, $v_1 + v_4$, $v_3 + v_4$, $2v_2 + v_4$, $v_1 + v_2$, $v_2 + v_3$, $3v_2$). In order to build the rovibrational Hamiltonian of tetrahedral molecules, the complete group of rotations O(3) or K_h must be employed, as well as the coupling coefficients of the chain of group O(3) $\supset T_d$.

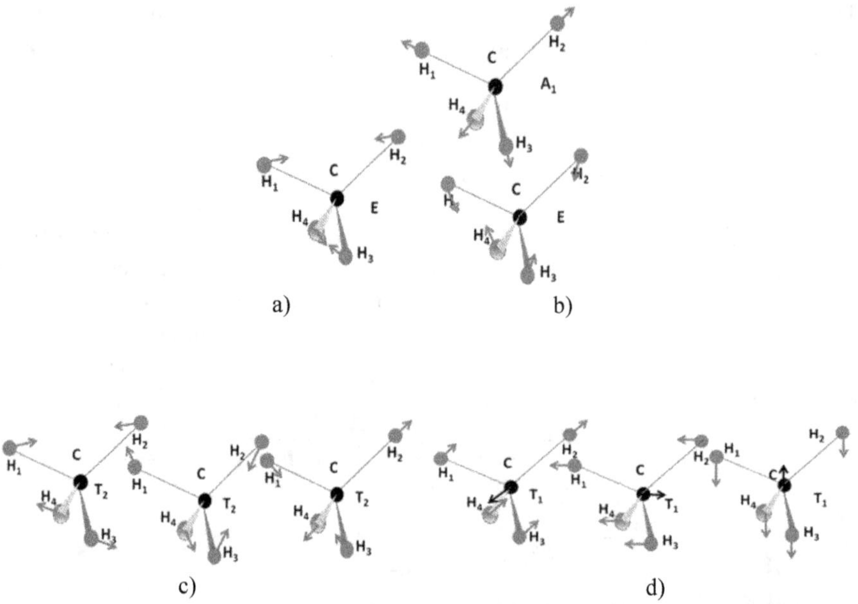

Figure 5.1. *Normal modes of vibration of CH_4: a) stretching ν_1 of type A_1; b) bending ν_2 of type E; c) stretching ν_3 of type T_2 (F_2); and d) bending ν_4 of type T_2 (F_2). For a color version of this figure, see www.iste.co.uk/dahoo/infrared.zip*

In the case of methane, an example of a spherical top, it can be shown from the Darling–Dennison Hamiltonian that $H_0^{(0)}$, the zeroth-order Hamiltonian is a sum of a vibrational Hamiltonian H_{vib}^0 (nine harmonic oscillators, one non-degenerate (ν_1), one doubly degenerate (ν_2) and two triply degenerate (ν_3, ν_4)) and of a rotational Hamiltonian H_{rot}^0 (rigid rotator).

The eigenfunctions of the rigid rotator depend on three quantum numbers J, K and M and are expressed as a function of Euler angles:

$$\langle \theta, \varphi, \chi | \Psi_R^0 \rangle = \Psi_R^0(\theta, \varphi, \chi) = \mathrm{N}_{JKM} \Theta_{JKM}(\theta) e^{iK\chi} e^{iM\varphi} \text{ (see equation [2.7])},$$

with:

$$\Theta_{JKM}(\theta) = t^{d/2}(1 - t)^{s/2}G_p(1 + d + s, 1 + d, t)$$

where G_p is a Jacobi polynomial and

$$t = \frac{1}{2}(1 - \cos\theta);\ d = |M - K|; s = |M + K|; p = J - \frac{1}{2}(d + s),$$

and

$$N_{JKM} = |(2p + d + s + 1)(p + d + s)!\,(p + d)!/4\pi p!\,d^2!\,(p + s)!|^{1/2}.$$

For each value of J, the quantum numbers K and M take $(2J + 1)$ positive, negative or zero integer absolute values, at most, equal to J.

The wave functions Ψ_1 of the non-degenerate oscillator are expressed as a function of Hermite polynomials $H_{v_1}(Q_1)$ that depend on the normal coordinate variable Q_1 and on the quantum number v_1:

$\Psi_1 = \left[\frac{1}{\sqrt{\pi}2^{v_1}v_1!}\right]^{\frac{1}{2}} e^{-\frac{Q_1}{2}} H_{v_1}(Q_1)$ (see section 2.2.2, Volume 1 and section 2.3, Volume 2 [DAH 17, DAH 19]).

The wave functions Ψ_2 of the doubly degenerate oscillator are expressed as a function of polar variables r_2 and χ_2 in the two-dimensional configuration space of normal coordinates Q_{21} and Q_{22}, such that $Q_{21} = r_2\cos(\chi_2)$ and $Q_{22} = r_2\sin(\chi_2)$:

$$\Psi_2 = N_{v_2,l_2} r_2^{l_2} e^{-\frac{r_2^2}{2}} L_{\tau_2}^{\frac{l_2}{2}}(r_2^2) e^{\pm i l_2 \chi_2},$$

where $\tau_2 = \frac{1}{2}(v_2 + l_2)$, N_{v_2,l_2} is a normalization coefficient and $L_{\tau_2}^{l_2}(r_2^2)$ is an associated Laguerre polynomial (section 2.3.2, Volume 2 [DAH 19]).

The wave functions Ψ_3 and Ψ_4 $(\Psi_s, s = 3, 4)$, of triply degenerate oscillators are expressed as a function of variables r_s, θ_s and φ_s in the three-dimensional configuration space of normal coordinates Q_{s1}, Q_{s2} and Q_{s3} such that:

$Q_{s1} = r_s\sin(\theta_s)\cos(\varphi_s)$, $Q_{s2} = r_s\sin(\theta_s)\sin(\varphi_s)$ and $Q_{s3} = r_s\cos(\theta_s)$,

with:

$$\Psi_s = N_{v_s,l_s} r_s^{l_s} e^{-\frac{r_s^2}{2}} L_{\tau_s+\frac{1}{2}}^{l_s+\frac{1}{2}}(r_s^2) P_{m_s}^{l_s}(\cos(\theta_s)) e^{\pm i m_s \varphi_s},$$

where $\tau_s = \frac{1}{2}(v_s + l_s)$, N_{v_s,l_s} is a normalization coefficient, $L_{\tau_s+\frac{1}{2}}^{l_s+\frac{1}{2}}(r_s^2)$ is an associated Laguerre polynomial and $P_{m_s}^{l_s}(cos(\theta_s))$ is a Legendre polynomial. The values of v_s (s = 1, 2, 3, 4) are positive integers or zero, l_s is a positive integer or zero, at most, equal to v_s and of the same parity as l_s and m_s is a relative integer or zero such that $m_s = -l_s, -l_s + 1, \ldots l_s - 1, l_s$ ($2l_s + 1$ values).

With the zeroth-order rovibrational wave function being the product of vibrational functions $\Psi_{\mathrm{v}}^0 = \Psi_1(\mathrm{v}_1)\Psi_2(\mathrm{v}_2,\mathrm{l}_2)\Psi_3(\mathrm{v}_3,\mathrm{l}_3,\mathrm{m}_3)\Psi_4(\mathrm{v}_4,\mathrm{l}_4,\mathrm{m}_4)$ and rotational functions $\Psi_R^0(\theta,\varphi,\chi) = \Psi_R^0(J,K,M)$, it depends on the rotational quantum numbers J, K and M and on the vibrational quantum numbers $\mathrm{v}_s, \mathrm{l}_s, \mathrm{m}_s$ (s = 1, 2, 3, 4 for v, s = 2, 3, 4 for l and s = 3, 4 for m).

5.3. Tensor formalism for the CH_4 molecule

As already underlined in the introduction, the tensor formalism developed by research teams in Dijon, France [CHA 92] stems from the initial works of Hecht [HEC 60a] and Moret-Bailly [MOR 61]. Hecht and Moret-Bailly express the vibration–rotation Hamiltonian operator $\widehat{H}_{VR}$ in the form of irreducible spherical tensors based on group theory tools. The diagonalization of the resulting operator leads to the wave functions and energy levels that are used to interpret the vibration–rotation spectra. Hecht [HEC 60b] extends the study of $\widehat{H}_{VR}$ through an expansion up to the third order, including the anharmonic vibrational terms. Moret-Bailly [MOR 61] builds a fourth-order quasi-diagonal formal transformed Hamiltonian with respect to vibrational quantum numbers v_s by developing the theory of spherical tensors adapted to the cubic symmetry. This approach involves building the operators and wave functions by appropriate coupling in the complete rotational group O(3). For example, the Moret-Bailly basic functions of the zeroth-order Hamiltonian can be used to introduce the coupling of rotational and vibrational angular momenta, which can be written as:

$$\Psi_{nC\sigma}^{(R_g)} = \left(\Psi_{Rot}^{(J_u)} \otimes \Psi_{Vib}^{(1_u)}\right)_{nC\sigma}^{(R_g)} \quad [5.1]$$

where $C = A_1, A_2, E, T_1\,(F_1), T_2\,(F_2)$, σ is the multiplicity index, i.e. the components of a degenerate representation basis (σ = 1, 2 or 3) and n (= 0, 1, 2, …) is the number of symbols that distinguish between the various bases of equivalent irreducible representations in a cubic tensor. For example, referring to Table 1.13 of Chapter 1, when $J = 4$, $n = 1$ and $\sigma = 1$ for $C = A_1$, $n = 1$ and $\sigma = 2$ for $C = E$ and $n = 1$ and $\sigma = 3$ for $C = T_1\,(F_1), T_2\,(F_2)$, $n = 0$ and $\sigma = 1$ for $C = A_2$. For $J = 5$, $n = 0$ for $C = A_1$ and A_2, $n = 1$ for $C = E$ and $T_2\,(F_2)$, and $n = 2$ for $C = T_1\,(F_1)$.

The symbols u and g relate to the parity in O(3) (the base level is denoted by g by convention and the excited level is denoted by u). R_g refers to the overall angular rotational moment of the molecule.

The tensor method developed by Moret-Bailly is applicable to the Hamiltonian transformation in such a way that it is diagonal with respect to the vibrational quantum numbers v_s. It can be used to obtain a diagonal or quasi-diagonal matrix form of the Hamiltonian. To solve the problem of interactions between vibrational bands, Champion [CHA 77, CHA 78, CHA 80] proposed a systematic general method for the construction of vibration–rotation Hamiltonian operators.

Using a non-symmetric coupling diagram in group T_d, he determined all the vibration–rotation operators of the Hamiltonian of molecules of type XY_4, including all the possible interactions, up to an arbitrary approximation order. A basis of Hamiltonian matrices was defined, whose matrix elements are only a function of coupling coefficients of the chain of group $O(3) \supset T_d$. A general formalism available for all the vibrational sub-levels irrespective of the symmetry type was developed. The tensor formalism in the overlapping group of the molecule, the tetrahedral group T_d for methane can thus be used to deal with the interactions between vibrational bands.

Using the same method for building tensor operators, Loëte [LOË 84] completes this work on the Hamiltonian, with the formal expansion of the dipole moment of tetrahedral molecules. A detailed description of the method can be found in reference [CHA 92] and the references therein. A summary of these models can be found in various references [ACE 00, BOU 04, CHE 99, NIK 03, NIK 13].

It is important to note that the eigenfunctions of $H_0^{(0)}$, similarly to the wave functions of H, depend on the vibrational quantum numbers v_s, l_s and m_s and on the rotational quantum numbers J, K and M. The central point of the tensor formalism developed in Dijon, France [CHA 92] is in the use of the symmetry properties within the framework of group theory, in order to write basis functions and operators. This involves the symmetrization of the angular momentum functions $|J_\alpha, \mathrm{m}_i\rangle$ ($\alpha = g$ or u the parity property defined from the inversion group (gerade or ungerade)) and the spherical tensors T_q^K of group SO(3) in the molecular point group.

If G is the point group of a molecule, a wider group, denoted by G' (G is a subgroup of G') is generally considered, for example the group SO(3) for the rotator. A chain of groups of type $G' \supset G$ is introduced. The tensor formalism adapted to the study of the spherical tops of type XY_4, such as methane CH_4, involves the use of the chain of groups $O(3) \supset T_d$. This requires:

– orienting the base of O(3) in the subgroup T_d (see section 1.7);

– building the tensor algebra adapted to the symmetry group T_d;

– symmetrizing the Hamiltonian operators and the transition moments;

– and finally, calculating their matrix elements in the appropriate bases.

Note that it is possible to only consider group T, given that group T_d can be deduced from T simply by adding the inversion operator. Consequently, a simpler approach is to carry out the study using the chain of group SO(3) ⊃ T.

5.3.1. *Orientation of SO(3) in* T$_d$

To determine the orientation of SO(3) within group T_d, the projector method can be used, starting from relation:

$$|J, nC\sigma\rangle = \frac{1}{\sqrt{N}} P_\sigma^{(C)} |J, M\rangle \qquad [5.2]$$

where:

– C is an irreducible representation of subgroup T_d, which, as shown in Chapters 1 and 2, refers to the irreducible representations of dimension 1, A_1 and A_2, of dimension 2, E and of dimension 3, T_1 and T_2 (or F_1 and F_2);

– n is the multiplicity index (of C in $D^{(J)}$);

– σ is a component of a degenerate basis.

This notation is the one used by Moret-Bailly [MOR 61].

Based on the definition of a projector (equation [1.26]):

$$P_{ij}^{\alpha} = \frac{l_\alpha}{h} \sum_{O_k \in G} \left(\Gamma_{ij}^{\alpha}(O_k)\right)^* O_k$$

projectors $P_\sigma^{(C)}$ can be built. They are expressed in the group SO(3), by the relation [HAM 62]:

$$P_\sigma^{(C)} = \frac{l_C}{h} \sum_{O_k \in T_d} \left(\Gamma_{ij}^{C}(O_k)\right)^* O_k \qquad [5.3]$$

where:

– l_C is the dimension of the irreducible representation C;

– h is the order of the subgroup;

– O_k is the symmetry element of the subgroup of SO(3) contained in T_d;

– $\Gamma_{ij}^{C}(O_k)$ is the matrix of the element O_k in the irreducible representation C of group T_d.

Since the tables of characters do not give the matrices $\Gamma_{ij}^{C}(O_k)$, but only the trace (sum of the diagonal elements) of these matrices (see equation [1.27]), a simplified projection operator is used. This operator is given as a function of characters as:

$$P^{(C)} = \frac{l_C}{h} \sum_{O_k \in T_d} \left(\chi^{(C)}(O_k)\right)^* O_k \qquad [5.4]$$

In the group SO(3), it can be written that:

$$P^{(C)} |J,M\rangle = \frac{l_C}{h} \sum_{O_k \in T_d} \left(\chi^{(C)}(O_k)\right)^* O_k |J,M\rangle \qquad [5.5]$$

with

$$\mathrm{O_k}|J,M\rangle = \sum_{M'} D^{J}_{M,M'}(\mathrm{O_k})\, |J,M'\rangle$$

where:

– $D^{J}_{M,M'}(O_k)$ is the Wigner matrix, i.e. the matrix of element O_k in SO(3);

– $\chi^{(C)}(O_k)$ is the trace of the element O_k in the irreducible representation C of subgroup T_d.

In group T_d, the following relation can be written:

$$P^{(C)} |\Psi,C\rangle = \frac{l_C}{h} \sum_{O_k \in T_d} \left(\chi^{(C)}(O_k)\right)^* O_k |\Psi,C\rangle \qquad [5.6]$$

with

$$\mathrm{O_k}|\Psi,C\rangle = \sum_{C} \left(\Gamma_{ij}^{(C)}(\mathrm{O_k})\right)^* |\Psi,C\rangle$$

where:

– $\left(\Gamma_{ij}^{(C)}(O_k)\right)^*$ is the matrix of element O_k in the irreducible representation C of group T_d;

– $\chi^{(C)}(O_k)$ is the trace of the element O_k in the irreducible representation C of subgroup T_d.

For example, in relation to Table 5.2 (see Table 2.11), which is given in this chapter for the sake of consistency, for a function $|\Psi, A_1\rangle = x^2 + y^2 + z^2$:

$T_d(M)$	E	(123)	(14) (23)	(1423)*	(23)*	Vectors	Tensors
Classes	1	8	3	6	6		
T_d	E	$8C_3$	$3C_2$	$6S_4$	$6\sigma_d$		
A_1	1	1	1	1	1		$x^2 + y^2 + z^2$ $\alpha_{xx} + \alpha_{yy} + \alpha_{zz}$
A_2	1	1	1	-1	-1		
E	2	-1	2	0	0		$(2z^2 - x^2 - y^2, x^2 - y^2)$ $(2\alpha_{zz} - \alpha_{xx} - \alpha_{yy}, \alpha_{xx} - \alpha_{yy})$
T_1	3	0	-1	1	-1	R_x, R_y, R_z	
T_2	3	0	-1	-1	1	T_x, T_y, T_z	(xy, xz, yz) $(\alpha_{yz}, \alpha_{zx}, \alpha_{xy})$

Table 5.2. *Table of characters of group T_d (M) for CH_4*

$$P^{(A_1)}|\Psi, A_1\rangle = \frac{1}{24}(1 \times |\Psi, A_1\rangle + 8 \times 1 \times |\Psi, A_1\rangle + 3 \times 1 \times |\Psi, A_1\rangle + 6 \times 1 \times |\Psi, A_1\rangle + 6 \times 1 \times |\Psi, A_1\rangle) = |\Psi, A_1\rangle \quad [5.7]$$

Hence, any general wave function $|\Psi\rangle$ can be decomposed in a sum of projections such that: $|\Psi\rangle = |\Psi, A_1\rangle + |\Psi, A_2\rangle + |\Psi, E\rangle + |\Psi, T_1\rangle + |\Psi, T_2\rangle$, where $|\Psi, C\rangle = P^{(C)}|\Psi\rangle$.

It is possible not to use the projection method, as it presents a problem for the definition of the multiplicity index, as seen in the Dijon group [REY 03]. The orientation matrix G is determined by the diagonalization of a rank 4 generic operator, which is completely symmetric in T_d.

5.3.2. *Vibrational tensor operators*

The vibration of a polyatomic molecule can be dealt with more generally, when the number of nuclei of the molecule becomes significant. This approach differs from the method used for diatomic and triatomic molecules, as described in the first two volumes [DAH 17, DAH 19]. The method adapted to the case of isotropic oscillators in a p-dimensional space can be found in section III of Chapter XII of the book by Messiah [MES 64]. Each of the vibrational spaces ε_i (i = 1, 2 ,..., p) is linked to canonical variables q_i and p_i so that the space in which the p harmonic oscillators evolve is given by the tensor product of spaces, such that $\varepsilon = \prod_{\otimes i} \varepsilon_i = \varepsilon_1 \otimes \varepsilon_2 \ldots \otimes \varepsilon_p$. In this case, a Hamiltonian H_i can be associated with each subspace so that $H = \sum_i H_i = \sum_i \frac{1}{2m}(p_i^2 + m^2\omega_i^2 q_i^2)$, where $|n_i\rangle$ are the eigenvectors of each Hamiltonian H_i with i given and $n_i = 0, 1, 2, \ldots, \infty$. The eigenvalue equation of such a Hamiltonian is written as:

$$H|n_1 \ldots n_p\rangle = (H_1 + \cdots + H_p)|n_1 \ldots n_p\rangle = \left(n_1 + \cdots + n_p + \frac{1}{2}p\right)\hbar\omega|n_1 \ldots n_p\rangle) \quad [5.8]$$

The eigenvalue only depends on the sum $n = n_1 + \cdots + n_p$ of p numbers. For a given n, there are $\frac{(n+p-1)!}{n!(p-1)!}$ possible distinct states so that the eigenvalue $\left(n + \frac{1}{2}p\right)\hbar\omega$ is $\frac{(n+p-1)!}{n!(p-1)!}$ degenerate. Hence, if a vibration is doubly degenerate, i.e. p = 2, the energy $(n+1)\hbar\omega$ corresponds to that of $(n+1)$ states. For a triply degenerate vibration, p = 3, the energy is $\left(n + \frac{3}{2}\right)\hbar\omega$ for $\frac{(n+1)(n+2)}{2}$ states.

Annihilation and creation operators (see section 5.7.2; equations [2.30] and [2.31], Volume 1) that verify the commutation relations (see section 5.7.2; equation [2.32], Volume 1) can be used, such that if denoted by a_i and $a_i^\dagger$, respectively, for the oscillator denoted by i, the following relation can be written as:

$$[a_i, a_j^\dagger] = \delta_{ij}\,, [a_i, a_j] = 0 \text{ and } [a_i^\dagger, a_j^\dagger] = 0 \quad [5.9]$$

Based on $a_i^\dagger|n_i\rangle = \sqrt{|n_i + 1}|n_i + 1\rangle$, $a_i|n_i\rangle = \sqrt{|n_i}|n_i - 1\rangle$, $a_i|0\rangle = 0|0\rangle$ and $|n_i\rangle = \frac{1}{\sqrt{n_i!}}\left(a_i^\dagger\right)^{n_i}|0\rangle$, it can be proved that:

$$|n_1 \ldots n_p\rangle = \frac{1}{\sqrt{n_1! \ldots n_p!}}\left(a_1^\dagger\right)^{n_1} \ldots \left(a_p^\dagger\right)^{n_p}|0\rangle \quad [5.10]$$

and that the observables $N_i = a_i^\dagger a_i$ with $i = 1, 2, \ldots, p$ each have a spectrum that comprises a sequence of non-negative integers. These integers are interpreted respectively as the number of quanta of type 1, 2,..., p. The sum $N = \sum_i N_i$ yields the total number of quanta. It can thus be written that $H = \left(N + \frac{p}{2}\right)\hbar\omega$. The N_i form a complete set of commuting observables (C.S.C.O.) and their basis is precisely the basis of the Hamiltonian operator *H*. However, N_i are not the only constant of motion (commutation with *H*) that form a complete set. Any operator of the form $a_i a_j^\dagger$ commutes with the Hamiltonian operator *H*. Overall, p^2 independent Hermitian operators can be formed by the linear combination of operators of type $a_i a_j^\dagger$ and their adjoints.

Among the functions of these constant of motion, there are several complete sets of observables that commute.

5.3.2.1. *Two-dimensional isotropic oscillator*

Considering a two-dimensional system, the Hamiltonian is written as:

$$H = \frac{1}{2m}(p_1^2 + m^2\omega_0^2 q_1^2) + \frac{1}{2m}(p_2^2 + m^2\omega_0^2 q_2^2) \quad [5.11a]$$

$$H = \hbar\omega(a_1^\dagger a_1 + a_2^\dagger a_2 + 1) = \hbar\omega(N_1 + N_2 + 1) \quad [5.11b]$$

Applying the results of the previous section, the eigenvectors and the energies, i.e. the eigenvalues of the first three levels and the most general level, can be obtained as written in Table 5.3.

Eigenvalues	Degeneracy	Eigenfunctions of N_1 and N_2
$\hbar\omega$	1	$\|00\rangle$
$2\hbar\omega$	2	$\|10\rangle$, $\|01\rangle$
$3\hbar\omega$	3	$\|20\rangle$, $\|11\rangle$, $\|02\rangle$
$(n+1)\hbar\omega$	n+1	$\|n0\rangle$, $\|n-11\rangle$, ... $\|n-s\ s\rangle$, ... $\|0n\rangle$

Table 5.3. *Eigenvalues of H and set of eigenvectors common to N_1 and N_2 that generate their eigenspaces*

The angular momentum operator *L* that is defined by:

$$L = \frac{1}{\hbar}(q_1 p_2 - q_2 p_1) = i(a_1 a_2^\dagger - a_1^\dagger a_2) \quad [5.12]$$

is also a constant of motion.

A further complete set of commuting observables can be built from $N = N_1 + N_2$ and *L*. In this case, the following operators are introduced [BIE 65, MES 64, SAK 94]:

$$A_{\pm} = \frac{1}{\sqrt{2}}(a_1 \mp a_2) \text{ and } A_{\pm}^{\dagger} = \frac{1}{\sqrt{2}}(a_1^{\dagger} \pm ia_2^{\dagger}) \qquad [5.13]$$

These operators verify commutation relations that are identical to those of the annihilation and creation operators (see equation [5.9]), such that:

$$[A_i, A_j^{\dagger}] = \delta_{ij}\,, [A_i, A_j] = 0 \text{ and } [A_i^{\dagger}, A_j^{\dagger}] = 0 \qquad [5.14]$$

where this time i = + or –, and j = + or –. A_+ and $A_+^{\dagger}$ can be interpreted as annihilation and creation operators of quanta of type + and A_- and $A_-^{\dagger}$ can be interpreted as annihilation and creation operators of type –. Consequently, the quantum number operator + and the quantum number operator – are defined as follows:

$$N_+ = A_+^{\dagger}A_+ \text{ and } N_- = A_-^{\dagger}A_- \qquad [5.15]$$

The space of eigenvectors common to these two operators can be built and the diagram of energy levels can be determined from the numbers n_+ and n_- such that

$$n_+ = 0,1,2 \ldots \text{ and } n_- = 0,1,2 \ldots \qquad [5.16]$$

These observables form a complete set of commuting observables such that each pair of quantum numbers (n_+, n_-) corresponds to only one common eigenvector. From relations $A_+|00\rangle = A_-|00\rangle = 0$, it can be deduced that the vector $|00\rangle$ is an eigenvector of the fundamental state such that $n_+ = 0$ and $n_- = 0$. The vectors:

$$|n_+n_-\rangle = \frac{1}{\sqrt{n_+!n_-!}}\left(A_+^{\dagger}\right)^{n_+}(A_-^{\dagger})^{n_-}|00\rangle \qquad [5.17]$$

form a complete set of orthonormal eigenvectors common to operators N_+ and N_- such that:

$$N_+|n_+n_-\rangle = n_+|n_+n_-\rangle \text{ and } N_-|n_+n_-\rangle = n_-|n_+n_-\rangle \qquad [5.18]$$

Two other sum and difference operators can be defined as follows:

$$N = N_+ + N_- \text{ and } L = N_+ - N_- \qquad [5.19]$$

which, similarly to N_+ and N_-, form a complete set of commuting observables.

Another complete set of orthonormal eigenvectors can thus be built for the problem of the two-dimensional isotropic harmonic oscillator. It can be written that:

$$H|n_+n_-\rangle = (n+1)\hbar\omega|n_+n_-\rangle \text{ and } L|n_+n_-\rangle = (n_+ - n_-)|n_+n_-\rangle \qquad [5.20]$$

Moreover, the relations involving the commutation of L with operators $A_\pm$ and $A_\pm^\dagger$ lead to:

$$[L, A_\pm^\dagger] = \pm A_\pm^\dagger \text{ and } [L, A_\pm] = \mp A_\pm \qquad [5.21]$$

This implies that the action of operators $A_\pm$ and $A_\pm^\dagger$ on the eigenvectors of L increases them ($A_+^\dagger$ and A_- on $|n_+n_-\rangle$) by one unit or reduces them ($A_-^\dagger$ and A_+ on $|n_+n_-\rangle$) by one unit, similarly to the action of ladder operators defined in the theory of angular momenta.

Given that the Hamiltonian of the two-dimensional isotropic oscillator is invariant, not only under a rotation in the space of configuration in the plane q_1-q_2 that can be put into correspondence with the physical space x-y (geometric symmetry), but also under a rotation that mixes the coordinates associated with variables of momentum p_i conjugate to the position variable q_i (dynamic symmetry), it can be shown that it is possible to define four constants of motion associated with operators that commute with the Hamiltonian H, out of which only three are independent.

This leads to the definition of the independent operators J_+, J_- and J_z, and of the Casimir operator J_0 corresponding to the Lie algebra SU(2) (Schwinger) generated by the operators J_+, J_- and J_z, such that [BIE 65, SAK 94]:

$$J_\pm = \hbar A_\pm^\dagger A_\mp \text{ and } J_z = \left(\frac{\hbar}{2}\right)\left(A_+^\dagger A_+ - A_-^\dagger A_-\right) \qquad [5.22a]$$

$$J_0 = \left(\frac{\hbar}{2}\right)\left(A_+^\dagger A_+ + A_-^\dagger A_-\right) \text{ and } H = \omega(2J_0 + 1) \qquad [5.22b]$$

It can be easily verified that these operators satisfy the commutation relations of angular momenta such that:

$$[J_z, J_\pm] = \pm\hbar J_\pm \text{ and } [J_+, J_-] = 2\hbar J_z \qquad [5.22c]$$

and

$$J_+|n_+n_-\rangle = \hbar A_+^\dagger A_-|n_+n_-\rangle = \hbar\sqrt{n_-(n_+ + 1)}|n_+ + 1 n_- - 1\rangle \qquad [5.23a]$$

$$J_-|n_+n_-\rangle = \hbar A_-^\dagger A_+|n_+n_-\rangle = \hbar\sqrt{n_+(n_- + 1)}|n_+ - 1 n_- + 1\rangle \qquad [5.23b]$$

$$J_z|n_+n_-\rangle = \frac{\hbar}{2}(N_+ - N_-)|n_+n_-\rangle = \frac{\hbar}{2}(n_+ - n_-)|n_+n_-\rangle|n_+n_-\rangle \qquad [5.23c]$$

These relations can be compared to the commutation properties of tensor operators, which satisfy the following relations (see equations [5.19] and [5.20]):

$$[J_z, T_q^{(k)}] = qT_q^{(k)} \text{ and } [J_\pm, T_q^{(k)}] = \sqrt{k(k+1) - q(q \pm 1)}T_{q\pm 1}^{(k)}$$

for which the operators associated with physical variables must be invariant under a rotation by an angle 2π, which requires k to be an integer greater or equal to zero, $k = 0,1,2, \ldots$ It is important to note in relations [5.22] the presence of a factor 1/2 in the definition of J_z, which implies that half-integral values are allowed for the oscillator. This leads to a Lie algebra of the rotational group *3D* whose values of j can be a half-integer.

The physical nature of the operators defined in [5.22] is not obvious. J_z and L are proportional, so that J_z is in relation to the orbital momentum related to the difference in energy between each *1D* oscillator. As for the two other operators, one of them has an effect on the phase between the two *1D* oscillators; the latter generates infinitesimal variations of the eccentricity of the elliptical orbit that maintains the sum of the semi-axes fixed. A circular orbit is modified into a line, then into an ellipse, the rotation changing its direction after a rotation by 2π, which requires a rotation by 4π for an orbit to find its initial position, which explains the possibility to define + states and – states.

Based on the definition of N, it can be shown that:

$$J^2 = J_z^2 + \frac{1}{2}(J_+J_- + J_-J_+) = \left(\frac{\hbar^2}{2}\right) N \left(\frac{N}{2} + 1\right) \quad [5.24a]$$

which corresponds to:

$$J^2 = J_z^2 + \frac{1}{2}(J_+J_- + J_-J_+) = \left(\frac{\hbar^2}{4}\right)(H^2 - 1) \quad [5.24b]$$

If a correspondence such that $n_+ = j + m$ and $n_- = j - m$ is established, then:

$$\left(\frac{\hbar^2}{2}\right)(n_+ + n_-)\left(\frac{(n_+ + n_-)}{2} + 1\right) = \hbar^2 j(j+1) \quad [5.25]$$

which corresponds to the eigenvalue of the operator J^2 as defined in [5.24a].

Based on the inverse relations $j = \frac{(n_+ + n_-)}{2}$ and $m = \frac{(n_+ - n_-)}{2}$ the simultaneous eigenvalues of operators J^2 and J_z can thus be characterized. Similarly, based on the properties of operators J_+ and J_-, it can be established that (see equation [5.17]):

$$|j,m\rangle = \frac{1}{\sqrt{(j+m)!(j-m)!}}\left(A_+^\dagger\right)^{j+m}(A_-^\dagger)^{j-m}|00\rangle \quad [5.26]$$

When $m = j$, which corresponds to the highest value of J_z, for J parallel to the axis Oz (Maximum value for the projection on the axis Oz), then:

$$|j,j\rangle = \frac{1}{\sqrt{(2j)!}}\left(A_+^\dagger\right)^{2j}|00\rangle \quad [5.27]$$

This state can be built from $2j$ particles of spin of half-integral value 1/2, all the spins being directed along the Oz axis. A complicated object of value j can be seen as being composed of primitive particles of spin 1/2, $j + m$ being directed upwards and $j - m$ downwards. It is possible to consider an object of angular momentum j as a composite system of $2j$ spins of half-integral value 1/2 from the perspective of the transformation properties in relation to rotations. These transformations show that a two-dimensional degenerate vibration can be studied by means of the theory used to study angular momenta.

In summary, the commutation relations show that the operators J_+, J_- and J_z are the generators of a Lie algebra of type SU(2). This algebra is the one applied in the theory of angular momenta (see sections 5.7.3 and 5.7.4). The possible eigenvalues of J^2 are $(h/2\pi)^2 j(j + 1)$, where j = 0, 1/2, 1, 3/2, ... At fixed j, the possible eigenvalues of J_i, where i = +, – and z, are m, with $m = -j, -j + 1, \ldots, j$. The degeneracy in j is $2j + 1$ which is also the degeneracy of the energy levels, since $(h/2\pi)^2 j(j + 1) = (1/4)(E^2-1)$ (see equation [5.25]). It can be deduced that $E_j = (h/2\pi)(2j + 1)$, where j = 0, 1/2, 1, 3/2, ... This method can be used to build only symmetric states, the primitive spins in this case being bosons. These various considerations show that a tensor approach to the study of vibrations is possible [CHA 92, IAC 81, IAC 94].

5.3.2.2. *Three-dimensional isotropic oscillator*

When a three-dimensional system is considered, the expression of the Hamiltonian is given by:

$$H = \frac{1}{2m}(p_1^2 + m^2\omega_0^2 q_1^2) + \frac{1}{2m}(p_2^2 + m^2\omega_0^2 q_2^2) + \frac{1}{2m}(p_3^2 + m^2\omega_0^2 q_3^2) \quad [5.28]$$

The eigenvalues are given by $\left(n + \frac{3}{2}\right)\hbar\omega$ for $\frac{(n+1)(n+2)}{2}$ states with n = 0, 1, 2,..., ∞. The observables $N_i = a_i^\dagger a_i$ with $i = 1,2,3$ can be defined, each of them having a spectrum that comprises a sequence of non-negative integers. These integers are interpreted, respectively, as the number of quanta of types 1, 2 and 3. The sum $N = \sum_i N_i$ gives the total number of quanta. It can thus be written that

$H = \left(n + \frac{3}{2}\right)\hbar\omega$. The N_i form a complete set of commuting observables and their basis is given by:

$$|n_1 n_2 n_3\rangle = \frac{1}{\sqrt{n_1! n_2! n_3!}} \left(a_1^\dagger\right)^{n_1} \left(a_2^\dagger\right)^{n_2} \left(a_3^\dagger\right)^{n_3} |000\rangle \quad [5.29]$$

with

$$a_1|000\rangle = a_2|000\rangle = a_3|000\rangle = 0 \quad [5.30]$$

This problem can be regarded as that of a particle in a central potential proportional to the square of the distance to the interaction center. In this case, an angular momentum $\vec{l} = \vec{r} \times \vec{p}$ can be defined. Similarly to the hydrogen atom, H, in this case l^2 *and* l_z form a complete set of commuting observables of common eigenvectors denoted by $|nlm\rangle$. The vectors $|nlm\rangle$ can be obtained from vectors $|n_1 n_2 n_3\rangle$ from a unitary transformation.

Similarly to the two-dimensional isotropic harmonic oscillator, the following transformations are applied:

$$A_1 = \frac{1}{\sqrt{2}}(a_1 - ia_2)\,,\ A_2 = \frac{1}{\sqrt{2}}(a_1 + ia_2) \text{ and } A_0 = a_3 \quad [5.31]$$

together with the conjugate Hermitian operators $A_1^\dagger, A_2^\dagger$ and $A_3^\dagger$. Similarly, $A_m^\dagger$ and A_m satisfy the commutation relations of [5.14] and can be interpreted as creation and annihilation operators of quanta of type m. The number of quanta of type m is represented by the operator $N_m = A_m^\dagger A_m$. In this case, N_1, N_0 and N_{-1} form a complete set of commuting observables, and the following relation can be written as:

$$H = \left(N_1 + N_0 + N_{-1} + \frac{3}{2}\right)\hbar\omega \text{ and } N = N_1 + N_0 + N_{-1} \quad [5.32]$$

Consequently, the eigenvalues $|n_1 n_0 n_{-1}\rangle$ are such that:

$$|n_1 n_0 n_{-1}\rangle = \frac{1}{\sqrt{n_1! n_0! n_{-1}!}} \left(A_1^\dagger\right)^{n_1} \left(A_0^\dagger\right)^{n_0} \left(A_{-1}^\dagger\right)^{n_{-1}} |000\rangle \quad [5.33]$$

and

$$H|n_1 n_0 n_{-1}\rangle = \left(n + \frac{3}{2}\right)\hbar\omega|n_1 n_0 n_{-1}\rangle \text{ and } n = n_1 + n_0 + n_{-1} \quad [5.34]$$

These vectors are, however, not the eigenvectors of l^2, but those of l_z such that:

$$l_z = (N_1 - N_0)\hbar \text{ and } m = n_1 - n_{-1} \quad [5.35]$$

In the subspace of eigenvectors of H corresponding to the eigenvalue $\left(n+\frac{3}{2}\right)\hbar\omega$, among the $\frac{(n+1)(n+2)}{2}$ eigenvectors $|n_1 n_0 n_{-1}\rangle$ that generate the subspace, there are $n = n_1 + n_0 + n_{-1}$ that form a set of orthonormal eigenvectors of l_z. The quantum number m can take the integral numbers from $-n$ to $+n$ as values. If the energy levels are labeled using the quantum numbers n, l and m, then for each n, l may have values from 0 to n–1, and for each l, m may have values from $-l$ to $+l$.

5.3.2.3. *Dijon tensor formalism*

The creation and annihilation operators operating in the vibrational Hilbert space are defined as a function of the dimensionless operators $q_{s\sigma}$ and $p_{s\sigma}$ by:

$$a_{s\sigma}^{+(C)} = \frac{1}{\sqrt{2}}(q_{s\sigma} - ip_{s\sigma}) \text{ and } a_{s\sigma}^{(C)} = \frac{1}{\sqrt{2}}(q_{s\sigma} + ip_{s\sigma}) \quad [5.36]$$

where s is related to various modes (s = 1, 2, 3, 4) and σ is the component of the operator (mode degeneracy equal to 2 for s = 2 with $C = E$, and equal to 3 for s = 3 and 4 with $C = T_2$ or F_2, depending on the preferred notation).

A first operator denoted by A is built from the two operators such that [CHA 92]:

$$A_{\{n_s\}\{m_s\}}^{\alpha_1\Gamma_1\alpha_2\Gamma_2(\Gamma)} = \left(\left(a_{n_s}^{+(\alpha_1\Gamma_1)}\right) \otimes \left(a_{m_s}^{(\alpha_2\Gamma_2)}\right)\right)^{(\Gamma)} \quad [5.37]$$

where

$a_{n_s}^{+(\alpha_1\Gamma_1)}$ and $a_{m_s}^{(\alpha_2\Gamma_2)}$ represent the couplings of n_s operators $a_{s\sigma}^{+(C)}$ and of m_s operators $a_{s\sigma}^{(C)}$ related to modes s, such that:

$$a_{n_s}^{+(\alpha_1\Gamma_1)} = \Bigg(\underbrace{a_{s\sigma}^{+(C)} \otimes a_{s\sigma}^{+(C)} \ldots \otimes a_{s\sigma}^{+(C)}}_{n_s}\Bigg)^{(\alpha_1\Gamma_1)} \quad [5.38a]$$

and such that:

$$a_{m_s}^{(\alpha_2\Gamma_2)} = \Bigg(\underbrace{a_{s\sigma}^{(C)} \otimes a_{s\sigma}^{(C)} \ldots \otimes a_{s\sigma}^{(C)}}_{m_s}\Bigg)^{(\alpha_2\Gamma_2)} \quad [5.38b]$$

α_1 and α_2 can be used to distinguish between the operators of the same symmetry Γ_1 and Γ_2 , where Γ satisfies the triangular relation $\Delta(\Gamma_1, \Gamma_2, \Gamma) = 0$.

The vibrational operators are built from the operator A and its adjoint such that:

$$^{\epsilon}V^{\alpha_1\Gamma_1\alpha_2\Gamma_2(\Gamma)}_{\{n_s\}\{m_s\}} = \frac{1}{N} e^{i\varphi}\left(\left(A^{\alpha_1\Gamma_1\alpha_2\Gamma_2(\Gamma)}_{\{n_s\}\{m_s\}}\right) + \left(\epsilon \times A^{\alpha_1\Gamma_1\alpha_2\Gamma_2(\Gamma)+}_{\{n_s\}\{m_s\}}\right)\right) \quad [5.39]$$

where $\epsilon = \pm 1$ and N is the normalization factor.

If $\epsilon = +1$, the operator V is odd with respect to $p_{s\sigma}$ and $e^{i\varphi} = 1$, and if $\epsilon = -1$, the operator V is even with respect to $p_{s\sigma}$ and $e^{i\varphi} = -1$.

For example, the only non-trivial vibrational operators that contribute to the energy of levels $v_2 = 1$ and $v_4 = 1$ are given by [PIE 77]:

$$V_2^{(\Gamma)} = \frac{1}{\sqrt{[\Gamma]}}\left(\left(a_2^{(E)}\right)\otimes\left(a_2^{(E)+}\right)\right)^{(\Gamma)} \text{ and } V_4^{(\Gamma)} = \frac{1}{\sqrt{[\Gamma]}}\left(\left(a_4^{(T_2)}\right)\otimes\left(a_4^{(T_2)+}\right)\right)^{(\Gamma)} \quad [5.40a]$$

where $[\Gamma]$ is the dimension of representation Γ.

The two first operators of interaction between levels $v_2 = 1$ and $v_4 = 1$ lead, in this case, to the following vibrational operators:

$$V_{24}^{(T_1)} = \frac{-i}{\sqrt{3}}\left(\left(a_4^{(T_2)} \otimes a_2^{(E)+}\right)^{(T_1)} \oplus \left(a_2^{(E)} \otimes a_4^{(T_2)+}\right)^{(T_1)}\right)^{(T_1)} \quad [5.40b]$$

$$V_{24}^{(T_2)} = \frac{1}{\sqrt{3}}\left(\left(a_4^{(T_2)} \otimes a_2^{(E)+}\right)^{(T_2)} \oplus \left(a_2^{(E)} \otimes a_4^{(T_2)+}\right)^{(T_2)}\right)^{(T_2)} \quad [5.40c]$$

5.3.3. *Rotational tensor operators*

Using Watson's method [WAT 67] (described in Chapter 2) the construction of the rotational Hamiltonian leads to the expression:

$$H_{rot} = \sum_{\substack{p+q+r \\ even}} h_{pqr}\left(J_x^p J_y^q J_z^r + J_z^r J_y^q J_x^p\right),$$

by determining the symmetric powers n of the representation of operator J in the complete rotation–reflection group, K_h, from $[\Gamma(J^2)]_{sym} = S_g + D_g$. This decomposition can be obtained from the fact that J is a vector operator and a spherical tensor of rank 1. To obtain the irreducible representation of J^2, the tensor product $D_1 \otimes D_1$ must be decomposed as follows:

$$D_2 \oplus D_1 \oplus D_0$$

and by only retaining the symmetric part, the following decomposition is obtained:

$$[\Gamma(\hat{J}^2)]_{sym} = S_g + D_g.$$

For J^4, the tensor product $D_1 \otimes D_1 \otimes D_1 \otimes D_1$ can be decomposed and the irreducible representations can be determined as follows:

$$D_4 \oplus 3D_3 \oplus 8D_2 \oplus 13D_1 \oplus 13D_0$$

If only the symmetric part is retained, then:

$$[\Gamma(\hat{J}^4)]_{sym} = S_g + D_g + G_g\,.$$

For J^6, the tensor product $D_1 \otimes D_1 \otimes D_1 \otimes D_1 \otimes D_1 \otimes D_1$ can be decomposed, which leads to $D_6 \oplus C_5^6 D_5 \oplus C_4^6 D_4 \oplus C_3^6 D_3 \oplus C_2^6 D_2 \oplus C_1^6 D_1 \oplus C_0^6 D_0$. Retaining only the symmetric part yields:

$$[\Gamma(\hat{J}^6)]_{sym} = S_g + D_g + G_g + I_g.$$

The numerical values of the coefficients giving the number of times that an irreducible representation appears in the decomposition (see Table 5.4) were omitted for $n = 5$ and $n = 6$; however, these values can easily be calculated.

n	$\underbrace{D_1 \otimes D_1 \ldots \otimes D_1}_{n} = \prod_n \otimes D_1$
2 ($D_1 \otimes D_1$)	$D_2 \oplus D_1 \oplus D_0$
3	$D_3 \oplus 2D_2 \oplus 3D_1 \oplus 3D_0$
4	$D_4 \oplus 3D_3 \oplus 8D_2 \oplus 13D_1 \oplus 13D_0$
5	$D_5 \oplus C_4^5 D_4 \oplus C_3^5 D_3 \oplus C_2^5 D_2 \oplus C_1^5 D_1 \oplus C_0^5 D_0$
6	$D_6 \oplus C_5^6 D_5 \oplus C_4^6 D_4 \oplus C_3^6 D_3 \oplus C_2^6 D_2 \oplus C_1^6 D_1 \oplus C_0^6 D_0$

Table 5.4. *Decomposition of tensor products into irreducible representations; only the values of the coefficients in the first three lines are mentioned; the other coefficients can be easily determined, for example* C_3^5*=18 and* C_1^5*=90*

As already underlined in Chapter 3, only the irreducible representations S_g, G_g and I_g (see Table 1.12) contain the fully symmetric representation of group T_d and only occur once. Consequently, there are two terms of fourth degree and three terms of sixth degree in the expansion of the rotational Hamiltonian. In Watson's formulation, $\hat{J}^2$, $\hat{J}_x^4 + \hat{J}_y^4 + \hat{J}_z^4$ and $\hat{J}_x^6 + \hat{J}_y^6 + \hat{J}_z^6$ are fully symmetric if axes (x, y, z) are along the axes of symmetry of fourth order (S_4). The two terms of fourth degree are given by $(\hat{J}^2)^2$, $\hat{J}_x^4 + \hat{J}_y^4 + \hat{J}_z^4$ and the terms of sixth degree are given by $(\hat{J}^2)^3$, $\hat{J}^2(\hat{J}_x^4 + \hat{J}_y^4 + \hat{J}_z^4)$ and $\hat{J}_x^6 + \hat{J}_y^6 + \hat{J}_z^6$. The rotational Hamiltonian is expressed in the form of spherical tensor operators Ω_i which are linear combinations of operators expressed in terms of J^2 [WAT 67]. The operators Ω_i [HEC 60a,HEC 60b,KIR 73] are transformed as a linear combination of components G_g (Ω_4) and Ig (Ω_6) in the symmetry operations of group O(3) (or K_h). Finally, Watson expresses the rotational Hamiltonian in the following form:

$$H_{rot} = B\hat{J}^2 - D(\hat{J}^2)^2 + D_{4t}\Omega_4 + H(\hat{J}^2)^3 + H_{4t}\hat{J}^2\Omega_4 + H_{6t}\Omega_6 \ldots \quad [5.41]$$

where the coefficients D, D_{4t}, H and H_{6t} are the real centrifugal constants. Chapters 10 and 11 of Hamermesch's book [HAM 62] describe the methods that can be used to build irreducible representations of tensor operators.

In the tensor formalism developed by the group in Dijon, the first rank primitive rotational operator is denoted by $R^{1(1_g)}$ in the group O(3) and its components are given by $2J_\alpha$ ($J_\alpha = \frac{P_\alpha}{\hbar}$ with $\alpha = x, y, z$), which leads to the relation $R_M^{1(1_g)} = 2J_M^{(1_g)}$, where $J_0^{(1_g)} = J_z$ and $J_{\pm 1}^{(1_g)} = \mp\frac{1}{\sqrt{2}}(J_x \pm iJ_y) = \mp\frac{1}{\sqrt{2}}J_\pm$. It is important to note that the components of O(3) are built as defined by Moret-Bailly [MOR 61] and can be obtained using the recurrence formula by Zhilinskii [ZHI 81], such that:

$$R_M^{\Omega(K_g)} = \left(R_0^{2(0)}\right)^{((\Omega-K_g)/2)} R_M^{K_g(K_g)} \quad [5.42]$$

and

$$R_M^{K_g(K_g)} = \left(R^{K_g-1(K_g-1)} \otimes R^{1(1)}\right)_M^{K_g} \quad [5.43]$$

with

$$R_0^{2(0)} = -\frac{4}{\sqrt{2}}J^2 \quad [5.44]$$

In the subgroup T_d, the operator is written in the form $R^{1(1_g,F_1)}$. An arbitrary rotational operator is Ω times the tensor product of the primitive operator $R^{1(1_g,F_1)}$ by itself. It is denoted by $R^{\Omega(K_g,n\Gamma)}$ such that:

$$R^{\Omega(K_g,n\Gamma)} = \Big(\underbrace{R^{1(1_g)} \otimes R^{1(1_g)} \ldots . \otimes R^{1(1_g)}}_{\Omega}\Big)^{(K_g,n\Gamma)} \qquad [5.45]$$

where:

– Ω is the degree of operator R;

– $K_g = \Omega - 2$, 0 or 1 is the rank of operator R;

– Γ is the symmetry of operator R in the group T_d given K_g;

– n is an index that can be used to distinguish between the operators of the same symmetry Γ.

It is important to note that several coupling schemes are possible to build the operator R. The justification of this approach within the framework of the group theory for building the vibration–rotation Hamiltonian of the molecules with tetrahedral symmetry was mentioned in [HIL 70]. To summarize, $R^{\Omega(K_g,n\Gamma)}$ is a polynomial in J_x, J_y and J_z of degree Ω; K_g is the rank with respect to O(3); and Γ is the irreducible representation of the point-group symmetry of the molecule, which, in this case, is the group T_d.

5.3.4. *Rovibrational tensor operators*

Within the framework of tensor formalism, the rovibrational operators are built by tensor product of a rotational operator and a vibrational operator [CHA 92], such that:

$$O_{RV}^{(\Gamma)} = \left(R^{\Omega(K_g,n\Gamma_R)} \otimes {}^{\epsilon}V_{\{n_s\}\{m_s\}}^{\Gamma_1\Gamma_2(\Gamma_V)} \right)^{(\Gamma)} \qquad [5.46]$$

5.3.5. *Expression of the rovibrational Hamiltonian*

According to the tensor formalism, the rovibrational Hamiltonian is built from sums of rovibrational operators, which in the group of symmetry T_d, belong to the irreducible representation of type A_1. The rovibrational operators can then be expressed in the following form [CHA 92]:

$$T_{\{n_s\}\{m_s\}}^{\Omega(K_g,n\Gamma_R)\Gamma_1\Gamma_2} = \left(R^{\Omega(K_g,n\Gamma_R)} \otimes {}^{\epsilon}V_{\{n_s\}\{m_s\}}^{\Gamma_1\Gamma_2(\Gamma_V)} \right)^{(A_1)} \quad [5.47]$$

Moreover, since the Hamiltonian must be Hermitian and invariant under a time-reversal operation, $\epsilon = +1$ if Ω is even and $\epsilon = -1$ if Ω is odd. Consequently, in the tensor formalism, the expression of the rovibrational Hamiltonian is given by:

$$H_{VR} = \sum_{indices} t_{\{n_s\}\{m_s\}}^{\Omega(K_g,n\Gamma)\Gamma_1\Gamma_2} T_{\{n_s\}\{m_s\}}^{\Omega(K_g,n\Gamma)\Gamma_1\Gamma_2} \quad [5.48]$$

where the coefficients t are the parameters of the rovibrational Hamiltonian. Depending on the rovibrational degrees of freedom, the order of expansion d is given by:

$$d = \Omega - 2 + \sum_s n_s + m_s \text{ with } \Omega \leq 2 \quad [5.49]$$

where each term of the equation [5.49] is, at most, of second degree with respect to the angular momentum, since, in fact, the zeroth-order term corresponds to:

$$H_0 = \frac{1}{2}\sum_{s\sigma} \hbar\omega_s \left(\frac{p_{s\sigma}^2}{\hbar^2} + q_{s\sigma}^2\right) + \frac{1}{2}\sum_{\alpha} \frac{J_{\alpha}^2}{I_{\alpha\alpha}^2} \quad [5.50]$$

Note that according to Watson's notations, $H_0 = H_{20} + H_{02}$.

5.3.6. *Expression of the vibration wave functions*

According to the tensor formalism, the vibrational wave functions are built by coupling the basis functions of the harmonic oscillators such that:

$$\psi_v^{C_v} = \left(\psi_{v_1}^{C_{v1}} \otimes \psi_{v_2}^{C_{v2}} \otimes \psi_{v_3}^{C_{v3}} \otimes \psi_{v_4}^{C_{v4}}\right)^{(C_v)} \quad [5.51]$$

where v is the set of the vibration quantum numbers v_s of the given state. C_v is the vibrational symmetry relative to the vibrational state v.

5.3.7. *Expression of rotational wave functions*

The rotational wave functions are the basis functions $\psi_R^{(J_g,nC_r)}$ of the rigid rotator oriented in T_d. By convention, the parity of these functions is g. Depending on the basis functions $\psi_R^{(j,m)} = \psi_m^j$ of the rigid rotator oriented in O(3), the following relation is obeyed:

$$\Psi_R^{(j,nC_r)} = \sum_m {}^{(j)}G_{nC_r}^m \Psi_m^j \qquad [5.52]$$

Components of T_P^J in T_d				Coefficients of matrices $^{(J)}G_P^m$ in Td for $m = -J \ldots J$								
J	n	Γ	σ	-4	-3	-2	-1	0	1	2	3	4
0	1	A_1	1					1				
1	1	T_1	1				1		-1			
			2				-i		-i			
			3					2				
2	1	E	1					2				
			2			1				1		
	1	T_2	1				-i		-i			
			2				1		-1			
			3			-i				i		
3	1	A_2	1			-8i				8i		
	1	T_1	1		5		-3		3		-5	
			2		5i		3i		3i		5i	
			3					16				
	1	T_2	1		-3		-5		5		3	
			2		3i		-5i		-5i		3i	
			3			8				8		
4	1	A_1	1	10				28				10
	1	E	1	14				20				14
			2			24				24		
	1	T_1	1		3i		21i		21i		3i	
			2		-3		21		-21		3	
			3			-24i				24i		
	1	T_2	1		-21i		3i		3i		-21i	
			2		-21		-3		3		21i	
			3			-24				24i		

Table 5.5. *Coefficients of the transition matrix $^{(j)}G_p^m$ for j = 1 to 4 [CHA 77b]; for each J, the coefficients must be divided by the sum of inputs for m=0. A square root must be applied on each positive real part of the coefficient; for example, –21i must be understood as $-(21/48)^{1/2}i$*

similarly to tensor operators:

$$T_{nC_r}^{(j)} = \sum_m {}^{(j)}G_{nC_r}^m T_m^{(j)} \quad [5.53]$$

$$|\Psi, j, A_1\rangle = \frac{1}{24}(1 \times |\Psi, A_1\rangle + 8 \times 1 \times |\Psi, A_1\rangle + 3\times 1 \times |\Psi, A_1\rangle$$
$$+6 \times 1 \times |\Psi, A_1\rangle + 6\times 1 \times |\Psi, A_1\rangle) = |\Psi, A_1\rangle \quad [5.54]$$

If the indices nC_r are grouped in only one index p [MOR 61], equations [5.52] and [5.53] can be rewritten in the form $\Psi_p^{(j)} = \sum_m {}^{(j)}G_p^m \Psi_m^{(j)}$ for the wave functions and $T_p^{(j)} = \sum_m {}^{(j)}G_p^m T_m^{(j)}$ for the operators, respectively, with $m = -J ,\ldots, J$. The matrix elements ${}^{(j)}G_p^m$ were calculated for $G = T_d$ by Moret-Bailly *et al.* [MOR 65] (numerical values are given in Table 1 of [MOR 65]) and Champion *et al.* [CHA 77b] (see Table 5.5).

5.3.8. *Expression of rovibrational wave functions*

The rovibrational basis wave functions $\psi_\sigma^{(C)}$ are built by the tensor product of the wave functions $\psi_v^{C_v}$ and $\psi_R^{(J_g, nC_r)}$ and are coupled in a similar way to the Hamiltonian operators, such that:

$$\psi_\sigma^{(C)} = \left(\psi_v^{C_v} \otimes \psi_R^{(J_g, nC_r)}\right)_\sigma^{(C)} \quad [5.55]$$

where $= C_r \otimes C_v$.

The matrix elements of the rovibrational operators are calculated using the Wigner–Eckart theorem.

5.4. Application to the CH_4 molecule

When a non-expert in the theoretical models developed in molecular spectroscopy, who is interested in the observations, uses the results of these models, they mainly focus on the topic pertaining to light–matter interaction. In fact, the spectral signature of the chemical species being observed, whether in the field of (vibration–rotation) IR or microwave (rotation) spectroscopy, depends on the selection rules authorized during this interaction that induces a transition between two energy levels (see Chapter 3). Two main operators are involved in this interaction, the dipole moment (first-order tensor) and the polarizability (second-order tensor), although magnetic dipole and electric quadrupole moments may also contribute to a transition which is generally weaker. It is sometimes necessary to take into account other factors in order to correctly interpret the profile of the observed spectrum; for example, such as when there is a coupling between vibration and rotation, or when the environment in which the molecule evolves is of cylindrical symmetry or is no longer isotropic of spherical symmetry as in free space, and which consequently leads to a degeneracy lifting of degenerate vibrations.

5.4.1. *Electric dipole transition moment*

In order to determine the selection rules in the case of an electric dipole moment operator within the framework of the Dijon tensor formalism, the results found in the thesis of Loëte [LOË 84] or in [COQ 94, COQ 95] are presented.

The electric dipole moment is a first rank tensor expressed as a function of spherical coordinates using the Wigner matrices such that:

$$\mu_m^{(1)} = \sum_k D_{k,m}^{(1)} \mu_k^{(1)} \qquad [5.56]$$

with k and m = −1, 0 and 1, and where $\mu_m^{(1)}$ are the spherical coordinates of the dipole moment in the laboratory reference frame, and $\mu_k^{(1)}$ are the spherical coordinates of the dipole moment in the molecular reference frame.

Wigner matrices are expressed as a function of Euler angles (see Chapter 1) and are related to the direction cosines $C_k^{(1)}$ by the relation:

$$D_{k,m}^{(1)} = (-1)^k C_{-k}^{(1)} \qquad [5.57]$$

such that:

$$\mu_m^{(1)} = \sum_k (-1)^k C_{-k}^{(1)} \mu_k^{(1)} = -\sqrt{3}\left(C^{(1)} \times \mu^{(1)}\right)_0^{(0)} \qquad [5.58]$$

Consequently, it suffices to establish a correspondence between the spherical components and the Cartesian components in the laboratory reference frame using the Stone coefficients [STO 75] given in Table 5.6 and written as:

$$\mu_\Gamma = \sum_k \langle 1; k|\Gamma\rangle \mu_k^{(1)} \qquad [5.59]$$

In the laboratory reference frame, the dipole moment, which is a polar vector, transforms following the irreducible representation $D^{(1u)}$ of O(3) (denoted ^{M}O(3)).

$\langle l; m\|\Theta\rangle$				
l	m	X	Y	Z
1	1	1/√2	-i/√2	0
1	0	0	0	1
1	-1	-1/√2	-i/√2	0

$\langle l; m\|\Theta_1\Theta_2\rangle$							
l	m	XX	YY	ZZ	XY	XZ	YZ
0	0	-1/√3	-1/√3	-1/√3	0	0	0
2	-2	1/2	-1/2	0	-i/2	0	0
2	-1	0	0	0	0	1/2	-i/2
2	0	-1/√6	-1/√6	√2/3	0	0	0
2	1	0	0	0	0	-1/2	-i/2
2	2	1/2	-1/2	0	i/2	0	0

Table 5.6. *Stone coefficients*

From the transformation of direction cosines as a function of spherical harmonics (hence, tensor operators) that transform according to the representation $D^{(1g)}$ of ${}^{M}O(3)$, equations [5.58] and [5.59] lead to:

$$\mu_{\Gamma}^{(0_u)} = -\sqrt{3}\sum_k \langle 1; k|\Gamma\rangle \left(C^{(1_g)} \times \mu^{(1_u)}\right)_0^{(0_u)} \qquad [5.60]$$

The irreducible representation $D^{(Ou)}$ of ${}^{M}O(3)$ can be reduced according to the irreducible representation A_2 of group T_d such that:

$$\left(C^{(1_g)} \times \mu^{(1_u)}\right)_0^{(0_u)} = \left(C^{(1_g)} \times \mu^{(1_u)}\right)_0^{(0_u, A_2)} \qquad [5.61]$$

Then, operators $C^{(1_g)}$ and $\mu^{(1_u)}$ in the group T_d must be oriented using isoscalar coefficients K. The representations $D^{(1_g)}$ and $D^{(1_u)}$ are reducible in the representations T_1 (or F_1) and T_2 (or F_1) of T_d. In this case:

$$\mu_{\Gamma}^{(0_u, A_2)} = \sqrt{3}\sum_k \langle 1; k|\Gamma\rangle \left(C^{(1_g, T_1)} \times \mu^{(1_u), T_2}\right)_0^{(0_u, A_2)} \qquad [5.62]$$

Given that the dipole moment depends only on normal coordinates, it can be expressed as a sum of purely vibrational symmetry operators T_2 (F_2), such that:

$$\mu_{\Gamma}^{(0_u,A_2)} = \sum_{indices} \mu_{\{n_s\}\{m_s\}}^{\alpha_1\Gamma_1\alpha_2\Gamma_2(T_2)} V_{\{n_s\}\{m_s\}}^{\alpha_1\Gamma_1\alpha_2\Gamma_2(T_2)} \qquad [5.63]$$

where the parameters $\mu_{\{n_s\}\{m_s\}}^{\alpha_1\Gamma_1\alpha_2\Gamma_2(T_2)}$ are expressed as a function of the partial derivatives of the dipole moment with respect to normal coordinates.

The transformed dipole moment must then be determined by applying the contact transformation, whose formal expression is written as:

$$\tilde{\mu}_{\Gamma}^{(0_u,A_2)} = \sqrt{3} \sum_k \langle 1; k|\Gamma\rangle \left\{C^{(1_g,T_1)} \times \tilde{\mu}^{(1_u,T_2)}\right\}_0^{(0_u,A_2)} \qquad [5.64]$$

where the term between curled brackets $\left\{C^{(1_g,T_1)} \times \tilde{\mu}^{(1_u),T_2}\right\}_0^{(0_u,A_2)}$ designates a symmetrized product. In the laboratory reference frame, a component of the dipole moment is written as:

$$\tilde{\mu}_{\Gamma}^{(0_u,A_2)} = \sqrt{3} \sum_k \langle 1; k|\Gamma\rangle \sum_i \tilde{\mu}_i \left\{C^{(1_g,T_1)} \times M^{(i,T_2)}\right\}_0^{(0_u,A_2)} \qquad [5.65]$$

with

$$M_{\alpha}^{(i,T_2)} = \left(R^{\Omega(K_r,n_r\Gamma_r)} \otimes V_{\{n_s\}\{m_s\}}^{\alpha_1\Gamma_1\alpha_2\Gamma_2(\Gamma_v)}\right)_{\alpha}^{(T_2)} \qquad [5.66]$$

where i takes into account the set of vibrational and rotational indices other than T_2. $\tilde{\mu}_i$ are the parameters of the transformed dipole moment that is written as:

$$\tilde{\mu}_{\mathrm{i}}^{(0_u,A_2)} = \sum_i \tilde{\mu}_i\, M_i^{(i,A_2)} \qquad [5.67]$$

Consequently, the calculation of the matrix element $\tilde{\mu}_{\mathrm{i}}^{(0_u,A_2)}$ requires that of the matrix element $M_i^{(i,A_2)}$ in the basis of eigenfunctions of the transformed Hamiltonian. Finally, this amounts to calculating the matrix element of various operators in the initial basis, which requires the application of the Wigner–Eckart theorem. The matrix element brings in the coupling symbol denoted by F [CHA 92, MOR 61] in the group T_d and the symbol *3j* in the expression of the reduced matrix element of the tensor of direction cosines.

Finally, the selection rules in a transition involving the dipole moment are written as:

$$F_{\sigma\sigma'}^{A_2CC'} \neq 0 \Rightarrow A_1 \leftrightarrow A_2, E \leftrightarrow E\ (\Delta\sigma \neq 0), T_1 \leftrightarrow T_2 (\Delta\sigma = 0) \qquad [5.68]$$

and

$$\begin{pmatrix} J' & 1 & J \\ M' & m & M \end{pmatrix} \neq 0 \Rightarrow |J' - J| \leq 1, m = M' - M = 0, \pm 1 \qquad [5.69]$$

It can be noted that only the reduced matrix element of the tensor of direction cosines depends on the secondary quantum numbers M and M' and on the choice of the component Γ or m of the electric dipole moment. The eigenfunctions are $(2J+1)$ times degenerate in M. The lifting of this degeneracy is possible in the presence of an external electric field.

5.4.2. *Polarizability*

Polarizability is a second rank tensor that can be decomposed according to the irreducible representations of O(3) which are $D^{(0_g)}$ and $D^{(2_g)}$. The spherical components $\alpha_m^{(L_g)}$ (L_g = 0 or 2, $|m| \leq L_g$) in the laboratory reference frame are related to the spherical components $\alpha_k^{(L_g)}$ in the molecular reference frame by:

$$\alpha_m^{(L_g)} = \sum\nolimits_k D_{k,m}^{(L_g)} \alpha_k^{(L_g)} \text{ with } |k| \leq L_g \qquad [5.70]$$

Introducing the expression [5.57] in $\alpha_m^{(L_g)}$ leads to:

$$\alpha_m^{(L_g)} = \sum\nolimits_k (-1)^k C_{-k}^{(L_g)} \alpha_k^{(L_g)} = (-1)^{L_g} \left(2L_g + 1\right) \left(C^{(L_g)} \times \alpha^{(L_g)}\right)^{(0_g)} \qquad [5.71]$$

It is now sufficient to link the spherical components to the Cartesian components in the laboratory reference frame using the Stone coefficients associated with second rank tensors [STO 75] to write:

$$\alpha_{\Gamma_1\Gamma_2} = \sum_{L_g,m} \left\langle L_g; m \middle| \Gamma_1\Gamma_2 \right\rangle \alpha_m^{(L_g)} \qquad [5.72]$$

which can be written in the spherical formalism as:

$$\alpha_{\Gamma_1\Gamma_2}^{(0_g)} = \sum_{L_g,m} \left\langle L_g; m \middle| \Gamma_1\Gamma_2 \right\rangle (-1)^{L_g} (2L_g + 1) \left(C^{(L_g)} \times \alpha^{(L_g)}\right)^{(0_g)} \qquad [5.73]$$

In the form of symmetrized tensor of ${}^{L}O(3)$, oriented in the group T_d, it was shown that:

$$\alpha^{(0_g,A_1)}_{\Gamma_1\Gamma_2} = \sum_{L_g,m} \langle L_g;m|\Gamma_1\Gamma_2\rangle \{\Gamma\}^{\frac{1}{2}} \left\{C^{(L_g,\Gamma)} \times \alpha^{(L_g,\Gamma)}\right\}^{(0_g,A_1)}$$

$$L_g = 0 \qquad \Gamma = A_1 \qquad [5.74]$$

$$L_g = 2 \qquad \Gamma = E, F_2$$

Similar to the case of the dipole moment, $\alpha^{(L_g,\Gamma)}$ is expressed as a sum of purely vibrational operators of symmetry Γ, weighted by parameters $\alpha^{(i,\Gamma)}$. In practice, the transformed polarizability tensor is used. This tensor is written as a sum of rovibrational operators weighted by transformed parameters $\tilde{\alpha}^{(i,\Gamma)}$:

$$\tilde{\alpha}^{(A_1)}_{\Gamma_1\Gamma_2} = \sum_{L_g,m} \langle L_g;m|\Gamma_1\Gamma_2\rangle \{\Gamma\}^{\frac{1}{2}} \left\{C^{(L_g,\Gamma)} \times \tilde{\alpha}^{(L_g,\Gamma)}\right\}^{(A_1)}$$

with:

$$= \sum_{L_g,m} \langle L_g;m|\Gamma_1\Gamma_2\rangle \{\Gamma\}^{\frac{1}{2}} \sum_i \tilde{\alpha}^{(i,\Gamma)} \left\{C^{(L_g,\Gamma)} \times P^{(i,\Gamma)}\right\}^{(A_1)}$$

$$P^{(i,\Gamma)} = \left(R^{\Omega(K_r,n_r\Gamma_r)} \otimes (-1)^{K_r} V^{\alpha_1\Gamma_1\alpha_2\Gamma_2(\Gamma_v)}_{\{n_s\}\{m_s\}}\right)^{(\Gamma)} \qquad [5.75]$$

For a given operator, i takes into account the set of vibrational and rotational indices other than Γ. Hence, the polarizability tensor is written as the sum:

$$\tilde{\alpha}^{(A_1)}_{\Gamma_1\Gamma_2} = \sum_i \langle 0;m|\Gamma_1\Gamma_2\rangle \tilde{\alpha}^{(i,A_1)} \left\{C^{(0,A_1)} \times P^{(i,A_1)}\right\}^{(A_1)}$$

$$+ \sum_{i,m} \sqrt{2} \langle 2;m|\Gamma_1\Gamma_2\rangle \tilde{\alpha}^{(i,E)} \left\{C^{(2,E)} \times P^{(i,E)}\right\}^{(A_1)} \qquad [5.76]$$

$$+ \sum_{i,m} \sqrt{3} \langle 2;m|\Gamma_1\Gamma_2\rangle \tilde{\alpha}^{(i,F_2)} \left\{C^{(2,F_2)} \times P^{(i,F_2)}\right\}^{(A_1)}$$

In these expressions, operators $P^{(i,A_1)}$ belong to the isotropic part of polarizability and operators $P^{(i,E)}$ and $P^{(i,F_2)}$ to the anisotropic part.

The calculation of the matrix element of $\tilde{\alpha}_{\Gamma_1\Gamma_2}^{(A_1)}$ between two eigenstates is written as:

$$\left\langle \tilde{\varphi}_{\sigma'}^{(J'C'\alpha')} \middle| \tilde{\alpha}_{\Gamma_1\Gamma_2}^{(A_1)} \middle| \tilde{\varphi}_{\sigma}^{(JC\alpha)} \right\rangle = \sum_{\substack{n,C_r,v,C_v \\ n',C_{r'},v',C_{v'}}} U_{n',C_{r'},v',C_{v'}}^{(J'C'\alpha')} U_{n,C_r,v,C_v}^{(JC\alpha)} \left\langle \Psi_{\sigma'}^{(C')} \middle| \tilde{\alpha}_{\Gamma_1\Gamma_2}^{(A_1)} \middle| \Psi_{\sigma}^{(C)} \right\rangle \quad [5.77]$$

This requires the calculation of the matrix element of operators $P^{(i,\Gamma)}$ in the basis of eigenfunctions of the transformed Hamiltonian. Then, the Wigner–Eckart theorem can be applied to relate this calculation to that of the matrix element in the initial basis.

The selection rules in a transition involving the polarizability tensor are written as:

$$F_{\sigma\sigma'}^{A_1CC'} \neq 0 \Rightarrow \Delta C = 0,\ \Delta\sigma = 0$$

$$C = A_1, A_2, E, F_1, F_2 \quad [5.78a]$$

and

$$\begin{pmatrix} J' & L_g & J \\ -M' & m & M \end{pmatrix} \neq 0 \Rightarrow |J' - J| \leq L_g$$

$$m = M' - M = 0, \pm 1, \pm 2 \quad [5.78b]$$

$$m = -L_g, +L_g$$

It can be shown that the Raman and IR activities obtained are those summarized in Table 5.7.

Vibrations	Type of symmetry	Active
ν_1 (stretching)	A_1	Raman
ν_2 (bending)	E	Raman
ν_3 (stretching)	T_2	IR/Raman
ν_4 (bending)	T_2	IR/Raman

Table 5.7. *IR and Raman transitions allowed by the selection rules for methane*

5.5. Rotational structure in the degenerate vibrational levels

The Dijon tensor formalism does not always allow the physical interpretation of the couplings between the various degrees of freedom of vibration and rotation. It is sometimes necessary to go back to less sophisticated models in order to identify the physical nature of couplings. In the case of methane, which is characterized by a non-degenerate vibration of type A_1, a doubly degenerate vibration of type E and two triply degenerate vibrations of type T_2 or F_2, the theory developed for two- and three-dimensional isotropic harmonic oscillators (see section 5.3.2) can be applied, and a physical interpretation of the operators used for building the vibration–rotation spectrum can be given.

The energy and the wave function that characterize a vibrational eigenstate are given by:

$$E_{\text{vib}}^0 = E_{v_1}^0 + E_{v_2}^0 + E_{v_3}^0 + E_{v_4}^0 = \hbar\omega_1\left(v_1 + \frac{1}{2}\right) + \hbar\omega_2(v_{21} + v_{22} + 1)$$

$$+\hbar\omega_3\left(v_{31} + v_{32} + v_{33} + \frac{3}{2}\right) + \hbar\omega_4\left(v_{41} + v_{42} + v_{43} + \frac{3}{2}\right) \quad [5.79a]$$

$$|v_1 v_{21} v_{22} v_{31} v_{32} v_{33} v_{41} v_{42} v_{43}\rangle = |v_1\rangle_0 |v_{21} v_{22}\rangle_0 |v_{31} v_{32} v_{33}\rangle_0 |v_{41} v_{42} v_{43}\rangle_0 \quad [5.79b]$$

This expression shows that the energy of an angular stretching mode ($\nu = 2$) depends only on the quantum number $v_2 = v_{21} + v_{22}$ that can take the values 0, 1, 2, 3, ... If $v_2 \neq 0$, the eigenvalues are degenerate. For example, the energy level $v_2 = 2$ has three independent eigenfunctions $|2_{21}0_{22}\rangle_0$, $|1_{21}1_{22}\rangle_0$ and $|0_{21}2_{22}\rangle_0$. According to the formula given in section 5.3.2 for $p = 2$, the degree of degeneracy of level v_2 is equal to $v_2 + 1$.

Similarly, the energy of modes ($\nu = s, s = 3 \; or \; 4$) depends only on the quantum number $v_s = v_{s1} + v_{s2} + v_{s3}$ that can take the values 0, 1, 2, 3, ... If $v_s \neq 0$, the eigenvalues are degenerate.

For example, in the case of the energy level $v_s = 2$, according to the formula given in section 5.3.2 for $p = 3$, there are six independent eigenfunctions:

$$|2_{31}0_{32}0_{33}\rangle_0, |1_{31}1_{32}0_{33}\rangle_0, |0_{31}2_{32}0_{33}\rangle_0, |1_{31}0_{32}1_{33}\rangle_0, |0_{31}0_{32}2_{33}\rangle_0$$

and $|0_{31}1_{32}1_{33}\rangle_0$

Referring to the functions given in section 5.2, functions other than Hermitian polynomials can be used as eigenfunctions. In this case, other quantum numbers are used to identify the energy levels:

$$|v_1 v_2 l_2 v_3 l_3 m_3 v_4 l_4 m_4\rangle = |v_1\rangle_0 |v_2 l_2\rangle_0 |v_3 l_3 m_3\rangle_0 |v_4 l_4 m_4\rangle_0 \qquad [5.79c]$$

5.5.1. *The degenerate vibrational level* $v_2 = 1$

For the degenerate oscillator of type E, two types of operators N_1 and N_2 can be built such that:

$$H = \hbar\omega\left(a_1^\dagger a_1 + a_2^\dagger a_2 + 1\right) = \hbar\omega(N_1 + N_2 + 1) \qquad [5.80a]$$

As described in section 5.3.2, a doubly degenerate vibration has $(n + 1)$ corresponding states denoted by $|n_1 n_2\rangle$ of energy $E = (n + 1)\hbar\omega$.

For the same degenerate oscillator of type E, two other types of operators, N_+ and N_-, can be built such that:

$$N_+ = A_+^\dagger A_+ \quad \text{and} \quad N_- = A_-^\dagger A_- \qquad [5.80b]$$

from:

$$A_\pm^\dagger = \frac{1}{\sqrt{2}}\left(a_{21}^\dagger \pm i a_{22}^\dagger\right) \quad \text{and} \quad A_\pm = \frac{1}{\sqrt{2}}(a_{21} \mp a_{22}) \qquad [5.80c]$$

where A_+ and $A_+^\dagger$ are the annihilation and creation operators of quanta of type +, and A_- and $A_-^\dagger$ are the annihilation and creation operators of quanta of type –.

N_+ becomes the operator, number of vibrons associated with states +, and N_- becomes the operator, number of vibrons associated with states –, and the following relations can be written as:

$$N_+|n_+ n_-\rangle = n_+|n_+ n_-\rangle \quad \text{and} \quad N_-|n_+ n_-\rangle = n_-|n_+ n_-\rangle \qquad [5.80d]$$

So that the space of eigenvectors common to these two operators leads to the diagram of energy levels from numbers n_+ and n_- such that:

$$n_+ = 0,1,2 \ldots \quad \text{and} \quad n_- = 0,1,2 \ldots \qquad [5.80e]$$

The states $|n_1 n_2\rangle$ are equivalent to states $|v_{21} v_{22}\rangle_0$ and the states $|n_+ n_-\rangle$ that combine the eigenvectors of mixed operators associated with each *1D* oscillator are linear combinations of states $|v_{21} v_{22}\rangle_0$.

The following operators can be used as a starting point:

$$N = N_+ + N_- \quad \text{and} \quad L = N_+ - N \qquad [5.81]$$

where L is the angular momentum operator, which in the space of configuration mixes the two vibrations and builds a complete set of orthonormal eigenvectors for the same two-dimensional isotropic harmonic oscillator. Given that the operator N is related to the total number of vibrational quanta and the operator L is related to the angular momentum, the eigenvectors common to N and L can be denoted by $|nl\rangle$. When n_+ increases by one unit, n_- decreases by one unit and vice versa, when n_+ decreases by one unit, n_- increases by one unit. Finally, the difference $(n_+ - n_-)$ varies by two units. If n denotes the number $n_+ + n_-$ and l denotes the number $(n_+ - n_-)$, then:

$$H|nl\rangle = (n+1)\hbar\omega|nl\rangle \quad \text{and} \quad L|nl\rangle = l|nl\rangle \qquad [5.82]$$

In this form, the equivalence $|nl\rangle \equiv |v_2 l_2\rangle_0$ can be written. The quantum numbers v_2 correspond to the number of vibration quanta and for $v_2 \neq 0$, $|l_2| = v_2, v_2 - 2, \dots, 0$. Note that in the space generated by the operators N and L, the operators $A_+^\dagger$ and A_- are the ladder operators of eigenvectors of L, which decrease the quantum number associated with L by one unit. Similarly, the operators $A_-^\dagger$ and A_+ are the ladder operators of the eigenvectors of L, which increase the quantum number associated with L by one unit. These operators are, however, not observables as they are not self-adjoint.

As mentioned in the notes of Schwinger [BIE 65, SAK 94], which refer to the works of Marojana [MAR 32], the operators J_+, J_- and J_z (see equation [5.22a]) can also be defined. It can be noted that J_z is proportional to the operator L and a three-dimensional Lie algebra is used. These operators satisfy the commutation relations of angular momenta (see equation [5.22c]) and it can be shown that: $J^2 = J_z^2 + \frac{1}{2}(J_+J_- + J_-J_+) = \left(\frac{\hbar^2}{2}\right)N\left(\frac{N}{2}+1\right) = \left(\frac{\hbar^2}{4}\right)(H^2-1)$ (see equation [5.24]) and establish a correspondence between the vibration operators and the spherical tensor operators.

5.5.2. *The degenerate vibrational level* $v_s = 1$ *(s = 3 or 4)*

For the degenerate oscillator of type T_2, three types of operators can be built, N_1, N_2 and N_3 such that (consider $s = 3$):

$$H = \hbar\omega\left(a_{31}^\dagger a_{31} + a_{32}^\dagger a_{32} + a_{33}^\dagger a_{33} + \tfrac{3}{2}\right) = \hbar\omega\left(N_1 + N_2 + N_2 + \tfrac{3}{2}\right) \qquad [5.83]$$

As described in section 5.3.2, for a triply degenerate vibration, there are $\frac{1}{2}(v_3+1)(v_3+2)$ states, denoted by $|v_{31}v_{32}v_{33}\rangle$, with the energy $E = \left(v_3+\frac{3}{2}\right)\hbar\omega$, where $v_3 = v_{31} + v_{32} + v_{33}$.

In the case of triply degenerate vibrations, the operators A_1, A_2 and A_0 and their conjugate as expressed below can be defined. They combine operators of two of the oscillators (consider $s = 3$):

$$A_1 = \frac{1}{\sqrt{2}}(a_{31} - ia_{32}),\ A_2 = \frac{1}{\sqrt{2}}(a_{31} + ia_{32}) \text{ and } A_0 = a_{33} \qquad [5.84a]$$

and the conjugate Hermitian operators $A_1^\dagger, A_2^\dagger$ and $A_3^\dagger$:

$$A_1^\dagger = \frac{1}{\sqrt{2}}\left(a_{31}^\dagger + ia_{32}^\dagger\right),\ A_2^\dagger = \frac{1}{\sqrt{2}}\left(a_{31}^\dagger - ia_{32}^\dagger\right) \text{ and } A_0 = a_{33}^\dagger \qquad [5.84b]$$

$A_m^\dagger$ and A_m satisfy the commutation relations [5.14] and can be interpreted as creation and annihilation operators of quanta of type m. The number of quanta of type m is represented by the operator $N_m = A_m^\dagger A_m$. In this case, N_1, N_0 and N_{-1} form a complete set of commuting observables, and

$$H = \left(N_1 + N_0 + N_{-1} + \frac{3}{2}\right)\hbar\omega \text{ with } N = N_1 + N_0 + N_{-1} \qquad [5.85a]$$

Consequently, the eigenvalues $|n_1 n_0 n_{-1}\rangle$ are such that:

$$|n_1 n_0 n_{-1}\rangle = \frac{1}{\sqrt{n_1! n_0! n_{-1}!}}\left(A_1^\dagger\right)^{n_1}\left(A_0^\dagger\right)^{n_0}\left(A_{-1}^\dagger\right)^{n_{-1}}|000\rangle \qquad [5.85b]$$

and

$$H|n_1 n_0 n_{-1}\rangle = \left(n + \frac{3}{2}\right)\hbar\omega|n_1 n_0 n_{-1}\rangle \text{ with } n = n_1 + n_0 + n_{-1} \qquad [5.86]$$

However, these vectors are not the eigenvectors of l^2, but of l_z and

$$l_z = (N_1 - N_0)\hbar \text{ and } m = n_1 - n_{-1} \qquad [5.87]$$

In the subspace of eigenvectors of H corresponding to the eigenvalue $\left(n+\frac{3}{2}\right)\hbar\omega$, among the $\frac{(n+1)(n+2)}{2}$ eigenvectors $|n_1 n_0 n_{-1}\rangle$ that generate the subspace, there are $n = n_1 + n_0 + n_{-1}$ that form a set of orthonormal eigenvectors of l_z. The values taken by the quantum number m are integers that span $-n$ to $+n$.

As this problem is analogous to that of a particle in a central potential, a Hilbert space can be built on the eigenvectors common to operators H, Hamiltonian sum of the kinetic energy and the central potential energy, l^2 the square of the angular momentum and l_z the projection of the angular momentum on the Oz axis. These operators are associated with commuting observables. If the energy levels are labeled using the quantum numbers n, l and m, the eigenvalue equations are written as follows:

$$H|nlm\rangle = \left(n + \frac{3}{2}\right)\hbar\omega|nlm\rangle \quad [5.88a]$$

$$l^2|nlm\rangle = l(l+1)\hbar^2|nlm\rangle \quad [5.88b]$$

$$l_z|nlm\rangle = m\hbar|nlm\rangle \quad [5.88c]$$

Each eigenvalue $\left(n + \frac{3}{2}\right)\hbar\omega$ of H has $\frac{1}{2}(n+1)(n+2)$ corresponding states of the angular momentum $|lm\rangle$ such that for each n, l can take values ranging from 0 to $n - 1$, and such that for each l, m can take values ranging from $-l$ to $+l$. The values of l are equal to $n, n - 2, \ldots, 0$ if $(-1)^n = 1$ for $(n + 2)/2$ states and equal to $n, n - 2, \ldots, 1$ if $(-1)^n = -1$ for $(n + 1)/2$ states.

The diagram of the state configurations can have the form presented in Figure 5.2, similar to the energy diagram of the hydrogen atom.

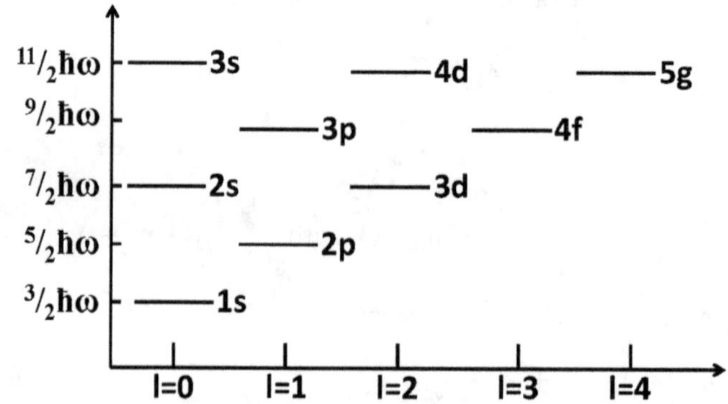

Figure 5.2. *Spectrum of the degenerate three-dimensional harmonic oscillator [MES 64]*

In an environment that modifies the spherical symmetry, the degeneracy lifting of these states can be identified by adding the perturbation generated by the coupling between the environment and the molecule. Symmetry lowering may lead to treating vibration as a vibration of double degeneracy if the symmetry becomes cylindrical, which partially lifts degeneracy or, if it is no longer symmetric, a complete degeneracy lifting. If the molecule can still have a rotational motion, the vibration–rotation coupling that involves the interaction between the internal angular momentum of vibration and the external angular momentum of rotation should be considered.

5.5.3. *Vibration–rotation Coriolis interaction*

If the rotational structure in the degenerate vibrational levels is considered: $v_3 = 1$ and $v_4 = 1$, the effects of distortion in the fundamental vibrational level, or in the excited vibrational level, can be neglected. On the contrary, the Hamiltonian of the molecule must include the Coriolis forces, which are present for the degenerate vibration modes. The Hamiltonian can be written in the form:

$$H_{VR} = H_{vib} + H_{rot} + H_{cor} \quad [5.89]$$

The interaction term H_{cor} results from the coupling between the vibrational angular moment and the total angular moment of the molecule, such that $\vec{J} = \vec{R} + \vec{l}$.

The calculation of the Coriolis interaction in the case of a triply degenerate vibration of a spherical top can be performed starting from the expression of the vibration–rotation Hamiltonian at first-order approximation and whose expression is given by equation [5.50] and written as:

$$H_{VR} = H_V + \sum_\alpha \frac{(J_\alpha - p_{v\alpha})^2}{2I^e_{\alpha\alpha}} \quad [5.90]$$

where J_α is the component α of the total angular momentum, sum of the angular momentum associated with the rotation and of the angular momentum associated with the vibration such that $\vec{J} = \vec{R} + \vec{p}_v$, with $\alpha = x, y, z$, corresponding to the axes related to the equilibrium configuration of the molecule (see Figure 5.3a and c). The expansion of the second term of equation [5.89] yields:

$$H_{VR} = \left(H_V + \sum_\alpha \frac{(p_{v\alpha})^2}{2I^e_{\alpha\alpha}}\right) + \sum_\alpha \frac{J_\alpha{}^2}{2I^e_{\alpha\alpha}} - \sum_\alpha \frac{J_\alpha p_{v\alpha}}{I^e_{\alpha\alpha}} = \widetilde{H}_V + H_{rot} + H_{cor} \quad [5.91]$$

In the specific case of spherical tops: $I^e_{xx} = I^e_{yy} = I^e_{zz} = I^e$, which implies that:

$$H_{cor} = -\frac{\vec{J}\cdot\vec{p}_v}{I^e}$$

which represents the coupling between rotation and vibration, i.e. the Coriolis interaction.

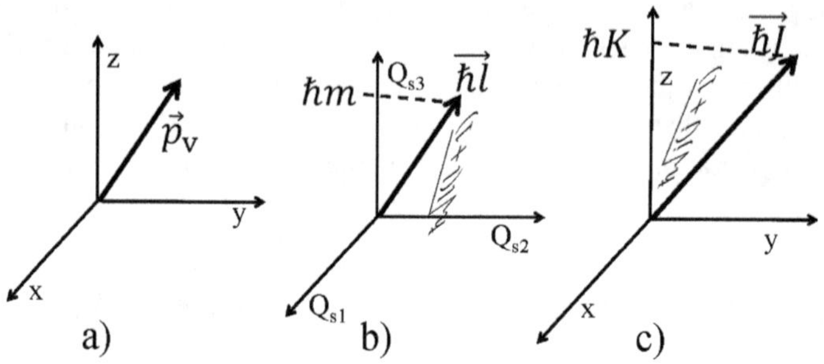

Figure 5.3. *a) Angular momentum in the physical space; b) angular momentum in the configuration space; c) total angular momentum in the physical space*

Considering $\vec{p}_v = \zeta\hbar l$, then:

$$\frac{H_{cor}}{hc} = -\zeta\hbar^2 \frac{\vec{J}\cdot\vec{l}}{hcI^e} = -2B^e\zeta_s.\vec{J}\cdot\vec{l}.$$

Consequently, for a triply degenerate state, the perturbation can be written in a first approximation as:

$$\hat{H}_{cor} = -2B\zeta_s\vec{p}.\vec{J} \ (s = 3 \text{ or } 4)$$

where in the configuration space (see Figure 5.3b), the components of the internal angular momentum are defined as follows:

$$l_x = Q_yP_z - Q_zP_y,\ l_y = Q_zP_x - Q_xP_z,\ l_z = Q_xP_y - Q_yP_x.$$

Based on the composition of angular momenta (see Figure 5.4a and b), moving from the decoupled representation $|JKlm\rangle$ to the coupled representation $|JKR\ K - m\rangle$ allows us to diagonalize the operator $\vec{J}\cdot\vec{l}$, by writing that:

$$\vec{J}\cdot\vec{l} = \frac{1}{2}(\vec{J}.\vec{J} + \vec{l}.\vec{l} - \vec{R}.\vec{R}).$$

The eigenvalue of the operator $\vec{J}\cdot\vec{l}$ in the coupled basis is given by:

$$\frac{1}{2}(J(J+1) + l(l+1) - R(R+1)) \text{ where } R = J + l, J + l - 1, \ldots, |J - l|.$$

Consequently, for a triply degenerate state and for a vibration quantum in a degenerate mode, then $l = 1$, and therefore $R = J - 1$, $R = J$ or $R = J + 1$.

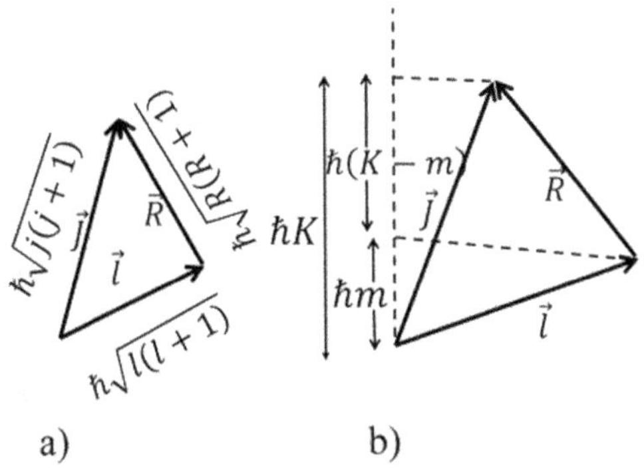

Figure 5.4. *a) Composition of angular momenta; b) projections of angular momenta*

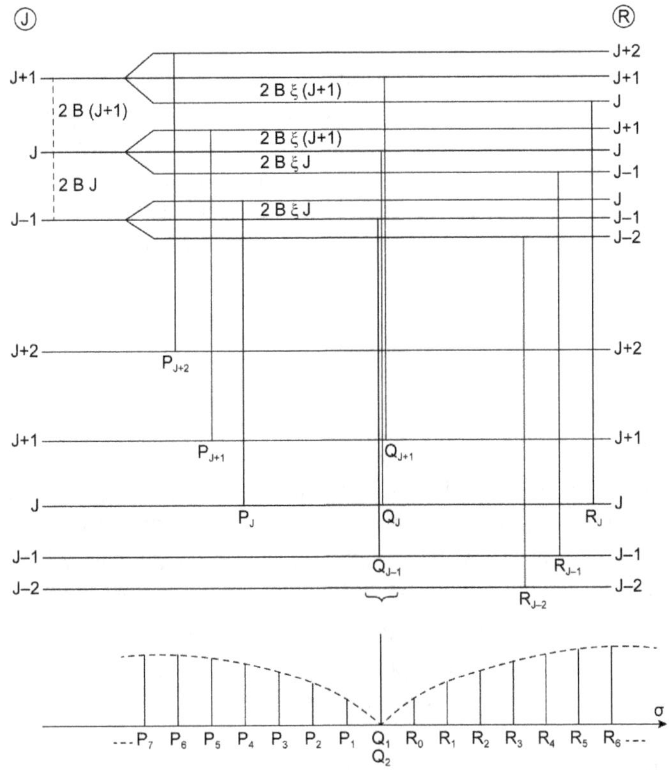

Figure 5.5. *Possible transitions of a cold band from the fundamental level to the first excited level of the triply degenerate level $v_3 = 1$, $l_3 = 1$ or $v_4 = 1$, $l_4 = 1$. (Branches P, Q and R)*

The first-order correction to the vibration–rotation energy is given by:

$$\frac{E}{hc} = -2B^e\zeta(1+J) - 2B^e\zeta, -2B^e\zeta(1-J(J+1))$$

for $R = J - 1$, J and $J + 1$, respectively. (The Hamiltonian matrix is necessarily diagonal, since $l = 1$). The rotational level splits into three sub-levels (see Figure 5.5).

For a triply degenerate vibration, a transition from the fundamental level to the first excited level (cold band) leads to an IR spectrum composed of branches P ($\Delta J = -1$), Q ($\Delta J = 0$) and R ($\Delta J = 1$). The lower level corresponds to $v_s = 0$ and $l_s = 0$ and the higher level to $v_s = 1$ and $l_s = 1$ ($s = 3$ or 4) with $R = J - 1$, J and $J + 1$. It is important to note that the Hamiltonian matrix is necessarily diagonal, since $l = 1$ and the rotational level splits into three sub-levels.

5.6. Conclusion

This chapter presents the theoretical model in the tensor formalism of the research group in Dijon, France. Applications to the electric dipole moment operators and to polarizability were presented as methods to determine the selection rules in IR and Raman spectroscopy. The vibration–rotation coupling of a triply degenerate vibration was also described in the coupled basis to determine the correction to vibration–rotation energy at first order that depends on the Coriolis interaction.

5.7. Appendices

5.7.1. *Appendix A: Quantum mechanics review*

In quantum mechanics (QM), the state space of a many-particle system is the tensor product of the state spaces of each particle ($\Omega = \Omega_1 \otimes \Omega_2 \otimes \Omega_3 \ \ldots \ \otimes \Omega_k \ldots$). This is not applicable if the particles are identical because the physical kets must be either symmetric ($\Omega_S \subset \Omega$, integral spin bosons: photon, meson, etc.) or antisymmetric ($\Omega_A \subset \Omega$, half-integral fermions: electron, positron, muon, etc.) in a permutation of particles. Consequently, only certain kets of the space of identical particles can describe these physical states.

In QM, the classical variables position vector and linear momentum vector are described by Hermitian operators that verify the commutation relation $[x, p] = i\hbar$ and operate over a space of quantum states that are square integrable functions (or kets $|\psi\rangle$ of a Hilbert space). The ket $|\psi\rangle$ and bra $\langle\psi|$ formalism, which is a notation independent of the representation (*x* or *p*), was introduced by Dirac in order to simplify the notation in QM. The states of the Hilbert space are determined by the stationary Schrödinger eigenvalue equation:

$$H|\psi_k\rangle = E_k|\psi_k\rangle \qquad [5.92]$$

where *H* is the Hamiltonian operator of the given physical system, sum of its kinetic energy *T* and its potential energy *V*. The eigenfunctions are mutually orthogonal. They are generally normed ($\langle\psi_i|\psi_k\rangle = \delta_{ik}$) and form a complete set. Each state vector $|\Psi\rangle$ of the physical system is expressed as a linear combination of $|\psi_k\rangle$, such that $|\Psi\rangle = \sum_{k=0}^{\infty} C_k|\psi_k\rangle$. Given that the C_k are determined by $C_k = \langle\psi_k|\Psi\rangle$, the state vector can be written as:

$$|\Psi\rangle = \sum_{k=0}^{\infty}|\psi_k\rangle\langle\psi_k|\Psi\rangle$$

where $P_k = |\psi_k\rangle\langle\psi_k|$ is the projection operator. This operator verifies the relation $P_k^2 = P_k$. The Hilbert space of a system Σ composed of several systems Σ_i, such that $\Sigma = \cup_i \Sigma_i$ (countable set), is the tensor product of Hilbert spaces of each subsystem such that $H_\Sigma = \prod_{\otimes i} H_{\Sigma_i}$.

For example, the Hilbert space associated with the vibrational, rotational and spin degrees of freedom of a molecule when it is in the fundamental electronic state is written as:

$$H_{VRS} = H_V \otimes H_R \otimes H_S,$$

at zeroth order.

In QM it is accepted that a given system can be prepared in a state described by a vector $|\Psi\rangle$ of the Hilbert space. In theory, if all the possible $|\Psi\rangle$ are determined and known, the same is true for the probabilities of all the possible results of a given measurement. Such states are termed pure states and the probability attached to each measurement is calculated from QM principles.

The evolution of the system is determined by the time-dependent equation:

$$\hat{H}|\psi\rangle = i\hbar\frac{\partial}{\partial t}|\psi\rangle$$

If $\hat{H}$ does not depend on time, then:

$$|\psi\rangle(t) = e^{-i\hat{H}t/\hbar}|\psi\rangle(0)$$

It can be expressed as:

$$|\psi\rangle(t) = \hat{U}(t)|\psi\rangle(0)$$

where $\hat{U}(t)$ is the time evolution operator of the physical system such that:

$$\hat{U}(t) = e^{-i\hat{H}t/\hbar}$$

When the knowledge on the state of the system is incomplete, the description of the quantum system requires the introduction of a density operator.

In QM, there are two postulates associated with the result of a measurement of an observable. They are as follows:

– Postulate 1: "a physical observable is represented by a Hermitian operator A and the result of its measurement is one of the eigenvalues of the operator with a certain probability".

– Postulate 2: "if the quantum system is in a state $|\Psi\rangle$, the average value of an observable is given by: $\langle\Psi|A|\Psi\rangle$". If p_k is the probability for the state $|\psi_k\rangle$ to be known, the average that takes into account both the quantum and statistical properties is expressed as:

$$\langle\hat{A}\rangle = \sum_k p_k \langle\psi_k|\hat{A}|\psi_k\rangle = Tr(\hat{\rho}\hat{A})$$

where Tr represents the trace (sum of diagonal elements) of the matrix $\hat{\rho}\hat{A}$

$$\hat{\rho} = \sum_k p_k |\psi_k\rangle\langle\psi_k|$$

is the density matrix of the system, where $|\psi_k\rangle\langle\psi_k| = P_k$ is the projection operator.

The density matrix is a description of a quantum system described by a statistical mixture of states. The mean of an operator associated with an observable comprises a quantum mean and a classical statistical mean. These mean values cannot be separated in the density matrix.

A pure state is represented by an eigenvector of a Hermitian operator in the Hilbert space, but a mixed state is a statistical state represented by a density matrix that is not a vector of the Hilbert space. In this case, the evolution of the density matrix is given by the time-dependent Von Neumann equation:

$$i\hbar\frac{\partial}{\partial t}\hat{\rho}=\left[\hat{H},\hat{\rho}\right]=\hat{H}\hat{\rho}-\hat{\rho}\hat{H} \quad [5.93]$$

Along with the space of quantum states of a physical system, square integrable functions (or kets $|\psi\rangle$) constitute a Hilbert space and the space of operators acting on the quantum states constitute the set of linear transformations $L(H_\Sigma)$ acting on the Hilbert space of quantum states. The construction of a complete basis of $L(H_\Sigma)$ can be achieved due to the properties of the tensor product in a vector space of infinite dimension and to the fact that an arbitrary operator of $L(H_\Sigma)$ is a representation of the group O(3).

5.7.2. *Appendix B: Creation and annihilation operators*

A real oscillator can be studied on the basis of the harmonic oscillator, by introducing Hermitian and conjugate creation and annihilation operators (second quantization), defined as:

$$\mathrm{a}^{\dagger}=\frac{1}{\sqrt{2}}(q-ip) \quad [5.94]$$

$$\mathrm{a}=\frac{1}{\sqrt{2}}(q+ip) \quad [5.95]$$

and verifying the commutation relation:

$$\left[\mathrm{a},\mathrm{a}^{\dagger}\right]=\mathrm{aa}^{\dagger}-\mathrm{a}^{\dagger}\mathrm{a}=1 \quad [5.96]$$

Inversely, the operators q and p are written as:

$$q = \frac{1}{\sqrt{2}}\left(\mathrm{a}^{\dagger} + \mathrm{a}\right) \quad [5.97]$$

$$p = \frac{i}{\sqrt{2}}\left(\mathrm{a}^{\dagger} - \mathrm{a}\right) \quad [5.98]$$

The matrix elements of the creation $\mathrm{a}^{\dagger}$ and annihilation a operators are given by:

$$_0\langle \mathrm{v}+1|\mathrm{a}^{\dagger}|\mathrm{v}\rangle_0 = \sqrt{\mathrm{v}+1} \quad [5.99]$$

$$_0\langle \mathrm{v}|\mathrm{a}|\mathrm{v}+1\rangle_0 = \sqrt{\mathrm{v}+1} \quad [5.100]$$

and their application to any eigenvector $|\mathrm{v}\rangle_0$ induces, respectively, the creation and annihilation of a vibrational photon such that:

$$\mathrm{a}^{\dagger}|\mathrm{v}\rangle_0 = \sqrt{\mathrm{v}+1}\,|\mathrm{v}+1\rangle_0 \quad [5.101]$$

$$\mathrm{a}|\mathrm{v}\rangle_0 = \sqrt{\mathrm{v}}\,|\mathrm{v}-1\rangle_0 \quad [5.102]$$

with

$$\mathrm{a}|0\rangle_0 = 0 \quad [5.103]$$

From the above, the Hamiltonian operator H_{vib}^{0} of the harmonic oscillator is written as:

$$H_{\mathrm{vib}}^{0} = \frac{1}{2}\hbar\omega_0\left(p^2 + q^2\right) = \hbar\omega_0\left(\mathrm{a}^{\dagger}\mathrm{a} + \frac{1}{2}\right) \quad [5.104]$$

This leads to:

$$H_{\mathrm{vib}}^{0}|\mathrm{v}\rangle_0 = E_{\mathrm{v}}^{0}|\mathrm{v}\rangle_0 \Rightarrow \hbar\omega_0\left(\mathrm{a}^{\dagger}\mathrm{a} + \frac{1}{2}\right)|\mathrm{v}\rangle_0 = \hbar\omega_0\left(\mathrm{v} + \frac{1}{2}\right)|\mathrm{v}\rangle_0 \quad [5.105]$$

This expression shows that the action of the Hermitian operator $N = a^{\dagger}a$ on an eigenvector $|v\rangle_0$ leads to the multiplication of the quantum number v by this eigenvector, hence:

$$N|v\rangle_0 = a^{\dagger}a|v\rangle_0 = v|v\rangle_0 \qquad [5.106]$$

This property is very useful in the calculations of perturbations that are applied to the harmonic oscillator.

5.7.3. *Appendix C: Clebsch–Gordan coefficients and Wigner* 3j*-symbols*

The coupled states of the spherical harmonics can be expanded from the closure relation in the decoupled basis such that

$$|JM\rangle = \sum_{m_1=-j_1}^{m_1=+j_1} \sum_{m_2=-j_2}^{m_2=-j_2} |j_1 m_1 j_2 m_2\rangle\langle j_1 m_1 j_2 m_2|JM\rangle \qquad [5.107]$$

where $\langle j_1 m_1 j_2 m_2|JM\rangle$ are the Clebsch–Gordan coefficients. They are sometimes written in the forms:

$$\langle j_1 j_2; m_1 m_2|JM\rangle \text{ or } C(j_1 j_2 J; m_1 m_2 M)$$

The quantum numbers j_1, j_2 and J verify the condition: $|j_1 - j_2| \leq J \leq j_1 + j_2$ and the magnetic quantum numbers m_1, m_2 and M verify the relation $m_1 + m_2 = M$.

A calculation code of the Clebsch–Gordan coefficients in FORTRAN language is presented in section 4.7.3.

Clebsch–Gordan coefficients can be expressed as a function of Wigner $3j$-symbols, easier to use considering their symmetry properties, as follows:

$$\begin{pmatrix} j_1 & j_2 & J \\ m_1 & m_2 & -M \end{pmatrix} = \frac{(-1)^{j_1-j_2+M}}{\sqrt{2J+1}} \langle j_1 m_1 j_2 m_2|JM\rangle \qquad [5.108]$$

where the exponent $j_1 - j_2 + M$ is an integer. The inverse relation is written as:

$$\langle j_1 m_1 j_2 m_2|JM\rangle = (-1)^{j_2-j_1-M} \sqrt{2J+1} \begin{pmatrix} j_1 & j_2 & J \\ m_1 & m_2 & -M \end{pmatrix} \qquad [5.109]$$

The sum of the arguments $j_1 + j_2 + J$ is an integer.

5.7.4. Appendix D: Tensor operators and the Wigner–Eckart theorem

A tensor operator denoted by $T_q^{(k)}$, where q = – k, …, k, is a set of operators that, after a rotation, are expressed linearly as a function of the operators of the set.

If J is the total angular momentum of the space in which these tensor operators act, the $2k + 1$ operators $T_q^{(k)}$ are the standard components of an irreducible operator of order k, if they satisfy the following relations:

$$\left[J_z, T_q^{(k)}\right] = qT_q^{(k)} \qquad [5.110]$$

and

$$\left[J_\pm, T_q^{(k)}\right] = \sqrt{k(k+1) - q(q \pm 1)}\,T_{q\pm 1}^{(k)} \qquad [5.111]$$

The operators associated with physical variables must be invariant under a rotation by an angle 2π. This requires k to be a non-negative integer, k = 0, 1, 2, ….

In a standard representation $(\alpha, J^2 J_z)$, the matrix elements of the operators $T_q^{(k)}$ verify a fundamental property known as the Wigner–Eckart theorem such that

$$\left\langle \alpha J M \middle| T_q^{(k)} \middle| \alpha' J' M' \right\rangle = \langle J' k M' q | J M \rangle \langle \alpha J || T^{(k)} || \alpha' J' \rangle \qquad [5.112]$$

In this relation, the Clebsch–Gordan coefficient contains all the geometric information (relative orientation of the states and of the operator through M, M' and q), the physical information (nature of the states and of the operator) being represented by the reduced matrix element.

If V_x, V_y and V_z are the Cartesian components of a vector operator, its standard components $V_1^{(1)}, V_0^{(1)}$ and $V_{-1}^{(1)}$ written as:

$$V_{\pm 1}^{(1)} = \mp \frac{1}{\sqrt{2}}\left(V_x \pm iV_y\right) \text{ and } V_0^{(1)} = V_z \qquad [5.\ 113]$$

More generally, it is written as:

$$\left[V_i, J_j\right] = i\hbar\, \epsilon_{ijk}\, V_k \qquad [5.\ 114]$$

where $\epsilon_{123} = \epsilon_{312} = \epsilon_{231} = 1$ and $\epsilon_{132} = \epsilon_{321} = \epsilon_{213} = -1$ (with 1 for x, 2 for y and 3 for z). This property can be used to define a vector operator.

The coupling of angular momenta and that of tensor operators can be analogously conducted. The transition from the decoupled basis to the coupled basis uses the following closure relation:

$$1 = \sum_{m_1,m_2} |j_1, m_1, j_2, m_2\rangle\langle j_1, m_1, j_2, m_2| \qquad [5.115]$$

This yields:

$$|J, M, j_1, j_2\rangle = \sum_{m_1,m_2} \langle j_1, m_1, j_2, m_2 | J, M, j_1, j_2\rangle \, |j_1, m_1, j_2, m_2\rangle \qquad [5.116]$$

Similarly, two tensor operators can be coupled such that:

$$T_q^{(k)} = \sum_{q_1=-k_1}^{q_1=k_1} \sum_{q_2=-k_2}^{q_2=k_2} \langle q_1, k_1, q_2, k_2 | k, q, k_1, k_2\rangle T_{q_1}^{(k_1)} T_{q_2}^{(k_2)} \qquad [5.117]$$

The first term in the sum is a Clebsch–Gordan coefficient that is non-zero if $q_1 + q_2 = q$ and if $|k_1 - k_2| \leq k \leq |k_1 + k_2|$.

5.7.5. *Appendix E: Hamiltonian as a function of dimensionless normal coordinates up to fourth order*

The Hamiltonian of a polyatomic molecule [AMA 57a, AMA 57b, AMA 58, AMA 71] can be written as a series expansion:

$$H = H_0 + H_1 + H_2 + \cdots \qquad [5.118]$$

with

$$H_0 = \frac{1}{2}\sum_\alpha \frac{J_\alpha^2}{I_{\alpha\alpha}^e} + \frac{\hbar}{2}\sum_{s\sigma} \omega_s \left(\frac{p_{s\sigma}^2}{\hbar^2} + q_{s\sigma}^2\right) \qquad [5.119]$$

$$H_1 = -\sum_\alpha \frac{J_\alpha p_\alpha}{I_{\alpha\alpha}^e} - \frac{1}{2}\sum_\alpha \sum_\beta \sum_{s\sigma} \left(\frac{\hbar}{\omega_{s\sigma}}\right)^{\frac{1}{2}} \frac{a_{s\sigma}^{\alpha\beta}}{I_{\alpha\alpha}^e I_{\beta\beta}^e} J_\alpha (J_\beta - 2p_\alpha)\, q_{s\sigma}$$

$$+hc \sum_{s\sigma} \sum_{s'\sigma'} \sum_{s''\sigma''} K_{ss's''} q_{s\sigma} q_{s'\sigma'} q_{s''\sigma''} \qquad [5.120]$$

and

$$H_2 = \frac{1}{2}\sum_\alpha \frac{p_\alpha^2}{I_{\alpha\alpha}^e} + \frac{1}{2}\sum_\alpha \sum_\beta \sum_{s\sigma} a_{s\sigma}^{\alpha\beta} \frac{J_\alpha p_\beta}{I_{\alpha\alpha}^e I_{\beta\beta}^e} q_{s\sigma}$$

$$-\frac{1}{2}\sum_{s\sigma}\sum_{s'\sigma'}\left(\frac{\hbar^2}{\omega_s\omega_{s'}}\right)^{\frac{1}{2}}\left\{\sum_{\alpha}\sum_{\beta}\left(A'^{\alpha\beta}_{s\sigma s'\sigma'}-\sum_{\gamma}\frac{a^{\alpha\gamma}_{s\sigma}a^{\gamma\beta}_{s'\sigma'}}{I^e_{\gamma\gamma}}\right)\frac{J_\alpha j_\beta}{I^e_{\alpha\alpha}I^e_{\beta\beta}}\right\}q_{s\sigma}q_{s'\sigma'}$$

$$+hc\sum_{s\sigma}\sum_{s'\sigma'}\sum_{s''\sigma''}\sum_{s'''\sigma'''}K_{ss's''s'''}q_{s\sigma}q_{s'\sigma'}q_{s''\sigma''}q_{s'''\sigma'''}\,..$$

[5.121]

where:

$$a^{\alpha\alpha}_{s\sigma}=2\sum_{i}^{3N}m_i^{1/2}\left(\beta^e_i\ell^\beta_{is\sigma}+\gamma^e_i\ell^\gamma_{is\sigma}\right) \qquad [5.122]$$

$$a^{\alpha\beta}_{s\sigma}=-\sum_{i}^{3N}m_i^{1/2}\left(\alpha^e_i\ell^\beta_{is\sigma}+\beta^e_i\ell^\gamma_{is\sigma}\right) \qquad [5.123]$$

$$A^{\alpha\alpha}_{s\sigma s'\sigma'}=\sum_{i}^{3N}\left(\ell^\beta_{is\sigma}\ell^\beta_{is'\sigma'}+\ell^\gamma_{is\sigma}\ell^\gamma_{is'\sigma'}\right) \qquad [5.124]$$

$$A^{\alpha\beta}_{s\sigma s'\sigma'}=-\sum_{i}^{3N}\ell^\alpha_{is\sigma}\ell^\beta_{is'\sigma'} \qquad [5.125]$$

$$A'^{\alpha\alpha}_{s\sigma s'\sigma'}=A^{\alpha\beta}_{s\sigma s'\sigma'}-\sum_{s''\sigma''}^{k}\zeta^\alpha_{s\sigma s''\sigma''}\zeta^\beta_{s'\sigma's''\sigma''} \qquad [5.126]$$

H_0 is the sum of the Hamiltonian of a rigid rotator of the inertia tensor $I^e_{\alpha\beta}$ and of $3N-6$ $(3N-5)$ independent harmonic oscillators for a nonlinear (linear) molecule in which there is a vibration–rotation separation.

While the translational motion can be rigorously separated from the vibrational and rotational motions by the Eckart–Sayvetz conditions, this is not true for the separation between vibration and rotation. The terms H_1 and H_2 contain the terms of interaction between rotation and vibration. The eigenstates of H_0 serve as a basis and H_1 and H_2 are treated as a perturbation.

– The first-order term H_1 is expressed as a function of the terms in $J_\alpha p_\alpha$ and $J_\alpha p^*_\beta q_{s\sigma}$ that represent the Coriolis interaction between the internal angular momentum and the total angular momentum; the term $J_\alpha J_\beta q_{s\sigma}$ corresponds to a first-order effect of the vibration on the value of the moments of inertia; the term in $q_{s\sigma}q_{s's}q_{s''s''}$ results from the anharmonicity of the vibrations.

– The second-order term H_2 comprises the term in p^2_α that represents the energy associated with the internal angular momentum, while the other terms represent the same effects as the terms of the second-order Hamiltonian H_1.

5.7.6. *Appendix F: Hamiltonian transformed using the contact method*

The contact transformation technique involves the use of the eigenfunctions of the zeroth-order Hamiltonian to calculate the zeroth-order energies. This is followed by the corrections to these energies that result from the other terms of the Hamiltonian, H'_1 and H'_2. The first-order corrective terms are obtained by a first-order perturbation calculation on H'_1; the second-order corrective terms are obtained by a second-order perturbation calculation on H'_1 and a first-order perturbation calculation on H'_2. A commonly used method in molecular physics is that of the unitary transformations proposed by Van Vleck. Its use relies on the physical properties of the molecule. An inspection of the vibrational states of a molecule makes it possible to group the states whose energies are too close to neglect their potential interactions. These groups of states are termed polyads of interacting states and must be simultaneously treated by the calculation. The transformation to be determined is such that the transformed Hamiltonian H'' is represented in the basis of the zeroth-order Hamiltonian H'_0, by a matrix that is decomposed into blocks according to the resonance diagram. The purpose of the contact transformation method is to eliminate, at a certain order of approximation, the matrix elements coupling the states that do not belong to the same polyad. More precisely, given H' the expanded Hamiltonian:

$$H' = H'_0 + \lambda H'_1 + \lambda^2 H'_2 + \cdots \qquad [5.127]$$

where λ designates an undefined parameter, and then H' is submitted to a contact transformation such that:

$$H'' = \mathrm{T}H'\mathrm{T}^{-1} = H''_0 + \lambda H''_1 + \lambda^2 H''_2 + \cdots \qquad [5.128]$$

where the matrix T is the unitary matrix of the form:

$$\mathrm{T} = e^{i\lambda_1 S_1} = 1 + i\lambda_1 S_1 - \frac{1}{2}\lambda_1^2 S_1^2 + \cdots \qquad [5.129]$$

and

$$\mathrm{T}^{-1} = e^{-i\lambda_1 S_1} = 1 - i\lambda_1 S_1 - \frac{1}{2}\lambda_1^2 S_1^2 + \cdots \qquad [5.130]$$

S is chosen such that $H''_0 = H'_0$ and such that the matrix of H''_1 is diagonal according to the resonance diagram

$$
\begin{aligned}
&H''_0 = H'_0 \\
&H''_1 = -\sum_{\alpha}\sum_{s\sigma}^{k} \zeta^{\alpha}_{s\sigma s\sigma'}\left(q'_{s\sigma}p'_{s\sigma'} - q'_{s\sigma'}p'_{s\sigma}\right)\frac{J_\alpha}{I^e_{\alpha\alpha}} \\
&H''_{2\mathrm{v}} = -hc\sum_{s\sigma}^{k}\frac{1}{\omega'_s}\left[\frac{3}{2}\left(p'^2_{s\sigma}q'^2_{s\sigma} + q'^2_{s\sigma}p'^2_{s\sigma}\right) + 1 + \frac{3}{2}q'^4_{s\sigma}\right] \\
&-\frac{hc}{2}\sum_{s\sigma}^{k}\sum_{s'\sigma'}^{k}\frac{k'^2_{sss'}}{\omega'_{s'}(4\omega'^2_s - \omega'^2_{s'})}\begin{bmatrix}(2\omega'^2_s - \omega'^2_{s'})q'^4_{s\sigma} + 4\omega_s\omega_{s'}q'^2_{s\sigma}q'^2_{s'\sigma'} + \\ 4\omega'^2_s(p'_{s\sigma}q'_{s\sigma} + q'_{s\sigma}p'_{s\sigma})\dfrac{p'_{s'\sigma'}q'_{s'\sigma'}}{\hbar^2} + 2\omega'^2_S q'^2_{s\sigma}\dfrac{p'^2_{s\sigma}}{\hbar^2}\end{bmatrix} \\
&-hc\sum_{s\sigma}^{k}\sum_{s'\sigma'}^{k}\frac{k'_{sss}k'_{ss's'}}{\omega'_s}\left[\frac{q'^2_{s'\sigma'}p'^2_{s\sigma}}{\hbar^2} + \frac{3q'^2_{s\sigma}p'^2_{s'\sigma'}}{\hbar^2}\frac{\omega'^2_{s'}}{4\omega^2_s - \omega'^2_{s'}} + q'^2_{s\sigma}q'^2_{s'\sigma'}\frac{5\omega'^2_{s'} - 2\omega'^2_s}{4\omega'^2_{s'} - 2\omega'^2_s}\right] \qquad [5.131] \\
&-\frac{hc}{2}\sum_{s\sigma}^{k}\sum_{s'\sigma'}^{k}\sum_{s''\sigma''}^{k}\frac{k'_{sss'}k'_{s''s''s'}}{\omega'_{s'}(4\omega'^2_s - \omega'^2_{s'})}\left[\frac{2\omega'^2_S}{\hbar^2}q'^2_{s''\sigma''}p'^2_{s\sigma} + (2\omega'^2_s - \omega'^2_{s'})q'^2_{s''\sigma''}q'^2_{s\sigma}\right] \\
&-\frac{hc}{2}\sum_{s\sigma<}^{k}\sum_{s'\sigma'<}^{k}\sum_{s''\sigma''}^{k}D_{ss's''}k^2_{ss's''}\begin{bmatrix}\omega'_s(\omega'^2_s - \omega'^2_{s'} - \omega'^2_{s''})q'^2_{s'\sigma'}q'^2_{s''\sigma''} \\ +\omega'_{s'}(\omega'^2_{s'} - \omega'^2_{s''} - \omega'^2_s)q'^2_{s''\sigma''}q'^2_{s\sigma} \\ +\omega'_{s''}(\omega'^2_{s''} - \omega'^2_s - \omega'^2_{s'})q'^2_{s\sigma}q'^2_{s\sigma} \\ -\dfrac{2}{\hbar^2}(\omega'^2_s\omega'^2_{s'}\omega'^2_{s''})^{1/2} \\ \times(p'_{s'\sigma'}q'_{s'\sigma'}p'_{s''\sigma''}q'_{s\sigma''} + q'_{s\sigma}p'_{s\sigma}q'_{s'\sigma'}p'_{s'\sigma'} + q'_{s\sigma}p'_{s\sigma}p'_{s''\sigma''}q'_{s''\sigma''})\end{bmatrix}
\end{aligned}
$$

The calculations can be conducted in this manner until the elimination, up to a given order of approximation, of all the operators coupling the states belonging to different polyads. Then, the Hamiltonian matrix appears in block diagonal form, and it is then sufficient to study each block individually relative to a polyad of interacting states.

At first-order approximation, the matrix H''_1 contains non-diagonal elements that take into account the resonance phenomena specific to each molecule: the Coriolis resonance interaction corresponds to very close frequencies ω'_s and $\omega'_{s'}$; the *anharmonic* resonance interactions concern the levels whose frequencies are such that $2\omega'_s \cong \omega'_{s'}$.

The calculation of the Hamiltonian H''_2 involves a great number of terms whose elements contribute to the second-order correction of the vibration energy. Two contact transformations, corresponding to a second-order approximation, are often necessary in order to eliminate non-diagonal terms in the Hamiltonian H''_2.

The vibration and rotation energies being calculated, the first- and second-order corrections due to Hamiltonians H''_1 and H''_2 must be added.

The first-order correction of Coriolis interaction must be added to the rotation energy; the second-order correction modifies the vibration energy. The basis functions Ψ_0 are used to calculate the contributions to the vibration energy due to H''_{2v} which contains the anharmonicity terms of the potential energy.

References

[ABO 73] ABOUAF-MARGUIN L., Étude du mouvement et de la relaxation vibrationnelle de l'ammoniac isolé en matrice à basse température, PhD thesis, Université Pierre et Marie Curie, Paris, 1973.

[AKI 19] AKIYAMA K. "The first M87 event horizon telescope results. The shadow of the supermassive black hole", *Astrophys. J.*, vol. 875, no. 1, 2019.

[ALI 76] ALIEV M.R., WATSON J.K.G., "Calculated sextic centrifugal distortion constants of polyatomic molecules", *J. Mol. Spec.*, vol. 61, pp. 29–52, 1976.

[ALI 85] ALIEV M.R., WATSON J.K.G., "Higher-order effects in the vibration-rotation spectra of semi-rigid molecules", in NARAHARI RAO K. (ed.), *Molecular Spectroscopy: Modern Reseach*, Academic Press, Orlando, pp. 1–67, 1985.

[ALI 86] ALIEV M.R., MIKHAILOV V.M., WATSON J.K.G., "Higher order dipole moments for pure rotational transitions of methane-type molecules", *J. Mol. Spec.*, vol. 118, pp. 544–547, 1986.

[AMA 57a] AMAT G., GOLDSMITH M., NIELSEN H.H., "Higher order rotation-vibration energies of polyatomic molecules II", *J. Chem. Phys.*, vol. 27, pp. 838–844, 1957.

[AMA 57b] AMAT G., NIELSEN H.H., "Higher order rotation-vibration energies of polyatomic molecules III", *J. Chem. Phys.*, vol. 27, pp. 845–850, 1957.

[AMA 58] AMAT G., NIELSEN H.H., "Higher order rotation-vibration energies of polyatomic molecules IV", *J. Chem. Phys.*, vol. 29, pp. 665–672, 1958.

[AMA 71] AMAT G., NIELSEN H.H., TARRAGO G., *Rotation-Vibration of Polyatomic Molecules*, Marcel Dekker, New York, 1971.

[ARI 75] ARIMA A., IACHELLO F., "Collective nuclear states as representations of a SU(6) group", *Phys. Rev. Lett.*, vol. 35, pp. 1069–1072, 1975.

[BAR 61] BARCHEWITZ P., *Spectroscopie infrarouge 1, Vibrations moléculaires*, Gauthier-Villars, Paris, 1961.

[BAR 88] BARBARA B., GIGNOUX D., VETTIER C., *Lectures on Modern Magnetism*, Science Press, Beijing and Springer-Verlag, Berlin/New York, 1988.

[BEA 11] BEAULIEU J.P., TINETTI G., KIPPING D. *et al.*, "Methane in the atmosphere of the transiting hot Neptune GJ436B", *Astrophys. J.*, vol. 731, pp. 16–27, 2011.

[BEN 57] BENEDICT W.S., PLYLER E.K., "Vibration-rotation bands of ammonia, II: The molecular dimensions and harmonic frequencies of ammonia and deuterated ammonia", *Can. J. Phys.*, vol. 35, pp. 1235–1241, 1957.

[BET 29] BETHE H.A., "Splitting of terms in crystals", *Annalen der Physik*, vol. 3, pp. 133–206, 1929.

[BIE 65] BIEDENHARN L.C., VAN DAM H., *Quantum Theory of Angular Momentum*, Academic Press, New York, 1965.

[BOI 85] BOISSEL P., Relaxation vibrationnelle en matrice: SF_6 et NH_3 en matrice de gaz rare et d'azote, PhD thesis, Université Paris Sud, Orsay, 1985.

[BOU 95] BOUDON V., Spectroscopie des hexafluorures à nombre impair d'électrons, PhD thesis, Université de Bourgogne, Dijon, 1995.

[BRA 72] BRADLEY C.J., CRACKNELL P., *The Mathematical Theory of Symmetry in Solids: Representation Theory for Point Groups and Space Groups*, Clarendon Press Oxford, Oxford, 1972.

[BRO 95] BROSSET P., Position et mouvement de la molécule d'ozone isolée en matrice inerte à basse température : spectre d'absorption infrarouge et modèle théorique, PhD thesis, Université Pierre et Marie Curie, Paris, 1995.

[BUC 67] BUCKINGHAM A.D., "Molecular moments and intermolecular forces", *Adv. Chem. Phys.*, vol. 12, p. 107, 1967.

[BUN 97] BUNKER P.R., SCHUTTE J.H., HOUGEN J.T. *et al.*, "Notations and conventions in molecular spectroscopy: Part 3. Permutation and permutation-inversion symmetry notation (IUPAC Recommendations 1997)", *Pure Appl. Chem.*, vol. 69, pp. 1651–1657, 1997.

[BUN 98] BUNKER P.R., JENSEN P., *Molecular Symmetry and Spectroscopy*, NRC Research Press, Ottawa, 1998.

[BUN 99] BUNKER P.R., JENSEN P., "Spherical top molecules and the molecular symmetry group", *Mol. Phys.*, vol. 97, pp. 255–264, 1999.

[CAM 85] CAMY-PEYRET C., FLAUD J.M., "Vibration-rotation dipole moment operator for asymmetric rotors", in NARAHARI RAO K. (ed.), *Molecular Spectroscopy: Modern Reseach*, Academic Press, Orlando, pp. 69–110, 1985.

[CHA 77a] CHAMPION J.P., "Développement complet de l'hamiltonien de vibration-rotation adapté à l'étude des interactions dans les molécules toupies sphériques. Application aux bandes ν_2 et ν_4 de $^{12}CH_4$", *Can. J. Phys.*, vol. 55, pp. 1802–1828, 1977.

[CHA 77b] CHAMPION J.P., PIERRE G., MICHELOT F. *et al.*, "Composantes cubiques normales des tenseurs sphériques", *Can. J. Phys.*, vol. 55, pp. 512–520, 1977.

[CHA 78] CHAMPION J.P., Développement complet de l'hamiltonien de vibration-rotation des molécules toupies sphériques. Application : étude des interactions vibrationnelles et analyse des bandes harmoniques et de combinaison, exemple du méthane CH_4, PhD thesis, Université de Dijon, Dijon, France, 1978.

[CHA 80] CHAMPION J.P., PIERRE G., "Vibration-rotation energies of harmonic and combination levels in tetrahedral XY_4 molecules", *J. Mol. Spec.*, vol. 79, pp. 255–280, 1980.

[CHA 92] CHAMPION J.P., LOËTE M., PIERRE G., *The Spectroscopy of the Earth's Atmosphere and Interstellar Medium*, Academic Press, New York, 1992.

[CHI 38] CHILDS W.H.J., JAHN H.A., "A new Coriolis perturbation in the methane spectrum III. Intensities and optical spectrum", *Proc. Roy. Soc. A*, vol. 169, no. 939, pp. 451–463, 1938.

[CHO 13] CHOUDHURY J., "Analysis of the vibrational spectra of polyatomic molecules using an algebraic model II. An application to AB_4", *Chin. J. Phys.*, vol. 51, pp. 500–510, 2013.

[CLA 75] CLARON HOSKINS L., "Pure rotational Raman spectroscopy of diatomic molecules", *J. Chem. Edu.*, vol. 52, pp. 568–572, 1975.

[CLA 81] CLAUDE M.L., VALENTIN A., "Tables for direct determination of spectral line parameters from spectrometric data – I. Pure Lorentz broadening", *J.Q.S.R.T.*, vol. 26, pp. 349–363, 1981.

[CLA 84] CLAUDE M.L., VALENTIN A., "Tables for direct determination of spectral line parameters from spectrometric data – II. Intermediate Voigt broadening", *J.Q.S.R.T.*, vol. 32, pp. 17–48, 1984.

[COQ 94] COQUARD M.P., Effet stark dans les molécules tétraédriques, PhD thesis, Université de Bourgogne, Dijon, 1994.

[COQ 95] COQUARD M.M., LOËTE M., "Stark effect in tetrahedral XY_4 molecules: Transition intensities", *J. Mol. Spec.*, vol. 172, pp. 485–502, 1995.

[COT 63] COTTON F.A., *Chemical Applications of Group Theory*, John Wiley & Sons, New York, 1963.

[COU 13] COUSTENIS A., BOUDON V., "Methane spectroscopy and its applications to planetary atmospheres, including the Earth's", *J. Mol. Spec.*, vol. 291, pp. 1–124, 2013.

[COU 86] COUDERT L., Étude des molécules possédant des potentiels à plusieurs minima, Application aux dérivés deutérés de l'ammoniac et du méthanol, PhD thesis, Université Pierre et Marie Curie, Paris, 1986.

[CRO 63] CROSS P.C., ALLEN H.C., *Molecular Vibration-Rotors*, John Wiley & Sons, New York, 1963.

[DAH 96] DAHOO P.R., Dynamique moléculaire en phase condensée, à basse température, HDR, Université de Versailles Saint-Quentin-en-Yvelines, Versailles, 1996.

[DAH 17] DAHOO P.R., LAKHLIFI A., *Infrared Spectroscopy of Diatomics for Space Observation*, ISTE Ltd, London and John Wiley & Sons, New York, 2017.

[DAH 19] DAHOO P.R., LAKHLIFI A., *Infrared Spectroscopy of Triatomics for Space Observation*, ISTE Ltd, London and John Wiley & Sons, New York, 2019.

[DAR 40] DARLING B.T., DENISSON D.M., "The water vapor molecule", *Phys. Rev.*, vol. 57, pp. 128–139, 1940.

[DEC 77] DECIUS J.C., HEXTER R.M., *Molecular Vibrations in Crystals*, McGraw-Hill, New York, 1977.

[DEN 32a] DENNISON D.M., HARDY J.D., "The parallel type absorption bands of ammonia", *Phys. Rev.*, vol. 39, pp. 938–947, 1932.

[DEN 32b] DENNISON D.M., UHLENBECK G.E., "The two-minima problem and the ammonia molecule", *Phys. Rev.*, vol. 41, pp. 313–321, 1932.

[DIN 97] DINELLI B.M., MILLER S., ACHILLEOS N. *et al.*, "UKIRT observations of the impact and consequences of Comet Shoemaker–Levy 9 on Jupiter", *Icarus*, vol. 126, pp. 107–125, 1997.

[DOR 72] DORNEY A.J., WATSON J.K.G., "Forbidden rotational spectra of polyatomic molecules: Stark effects and ΔJ=0 transitions of T_d molecules", *J. Mol. Spec.*, vol. 42, pp. 135–148, 1972.

[DUB 75] DUBOST H., Étude des mouvements moléculaires, de la relaxation et des transferts d'énergie vibrationnelle de l'oxyde de carbone isolé en matrice de gaz rare à basse température, PhD thesis, Université Paris Sud, Orsay, 1975.

[ECK 35] ECKART C., "Some studies concerning rotating axes and polyatomic molecules", *Phys. Rev.*, vol. 47, pp. 552–558, 1935.

[EDM 74] EDMONDS A.R., *Angular Momentum in Quantum Mechanics*, Princeton University Press, Princeton, 1974.

[EIN 15] EINSTEIN A., "The field equations of gravitation", *Sitzungsberichte der Preussischen Akademie der Wissenschaften*, pp. 844–847, Berlin, 25 November, 1915.

[FLA 81] FLAUD J.M., CAMY-PEYRET C., TOTH R.A., *Water Vapour Line Parameters from Microwave to Medium Infrared: An Atlas of* $H_2{}^{16}O$, $H_2{}^{17}O$ *and* $H_2{}^{18}O$ *Line Positions and Intensities between 0 and 4350* cm^{-1}, Pergamon Press, Oxford, 1981.

[FLA 90] FLAUD J.M., CAMY-PEYRET C., RINSLAND C.P. *et al.*, *Atlas of Ozone Spectral Parameters from Microwave to Medium Infrared*, Academic Press, Boston, 1990.

[GAB 96] GABART T., Étude des effets collisionnels dans les molécules tétraédriques. Applications au méthane perturbé par l'argon, PhD thesis, Université de Bourgogne, Dijon, 1996.

[GAU 80] GAUTHIER ROY B., Relaxation de l'énergie vibrationnelle du fluorure de méthyle en matrice inerte à basse température, PhD thesis, Université Paris Sud, Orsay, 1980.

[GEB 96] GEBALLE T.R., KULKARNI S.R., WOODWARD C.E. *et al.*, "The near infrared spectrum of the brown dwarf gliese", *Astrophys. J.*, vol. 467, pp. 101–104, 1996.

[GHE 79] GHERISSI S., Étude à haute résolution des bandes de la molécule $^{12}CH_4$. Calcul de l'intensité des raies de vibration-rotation et des constantes moléculaires, PhD thesis, Université Pierre et Marie Curie, Paris, 1979.

[GIL 74] GILMORE R., *Lie Groups, Lie Algebras and Some of Their Applications*, Dover Publications, Mineola/New York, 1974.

[GOL 56] GOLDSMITH M., AMAT G., NIELSEN H.H., "Higher order rotation-vibration energies of polyatomic molecules II", *J. Chem. Phys.*, vol. 24, pp. 838–844, 1956.

[GRA 79] GRAY D.L., ROBIETTE A.G., "The anharmonic force field and equilibrium structure of methane", *Mol. Phys.*, vol. 37, pp. 1901–1920, 1979.

[GRI 64] GRIFFITH J.S., *The Theory of Transitions-Metal Ions*, Cambridge University Press, London, 1964.

[HAM 62] HAMERMESH M., *Group Theory and Its Application to Physical Problems*, Wesley Publishing, Boston, 1962.

[HEC 60a] HECHT K.T., "The vibration-rotation energies of tetrahedral XY_4 molecules: Part I. Theory of spherical top molecules", *J. Mol. Spec.*, vol. 5, pp. 355–389, 1960.

[HEC 60b] HECHT K.T., "The vibration-rotation energies of tetrahedral XY_4 molecules: Part II. The fundamental ν_3 of CH_4", *J. Mol. Spec.*, vol. 5, pp. 390–404, 1960.

[HER 45] HERZBERG G., *Molecular Spectra and Molecular Structure II: Infrared and Raman Spectra of Polyatomic Molecules*, vol. 2, D. Van Nostrand, New York, 1945.

[HIL 70] HILICO J.C., "Expression tensorielle de l'hamiltonien de vibration-rotation des molécules à symétrie tétraédrique", *Journal de Physique*, vol. 31, pp. 15–20, 1970.

[HIL 76] HILICO J.C., CHAMPION J.P., PIERRE G., "The ground vibronic state of tetrahydrides", *J. Mol. Spec.*, vol. 60, pp. 422–425, 1976.

[HOU 62] HOUGEN J.T., "The classification of rotational energy levels for symmetric-top molecules", *J. Chem. Phys.*, vol. 37, pp. 1433–1441, 1962.

[HOU 63] HOUGEN J.T., "The classification of rotational energy levels II", *J. Chem. Phys.*, vol. 39, pp. 358–365, 1963.

[HOU 65] HOUGEN J.T., "The vibrational problem in molecules with nearly free internal rotation", *Pure Appl. Chem.*, vol. 11, pp. 481–496, 1965.

[HOU 71] HOUGEN J.T., "Interpretation of molecular-beam radiofrequeny ortho-para transitions in methane", *J. Chem. Phys.*, vol. 55, pp. 1122–1127, 1971.

[HOU 76] HOUGEN J.T., "Methane symmetry operations", in RAMSAY D.A. (ed.), *MTP International Reviews of Science, Physical Chemistry*, Butterworths, London, pp. 75–125, 1976.

[HOU 86] HOUGEN J.T., "Symmetry beyond point groups in molecular spectroscopy", *Phys. Chem.*, vol. 90, pp. 562–568, 1986.

[IAC 81] IACHELLO F., "Algebraic methods to molecular rotation-vibration spectra", *Chem. Phys. Lett.*, vol. 78, pp. 581–585, 1981.

[IAC 82] IACHELLO F., LEVINE R.D., "Algebraic approach to molecular rotation-vibration spectra. I. Diatomic molecules", *J. Chem. Phys.*, vol. 77, pp. 3036–3055, 1982.

[IAC 93] IACHELLO F., "Algebraic methods in quantum mechanics with applications to nuclear and molecular structure", *Nucl. Phys. A*, vol. 560, pp. 23–24, 1993.

[IAC 94] IACHELLO F., LEVINE R.D., *Algebraic Theory of Molecules*, Oxford University Press, Oxford, 1994.

[JAH 38a] JAHN H.A., "A new Coriolis perturbation in the methane spectrum: Part I. Vibrational-rotational Hamiltonians and wave functions", *Proc. Roy. Soc. A*, vol. 168, pp. 469–495, 1938.

[JAH 38b] JAHN H.A., "A new Coriolis perturbation in the methane spectrum: Part II. Energy levels", *Proc. Roy. Soc. A*, vol. 168, pp. 495–518, 1938.

[KIM 81a] KIM S.K., "A new method of constructing the symmetry coordinates of molecular vibrations based on the correspondence theorem", *J. Maths Phys.*, vol. 22, pp. 2203–2306, 1981.

[KIM 81b] KIM S.K., "A new method of constructing the symmetry-adapted linear combinations for a finite group: I. Salc's of equivalent invariant functions", *Physica A.*, vol. 105, pp. 577–592, 1981.

[KIM 81c] KIM S.K., "A new method of constructing the symmetry-adapted linear combinations for a finite group: II. Correspondence theorem and basis operators", *Physica A.*, vol. 106, pp. 521–538, 1981.

[KIR 73] KIRSCHNER S.M., WATSON J.K.G., "Sextic centrifugal distortion of tetrahedral molecules", *J. Mol. Spec.*, vol. 47, pp. 347–348, 1973.

[LAK 87] LAKHLIFI A., Étude théorique de la molécule d'ammoniac piégée en matrice. Interactions, mouvements et phénomènes relaxationnels, PhD thesis, Université Franche-Comté, Besançon, 1987.

[LAN 75] LANDAU L., LIFCHITZ E., PITAEVSKI L., *Physique Théorique : Mécanique Quantique*, vol. 3, Mir, Moscow, 1975.

[LIG 85] LIGHT J.C., HAMILTON I.P., LILL J.V., "Generalized discrete variable approximation in quantum mechanics", *J. Chem. Phys.*, vol. 82, pp. 1400–1409, 1985.

[LIP 65] LIPKIN J.H., *Lie Groups for Pedestrians*, North-Holland Publishing Company, Amsterdam, 1965.

[LOE 84] LOËTE M., Théorie des intensités absolues des transitions de vibration-rotation des molécules XY_4 application au méthane, PhD thesis, Université de Dijon, Dijon, 1984.

[LON 63] LONGUET-HIGGINS H.C., "The symmetry groups of non-rigid molecules", *Mol. Phys.*, vol. 6, pp. 445–460, 1963.

[MAJ 32] MAJORANA E., "Atomi orientate in campo magnetico variabile", *Nuovo Cimento*, vol. 9, pp. 43–50, 1932.

[MAN 35] MANNING M.F., "Energy levels of a symmetrical double minima problem with applications to the NH_3 and ND_3 molecules", *J. Chem. Phys.*, vol. 3, pp. 136–138, 1935.

[MAR 83] MARUANI J., SERRE J., *Symmetries and Properties of Non-Rigid Molecules: A Comprehensive Survey*, Elsevier Scientific Publishing, Amsterdam, 1983.

[MAR 92] MARTIN J.M.L., LEE T.J., TAYLOR P.R., "An accurate *ab initio* quartic force field for ammonia", *J. Chem. Phys.*, vol. 97, pp. 8361–8371, 1992.

[MES 64] MESSIAH A., *Mécanique quantique*, Dunod, Paris, 1964.

[MIZ 72] MIZUSHIMA M., *Theoretical Physics: From Classical Mechanics to Group Theory of Microparticles*, John Wiley & Sons, New York, 1972.

[MOR 61] MORET-BAILLY J., Sur l'interprétation des spectres de vibration-rotation des molécules à symétrie tétraédrique ou octaédrique, PhD thesis, Université de Paris, Paris, 1961.

[MOR 65] MORET-BAILLY J., GAUTIER L., MONTAGUTELLI J., "Clebsch-Gordan coefficients adapted to cubic symmetry", *J. Mol. Spec.*, vol. 15, pp. 355–377, 1965.

[NEW 48] NEWTON P.R., THOMAS L.H., "Internal molecular motions of large amplitude illustrated by the symmetrical vibration of ammonia", *J. Chem. Phys.*, vol. 16, pp. 310–323, 1948.

[NIE 51] NIELSEN H.H., "The vibration-rotation energies of molecules", *Rev. Mod. Phys.*, vol. 23, pp. 90–136, 1951.

[NIK 13] NIKITIN A.V., BOUDON V., WENGER C. *et al.*, "High resolution spectroscopy and the first global analysis of the Tetradecad region of methane $^{12}CH_4$", *Phys. Chem. Chem. Phys.*, vol. 15, pp. 10071–10093, 2013.

[OPE 40] OPECHOWSKI W., "The double crystallographic groups", *Physica*, vol. 7, pp. 552–562, 1940.

[PAP 83] PAPOUSEK D., ALIEV D., *Molecular Vibration-Rotation Spectra*, Elsevier, Amsterdam, 1983.

[PER 96] PERMOGOROV D., Spectroscopie intracavité laser des transistions du méthane, silane, germane et de leurs dérivés deutérés dans les domaines visible et proche infrarouge, PhD thesis, Université Joseph Fourier, Grenoble, 1996.

[PIE 77] PIERRE G., CHAMPION J.P., "Analyse de la bande ν_4 de $^{28}SiH_4$ en interaction avec la bande ν_2", *Journal de Physique Lettres*, vol. 38, pp. 311–316, 1977.

[REY 03] REY M., BOUDON V., WENGER C. *et al.*, "Orientation of $O(3)$ and $SU(2)\otimes C_I$ representations in cubic point groups (O_h, T_d) for application to molecular spectroscopy", *J. Mol. Spec.*, vol. 219, pp. 313–325, 2003.

[ROO 83] ROOSMALEN VAN O.S., "Algebraic approach to molecular rotation-vibration spectra II. Triatomic molecules", *J. Chem. Phys.*, vol. 79, pp. 2515–2536, 1983.

[ROS 67] ROSE M.E., *Elementary Theory of Angular Momentum*, Wiley, New York, 1967.

[SAK 67] SAKHAROV A., "Violation of Cp invariance, C asymmetry, and baryon asymmetry of the universe", *JETP Lett.*, vol. 5, pp. 24–27, 1967.

[SAK 94] SAKURAI J.J., *Modern Quantum Mechanics*, Addison Wesley Publishing, Reading, 1994.

[SCH 39a] SCHAFFER W.H., NIELSEN H.H., THOMAS L.H., "The rotation-vibration energies of tetrahedrally symmetric pentatomic molecules I", *Phys. Rev.*, vol. 56, pp. 895–907, 1939.

[SCH 39b] SCHAFFER W.H., NIELSEN H.H., THOMAS L.H., "High resolution spectroscopy and the first global analysis of the tetradecad region of methane $^{12}CH_4$", *Phys. Rev.*, vol. 57, pp. 128–10093, 1939.

[STO 75] STONE A.J., "Transformation between cartesian and spherical tensors", *Mol. Phys.*, vol. 29, pp. 1461–1471, 1975.

[SUG 65] SUGDEN T.M., KENNEY C.N., *Microwave Spectroscopy of Gases*, D. Van Nostrand, New York, 1965.

[SWA 62] SWALEN J.D., IBERS J.A., "Potential function for the inversion of ammonia", *J. Chem. Phys.*, vol. 36, pp. 1914–1918, 1962.

[SWA 08] SWAIN M.R., VASISHT G., TINETTI G., "The presence of methane in the atmosphere of an extrasolar planet", *Nature*, vol. 452, pp. 329–331, 2008.

[TAR 65] TARRAGO G., Fréquences des transitions de vibration-rotation des molécules possédant un axe de symétrie ternaire. Calcul des corrections jusqu'au quatrième ordre, PhD thesis, Université de Paris, Paris, 1965.

[TIN 03] TINKHAM M., *Group Theory and Quantum Mechanics*, Dover Publications, New York, 2003.

[TOW 75] TOWNES C.H., SCHAWLOW A.L., *Microwave Spectroscopy*, Dover Publications, New York, 1975.

[VAN 29] VAN VLECK J.H., "On sigma-type doubling and electron spin in the spectra of diatomic molecules", *Phys. Rev.*, vol. 33, pp. 467–506, 1929.

[VER 92] VERNOV A., STEELE W.A., "The electrostatic field at a graphite surface and its effect on molecule-solid interactions", *Langmuir*, vol. 8, pp. 155–159, 1992.

[VOT 84] VOTER A.F., DOLL J.D., "Transition state theory description of surface self-diffusion: Comparison with classical trajectory results", *J. Chem. Phys.*, vol. 80, pp. 5832–5838, 1984.

[WAN 29] WANG S.C., "On the asymmetrical top in quantum mechanics molecules", *Phys. Rev.*, vol. 34, pp. 243–252, 1929.

[WAT 66] WATANABE H., *Operator Methods in Ligand Field Theory*, Prentice-Hall, Englewood Cliffs, 1966.

[WAT 67] WATSON J.K.G., "Determination of centrifugal distortion coefficients of asymmetric-top molecules", *J. Chem. Phys.*, vol. 46, pp. 1935–1948, 1967.

[WAT 68a] WATSON J.K.G., "Simplification of the molecular vibration-rotation Hamiltonian", *Mol. Phys.*, vol. 15, pp. 479–490, 1968.

[WAT 68b] WATSON J.K.G., "Determination of centrifugal-distortion coefficients of asymmetric-top molecules II. Dreizler, Dendl, and Rudolph's results", *J. Chem. Phys.*, vol. 48, pp. 181–185, 1968.

[WAT 68c] WATSON J.K.G., "Determination of centrifugal distortion coefficients of asymmetric-top molecules III. Sextic coefficients", *J. Chem. Phys.*, vol. 48, pp. 4517–4524, 1968.

[WAT 71] WATSON J.K.G., "Forbidden rotational spectra of polyatomic molecules", *J. Mol. Spec.*, vol. 40, pp. 536–544, 1971.

[WAT 75] WATSON J.K.G., "Higher-degree centrifugal distortion of tetrahedral molecules", *J. Mol. Spec.*, vol. 55, pp. 498–499, 1975.

[WIG 59] WIGNER E., *Group Theory and Its Application to the Quantum Mechanics of Atomic Spectra*, Academic Press, Cambridge, 1959.

[WIL 36] WILSON E.B., HOWARD J.B., "The vibration-rotation energy levels of polyatomic molecules I. Mathematical theory of semirigid asymmetrical top molecules", *J. Chem. Phys.*, vol. 4, pp. 262–267, 1936.

[WIL 55] WILSON E.B., DECIUS J.C., CROSS P.C., *Molecular Vibrations: The Theory of Infrared and Raman Vibrational Spectra*, McGraw-Hill, New York, 1955.

[WIL 80] WILSON E.B., DECIUS J.C., CROSS P.C., *Molecular Vibrations: The Theory of Infrared and Raman Vibrational Spectra*, Dover Publications, New York, 1980.

[YUR 13] YURCHENKO N.S., TENNYSON J., BARBER R.J., *et al.*, "Vibrational transition moments of CH_4 from first principles", *J. Mol. Spec.*, vol. 291, pp. 69–76, 2013.

[ZHI 81] ZHILINSKII B.I., "Reduction of rotational operators to standard form", *Opt. Spec.*, vol. 51, pp. 262–263, 1981.

Index

G, H, I

K, L, M

N, O, P

Q, R

S

T

V, W, X

Other titles from

ISTE

in

Waves

2020

DANIELE Vito G., LOMBARDI Guido
Scattering and Diffraction by Wedges 1: The Wiener-Hopf Solution - Advanced Applications (Waves and Scattering Set – Volume 1)

DANIELE Vito G., LOMBARDI Guido
Scattering and Diffraction by Wedges 2: The Wiener-Hopf Solution - Advanced Applications (Waves and Scattering Set – Volume 2)

SAKHO Ibrahima
Introduction to Quantum Mechanics 2: Wave-Corpuscle, Quantization & Schrödinger's Equation

2019

BERTRAND Pierre, DEL SARTO Daniele, GHIZZO Alain
The Vlasov Equation 1: History and General Properties

DAHOO Pierre-Richard, LAKHLIFI Azzedine
Infrared Spectroscopy of Triatomics for Space Observation (Infrared Spectroscopy Set – Volume 2)

RÉVEILLAC Jean-Michel
Electronic Music Machines: The New Musical Instruments

ROMERO-GARCIA Vicente, HLADKY-HENNION Anne-Christine
Fundamentals and Applications of Acoustic Metamaterials: From Seismic to Radio Frequency
(Metamaterials Applied to Waves Set – Volume 1)

SAKHO Ibrahima
Introduction to Quantum Mechanics 1: Thermal Radiation and Experimental Facts Regarding the Quantization of Matter

2018

SAKHO Ibrahima
Screening Constant by Unit Nuclear Charge Method: Description and Application to the Photoionization of Atomic Systems

2017

DAHOO Pierre-Richard, LAKHLIFI Azzedine
Infrared Spectroscopy of Diatomics for Space Observation
(Infrared Spectroscopy Set – Volume 1)

PARET Dominique, HUON Jean-Paul
Secure Connected Objects

PARET Dominque, SIBONY Serge
Musical Techniques: Frequencies and Harmony

RÉVEILLAC Jean-Michel
Analog and Digital Sound Processing

STAEBLER Patrick
Human Exposure to Electromagnetic Fields

2016

ANSELMET Fabien, MATTEI Pierre-Olivier
Acoustics, Aeroacoustics and Vibrations

BAUDRAND Henri, TITAOUINE Mohammed, RAVEU Nathalie
The Wave Concept in Electromagnetism and Circuits: Theory and Applications

PARET Dominique
Antennas Designs for NFC Devices

PARET Dominique
Design Constraints for NFC Devices

WIART Joe
Radio-Frequency Human Exposure Assessment

2015

PICART Pascal
New Techniques in Digital Holography

2014

APPRIOU Alain
Uncertainty Theories and Multisensor Data Fusion

JARRY Pierre, BENEAT Jacques N.
RF and Microwave Electromagnetism

LAHEURTE Jean-Marc
UHF RFID Technologies for Identification and Traceability

SAVAUX Vincent, LOUËT Yves
MMSE-based Algorithm for Joint Signal Detection, Channel and Noise Variance Estimation for OFDM Systems

THOMAS Jean-Hugh, YAAKOUBI Nourdin
New Sensors and Processing Chain

TING Michael
Molecular Imaging in Nano MRI

VALIÈRE Jean-Christophe
Acoustic Particle Velocity Measurements using Laser: Principles, Signal Processing and Applications

VANBÉSIEN Olivier, CENTENO Emmanuel
Dispersion Engineering for Integrated Nanophotonics

2013

BENMAMMAR Badr, AMRAOUI Asma
Radio Resource Allocation and Dynamic Spectrum Access

BOURLIER Christophe, PINEL Nicolas, KUBICKÉ Gildas
Method of Moments for 2D Scattering Problems: Basic Concepts and Applications

GOURE Jean-Pierre
Optics in Instruments: Applications in Biology and Medicine

LAZAROV Andon, KOSTADINOV Todor Pavlov
Bistatic SAR/GISAR/FISAR Theory Algorithms and Program Implementation

LHEURETTE Eric
Metamaterials and Wave Control

PINEL Nicolas, BOURLIER Christophe
Electromagnetic Wave Scattering from Random Rough Surfaces: Asymptotic Models

SHINOHARA Naoki
Wireless Power Transfer via Radiowaves

TERRE Michel, PISCHELLA Mylène, VIVIER Emmanuelle
Wireless Telecommunication Systems

2012

LALAUZE René
Chemical Sensors and Biosensors

LE MENN Marc
Instrumentation and Metrology in Oceanography

LI Jun-chang, PICART Pascal
Digital Holography

2011

BECHERRAWY Tamer
Mechanical and Electromagnetic Vibrations and Waves

BESNIER Philippe, DÉMOULIN Bernard
Electromagnetic Reverberation Chambers

GOURE Jean-Pierre
Optics in Instruments

GROUS Ammar
Applied Metrology for Manufacturing Engineering

LE CHEVALIER François, LESSELIER Dominique, STARAJ Robert
Non-standard Antennas

2010

BEGAUD Xavier
Ultra Wide Band Antennas

MARAGE Jean-Paul, MORI Yvon
Sonar and Underwater Acoustics

2009

BOUDRIOUA Azzedine
Photonic Waveguides

BRUNEAU Michel, POTEL Catherine
Materials and Acoustics Handbook

DE FORNEL Frédérique, FAVENNEC Pierre-Noël
Measurements using Optic and RF Waves

FRENCH COLLEGE OF METROLOGY
Transverse Disciplines in Metrology

2008

FILIPPI Paul J.T.
Vibrations and Acoustic Radiation of Thin Structures

LALAUZE René
Physical Chemistry of Solid-Gas Interfaces

2007

KUNDU Tribikram
Advanced Ultrasonic Methods for Material and Structure Inspection

PLACKO Dominique
Fundamentals of Instrumentation and Measurement

RIPKA Pavel, TIPEK Alois
Modern Sensors Handbook

2006

BALAGEAS Daniel *et al.*
Structural Health Monitoring

BOUCHET Olivier *et al.*
Free-Space Optics

BRUNEAU Michel, SCELO Thomas
Fundamentals of Acoustics

FRENCH COLLEGE OF METROLOGY
Metrology in Industry

GUILLAUME Philippe
Music and Acoustics

GUYADER Jean-Louis
Vibration in Continuous Media

www.ingramcontent.com/pod-product-compliance
Lightning Source LLC
LaVergne TN
LVHW011751200926
848484LV00003B/672